The New
SPACE
Encyclopaedia

The New SPACE Encyclopaedia

A Guide to Astronomy and Space Exploration

New, revised Edition

A SUNRISE BOOK

E. P. DUTTON & CO., INC.

NEW YORK

FIRST IMPRESSION, SEPTEMBER 1969
SECOND IMPRESSION, DECEMBER 1969
SECOND EDITION, NOVEMBER 1973

ISBN 0-85141-311-0 (U.K. EDITION)
ISBN 0-87690-108-9 (U.S.A. EDITION)

LIBRARY OF CONGRESS CATALOG CARD NUMBER: 73-12348

Made and printed in Great Britain by
Redwood Press Limited, Trowbridge, Wiltshire

CONTRIBUTORS

Sir Bernard Lovell, O.B.E., F.R.S.
Professor of Radio Astronomy, Manchester University.

Sir Harold Spencer Jones, K.B.E., F.R.S.
Late Astronomer Royal.

Zdeněk Kopal, D.Sc.
Professor of Astronomy, Manchester University.

Homer E. Newell, Jr.
Late Deputy Administrator, N.A.S.A.

R. H. Garstang, Ph.D.
Professor of Astrophysics, University of Colorado.

Patrick Moore, F.R.A.S.

R. Griffin, Ph.D., F.R.A.S.
Cambridge University Observatories.

Valerie Myerscough, Ph.D.
Queen Mary College, University of London.

Peter J. Smith, Ph.D.
The Open University.

Squadron Leader C. W. V. McCleery, R.A.F.

J. G. Porter, Ph.D.
Royal Greenwich Observatory.

P. R. Owen, Ph.D., F.R.A.S.
J. A. J. Whelan, M.Sc.
T. M. H. Petersen, B.Sc.
G. T. Bath, B.A.
P. Lancaster Brown, B.Sc.
Hubertus Strughold, Ph.D.
Winston R. Withey, Ph.D., F.R.A.S.
Peter Martin, *Barrister-at-Law.*

General Editor:
M. T. Bizony, M.A.

The Publishers' thanks are also due for material and information received from the following:

The Royal Society, Royal Astronomical Society, N.A.S.A., Ministry of Defence, Ministry of Supply, California Institute of Technology, Harvard University Observatory, Royal Aircraft Establishment (Space Department and Dr. D. G. King-Hele), Novosti (Moscow), Krakov University Observatory, Sperry Rand Corporation, Glenn Martin Company, Douglas Aircraft Corporation, Lockheed Aircraft Corporation, the British Interplanetary Society and many others who have supplied photographs, including the U.S. Embassy, London.

On July 21, 1969, Man first set foot on the Moon and was watched by millions on Earth as he did so.

Ten days later he began the detailed mapping of Mars.

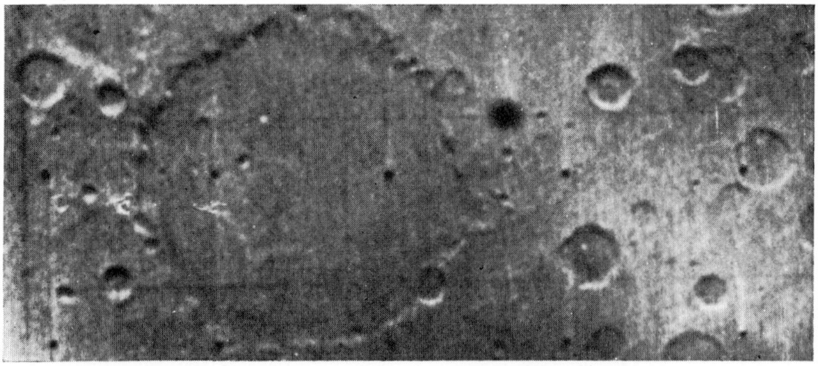

'A GREEK it was whose lively force of mind first won its way, so that he passed far beyond the flaming ramparts of the world, and in mind and spirit traversed the boundless whole.'

LUCRETIUS (99–55 B.C.), *De Rerum Natura.*

'THERE will certainly be no lack of human pioneers when we have mastered the art of soaring. Let us create vessels and sails adjusted to the heavenly ether, and men will present themselves who are unafraid of the boundless voids. In the meantime we shall prepare, for the brave sky-travellers, maps of the celestial bodies – I shall do it for the Moon, and you, Galileo, for Jupiter.'

JOHANNES KEPLER, 1610.

Words printed in **bold type** indicate those articles which it may be useful to consult for additional information.

The articles on Astronomy, Cosmology and Radio Astronomy, and those on Artificial Satellites, Planetary Probe and Rocketry are recommended as general introductions to their respective subjects. Individual missiles and space vehicles are not, as a rule, listed under their own names; **Apollo** is the major exception.

A

Å. See **Ångström Unit.**

ABERRATION OF LIGHT is responsible for a small apparent displacement of stars as seen by an observer moving with the Earth.

If a pellet is dropped down the middle of a vertical tube, and if the tube is moving sideways while the pellet is falling, the pellet will not fall along the centre line of the tube. To an observer moving with the tube, the pellet will appear to be slanting towards the wall of the tube. In the same way, light traversing a telescope is apparently deflected, since the telescope is moving with the Earth at nearly 20 miles per second while the light is travelling through it. This will affect the observed position of stars by up to 20″.5.

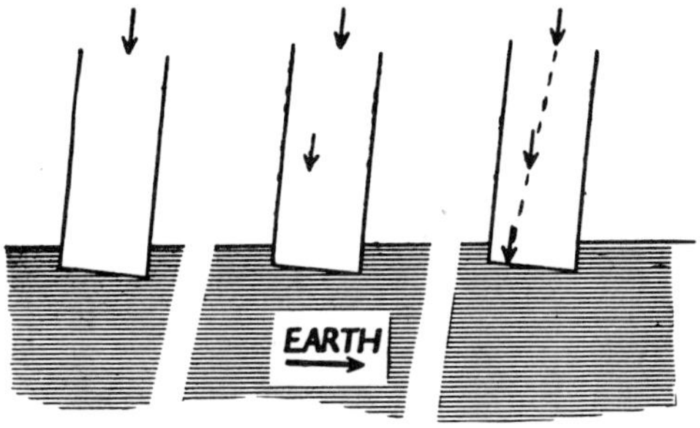

Three rays of light from a distant star enter a telescope along its mid-line. Moving at some 18 miles per second, the Earth carries the telescope across the line of sight, and the rays appear to be coming in from a direction which differs slightly from the true direction of the star.

The speed of light is reduced by about $\frac{1}{4}$ when it passes through water. In 1871 the English astronomer Airy filled a telescope with water to measure the effect on the aberration. Since the light would now take longer to traverse the telescope, the aberration should have been noticeably increased. Contrary to all expectation, it remained unchanged. This surprising result was incompatible with the old idea of the aether, and did not find its explanation until Einstein formulated his Theory of **Relativity,** for which it is still one of the most direct pieces of evidence.

ABORT. Failure or deliberate termination of a rocket launch or mission during or after count-down. (See also **Back-Out** and **Break-Up System.**)

ABSOLUTE TEMPERATURE is temperature expressed in degrees Kelvin (° K.) above **absolute zero.** A degree Kelvin equals a degree Centigrade in range, so that

$$\textit{absolute temp.} = \textit{Centigrade temp.} + 273{\cdot}16$$

ABSOLUTE ZERO. The lowest possible temperature; it is the starting point of the absolute scale of temperature, and equals −273·16° C.

Cold is merely the absence of heat. Heat may be defined as the energy inherent in the *random* motion of the molecules in a substance. As a body cools, so this motion decreases, but some of it remains even in a body which 'feels' ice-cold to the touch. When there is *no* random molecular motion left, the body is at absolute zero.

Temperatures less than a thousandth of a degree above absolute zero have been attained in the laboratory, but it can never be reached completely. Its precise value relative to the Centigrade scale is known from theoretical calculations.

ABSORPTION, GALACTIC. See **Galactic Absorption.**

ABSORPTION SPECTRA. Spectra consisting of dark lines, dark bands or a dark continuum superposed on a bright background and caused by the absorption of light by a substance placed in front of a bright source. For example, the dark lines of the solar spectrum are produced by radiation from the deeper layers being absorbed in the cooler, outermost layers of the Sun. (See **Spectroscopy.**)

ABUNDANCES OF THE ELEMENTS. See **Nucleogenesis.**

ACCELERATION. Rate of change of velocity. A body which at one moment moves with a velocity of 10 miles per second, and which one second later moves at 12 miles per second, has undergone an acceleration of 2

RS-2 ROCKET SLED travelling on its rails at 1,750 m.p.h. It can carry 2,000-pound payloads of missile guidance systems for testing under high accelerations.

miles per second per second, or 2 miles/sec.^{2}. If the velocity had decreased, the acceleration would be negative, i.e. it would be a *retardation.*

A body falling under the influence of gravity gathers speed, i.e. it is accelerated, and apart from air resistance this acceleration is exactly the same for a light body as for a heavy one. Thus the effect of gravity may be expressed in terms of the acceleration it produces, usually in cm./sec.^{2} or feet/sec.^{2}.

In the case of gravity, acceleration is uniform (or linear) over a short distance. In other cases it may not be; e.g. a stone propelled by an ordinary catapult has maximum acceleration at the moment of release, when the accelerating force of the tension is greatest but the stone's speed is still zero; when the elastic is relaxed and the catapult has spent its driving force, the acceleration will have sunk to zero, and the speed will have reached its maximum.

It must always be remembered that it requires just as much energy to accelerate a given body to a certain speed as to bring it to rest once it has attained that speed. In most everyday examples this energy is supplied by the action of the force of gravity, which slows down a stone that has been accelerated upwards, or by friction, which will slow down an object sliding on a horizontal surface. In the absence of such forces, a body once in motion will continue to move in a straight line. The work that has to be done to stop it is equal to the body's kinetic energy, i.e. the product of its mass with half the square of its velocity. (See also **Energy** and **Gravity.**) (M.T.B.)

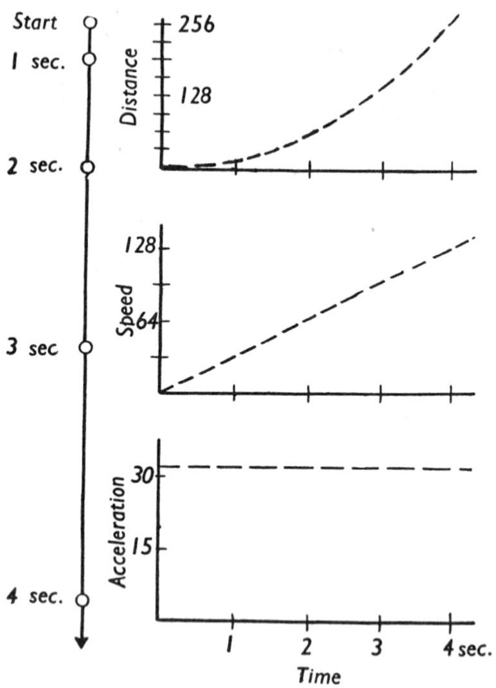

THREE GRAPHS for a body falling from rest near the surface of the Earth. *Bottom:* the acceleration due to gravity is constant at 32 feet per second^2. *Middle:* the speed of the falling body increases uniformly; it is plotted in feet per second. *Top:* the total distance fallen (in feet) is here plotted against time (in seconds); this graph should be compared with the diagram on the left, in which the falling body is seen in the positions it occupies after 1, 2, 3 and 4 seconds.

ACCELEROMETER. An instrument for measuring **acceleration.** It usually relies upon the fact that

$$force = mass \times acceleration.$$

In its simplest form it is a pendulum free to swing in a certain direction. The angle which it assumes to the vertical is a measure of the acceleration in that direction.

Accurate accelerometers, usually of the integrating gyroscopic type, are of fundamental importance in **inertial guidance** systems, and are also important components of aircraft and missile autopilots.

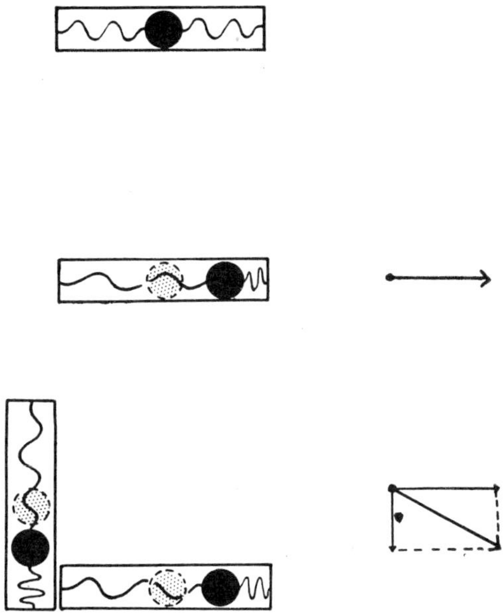

PRINCIPLE OF A SPRING ACCELEROMETER. *Top:* resting position. *Middle:* acceleration to the left. *Bottom:* acceleration to the upper left, resolved into vertical and horizontal components. The arrows show displacement of the weight (in the *opposite* sense to acceleration).

Where great accuracy is not required direct-reading spring-type linear accelerometers are quite adequate, but for most control system applications a servo instrument is used. This is basically a spring accelerometer with a device to drive the spring deflection back to a zero displacement. The force required to counterbalance the acceleration can be accurately converted into an electrical signal which may then be used to activate a control system or be fed into a navigational computer. Accelerometers are normally mounted in gimbals.

ACHILLES. One of the Trojan group of **asteroids.** Diam. 150 miles.

ADIABATIC CHANGE. A change in matter not involving any transfer of heat. A thermally insulated mass of gas may be heated (or cooled) by compressing (or expanding) it adiabatically.

ADONIS. A small **asteroid** with a diameter of a few miles. Owing to the eccentricity of its orbit it has come quite close to the Earth ($1\frac{1}{2}$ million miles in 1936).

AERIA. An ochre-coloured 'desert' region on **Mars.**

AERIAL. See **Antenna.**

AGENA. The Lockheed *Agena* is a highly reliable restartable upper stage rocket which has been widely used in the U.S. space programme. It was the first propulsion system to be 'docked' in orbit with a manned spacecraft and subsequently fired while still attached to the Gemini capsule.

AIR BREAK-UP. A method of recovery of instruments from a high-altitude rocket.

If a heavy, streamlined missile is allowed to fall to earth from a great height it can bury itself so deeply that recovery of instruments may be extremely difficult, and damage by the impact will be severe. This is sometimes avoided by firing a cartridge inside the missile by means of a time switch or radio signal while it is in mid-air, causing it to break up into separate sections. Those containing important instruments will now be of poor

An AGENA target vehicle photographed from a Gemini space craft in orbit.

aerodynamic form and will therefore be greatly retarded by air resistance, so that together with parachutes a relatively gentle landing becomes possible.

AIRGLOW. Even when the Moon is not in the night sky perfect blackness of the sky is not obtained. Some of the remaining light (about $\frac{1}{6}$) comes from the stars and nebulae, and from interplanetary material reflecting the solar rays. But a large proportion of the total arises within the Earth's atmosphere itself. During the daytime the solar rays dissociate molecules and ionize atoms in the upper atmosphere. At night the energy absorbed by the processes is released, and some of it appears as visible light, thus giving rise to the airglow. Unlike the **aurora polaris,** the airglow shows no structures such as arcs and it is emitted from the entire sky at all latitudes and at all times.

Airglow has been simulated by releasing nitric oxide gas or sodium vapour from rockets at varying heights and from the first Russian artificial planet at 71,000 miles. In each case a patch of light was observed which slowly spread and faded away. The observations gave information on winds at high altitudes, the nature of airglow, and the luminescence of comets, and allowed a visual check to be made on the position of the artificial planet at the time of release of vapour.

Airglow is very weak in the visible region of the **electromagnetic spectrum,** but is about a thousand times as strong in the infra-red. It limits the faintest objects which can be photographed from terrestrial observatories, and this can be overcome only by establishing **orbiting observatories** beyond the atmosphere.

AIR RESISTANCE. The retarding force acting on a body moving through air. A more sophisticated term is *aerodynamic drag*. At low speeds the air flow past a moving object is *streamline*, i.e. the layers of air near the body slip smoothly over one another; at high speeds turbulence sets in and the air eddies past the object. Various factors combine to give the total air resistance: skin, form, interference and wave drags.

SKIN DRAG occurs because of the motion between different layers of air near the moving body; the air immediately in contact with the surface moves with the body. As the air resistance is much greater for turbulent than streamline flow, designers aim to preserve the latter as far as possible.

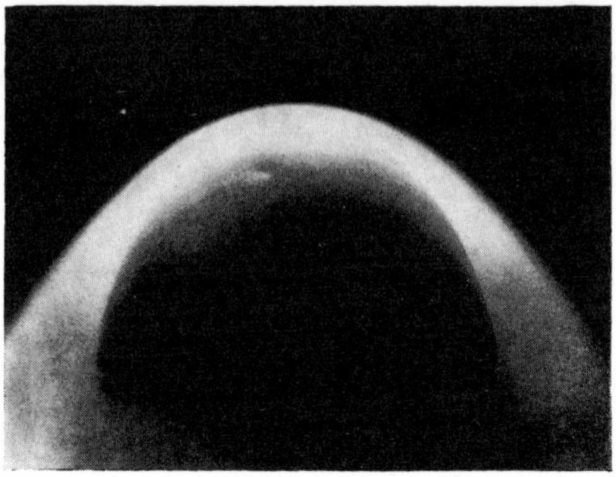

MISSILE NOSE CONE surrounded by incandescent glow while under test in a Hotshot wind tunnel. The dent in the cone was caused by a piece of metal torn off the tunnel by the test wind which reached 32,400 m.p.h.

FORM DRAG is occasioned by variations of pressure round a moving body. An extreme case is that of a flat plate, where the pressure is greatly increased on the forward face and greatly reduced behind the plate. At speeds less than that of sound, form drag may be reduced by using an elongated shape with a rounded nose and pointed tail.

INTERFERENCE DRAG is caused by projections which disturb the air flow round the object, for instance the wings and tail of an aircraft. It is minimized by rounding all projections and interstices.

WAVE DRAG occurs at supersonic speeds. The air in front of a body moving faster than sound cannot get out of the way quickly enough, and is suddenly compressed when the object arrives; shock waves are formed at the front and back of the object.

The compression of air in shock waves leads to adiabatic heating, and friction in the air layers near a moving surface also generates heat. At sufficiently great speeds, the temperature of the surface of, say, an aircraft becomes high enough to cause the structural materials to *creep*, and ultimately melt. Stainless steel and titanium structures and surface refrigeration are being used in an effort to overcome this so-called 'heat barrier'. (See **Wind Tunnel.**)

ALBEDO. The fraction of the total incident sunlight which is reflected back in all directions by a planet, satellite or asteroid, or any part of their surfaces. Albedos observed astronomically may be compared with those of known substances measured in the laboratory, and can give valuable clues as to the nature of the surfaces of the celestial body. Some typical albedos are:

Earth	0·39
Moon	0·07
Mars	0·15
Venus	0·59
Asteroids	0·1 (approx.)

ALBEDO. The albedo of a planet can be markedly influenced by clouds; in this picture, the clouds reflect far more sunlight than either the mountain top or the ground below.

Thus the Moon reflects less than 1/14th of the total sunlight it receives, and therefore has a fairly absorbent surface; Venus, with its atmosphere of dense white cloud, reflects more than half.

ALBERT. An **asteroid** (No. 719) with a diameter of about 3 miles. It has a very eccentric orbit, its perihelion being within 20 million miles of the Earth's orbit while its aphelion is nearly as far from the Sun as Jupiter.

ALCOR, or *The Test,* is a faint star in the **Great Bear** constellation. It lies close to the much brighter **Mizar,** with which it forms a binary system, and has for centuries been used as an eyesight test: a person who can distinguish it with the naked eye from Mizar has normal eyesight.

ALGOL (β Persei). An eclipsing double star. Its name means 'Demon Star', and its cyclic changes in magnitude were probably known to the Arab astronomers. During its 69-hour cycle it loses two-thirds of its brightness in 5 hours, and the change is easily observed with the naked eye. To astrologers it was the most ominous of all stars. (See **Binary Star.**)

ALGOL BINARY. A double star consisting, like **Algol,** of a **main sequence** star and a subgiant which eclipse each other. They were probably formed by splitting at a fairly recent time, but in that case it is hard to explain why the larger component should be the fainter. One interesting theory is that slightly uneven development of the two stars causes the larger and initially brighter one to deviate to the right of the main sequence; it expands until the other star is engulfed in it, loses material and so decreases in size without gaining in luminosity. Its central portion, still large, would then evolve naturally until it is once more distinct from its companion. (See **Hertzsprung – Russell Diagram** and **Binary Star.**)

ALPHA CENTAURI. A very bright multiple star in the constellation *Centaurus.* Its brightest component closely resembles the Sun. It is 4·3 light years away and the second-nearest star to the Earth, the only nearer star being a much fainter component of the same system called *Proxima Centauri.*

ALPHA PARTICLE. The nucleus of a helium **atom,** consisting of two protons and two neutrons, i.e. doubly ionized helium.

ALTITUDE. Height. In Astronomy, the angle between a star or object, the observer, and the point on the observer's horizon immediately below the star. It is often measured with a sextant, and has to be corrected for the dip of the horizon and atmospheric refraction to convert *observed* into *true* altitude.

True altitude + zenith distance = 90°.

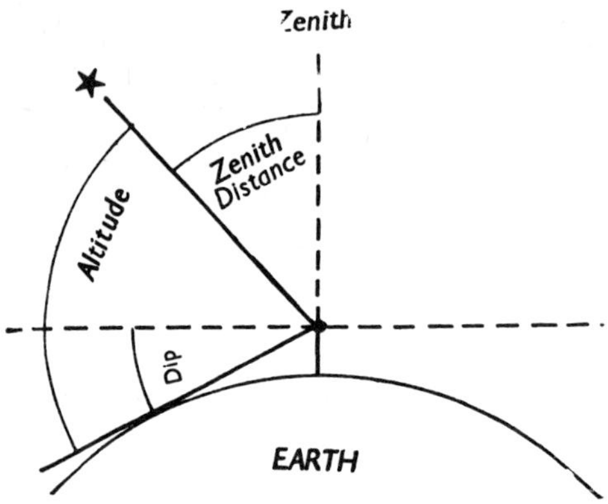

AMALTHEA. The unofficial name of one of **Jupiter's** moons. It is near enough to its planet to be deformed by the enormous gravitational strain, and is being drawn towards Jupiter at a rate of one or two inches a year. This will cause its destruction in about seventy million years. Its diameter is some 150 miles.

AMERICAN ROCKET SOCIETY. Founded in 1930 as the 'American Interplanetary Society', it has devoted itself to the development of propulsion methods. Its monthly journal *Jet Propulsion* maintains a high technical level. *Address:* Northampton Street, Easton, Pa., U.S.A.

AMMONIA, NH_3, is a colourless, pungent and strongly alkaline gas. Its melting point is $-78°$ C., and boiling point $-33°$ C. at normal pressure. It is poisonous except in low concentration. It forms a large part of the atmosphere of **Jupiter,** where it occurs both as fine crystals and as vapour, and exists in the solid form on **Uranus** and **Neptune.**

Ordinary 'household ammonia' is a solution of the gas in water.

AMOR. An **asteroid** with a diameter of ten miles. Its fairly eccentric orbit can bring it within 10 million miles of the Earth.

ANDROMEDA GALAXY (Messier 31). This great spiral galaxy in many ways resembles our own Milky Way system, and belongs to the same local group of about 20 galaxies. It is visible to the naked eye as a faint blur in the constellation *Andromeda*; but if the human eye were greatly more sensitive, then even without magnification this nebula would be one of the most splendid objects in the sky, with an apparent diameter six times that of the Sun.

The distance of M 31 is about 575,000 **parsecs,** and its diameter over 30,000 parsecs. The spiral arms contain **supergiants, Cepheids,** frequent **novae** (about two a month), diffuse and dark **nebulae,** and open clusters. The nucleus has been resolved into individual stars; it is by far the brightest part of the galaxy.

The spiral rotates, but not uniformly: the centre turns like a wheel, but the more outlying portions lag behind, so that the arms are becoming more 'tightly wound'. They already describe two to three turns about the centre.

Like our own galaxy, M 31 is surrounded by about two hundred **globular clusters,** each containing perhaps a hundred thousand stars similar to those in the spiral itself. It also has at least two elliptical companion galaxies, Messier 32 superposed on the spiral arm, and NGC 205 a small distance away.

ANDROMEDIDS. A **meteor** shower associated with the comet **Biela,** and hence also called *Bielids.* Showers of Andromedids were recorded in 1741 and have been traced with some probability to A.D. 524. There were great storms from this centre in 1872 and 1885, and the occurrence of these showers appeared at the time to have some connection with the disruption of the comet, which had actually been witnessed in 1846 (see **Comet**). Showers of Andromedids in 1798, 1830 and 1838, however, must have radiated from a point on the orbit in front of the comet. The orbit suffers a rapid regression of the **nodes,** as a result of which the shower occurs one day earlier every 7 years. At present it falls on November 14, but shows little activity.

ÅNGSTRÖM UNIT, abbreviated A. or Å, is a very small unit of length employed mainly for the measurement of the **wavelength** of light. It is equal to one hundred-millionth of a centimetre.

ANGULAR MOMENTUM. This is a quantity which is defined for any body which is rotating, or travelling in a curved path, around some axis; it is the product of the body's

THE GREAT ANDROMEDA GALAXY Messier 31, and its two elliptical satellites.
(*Ritchey and Pease*)

moment of inertia and its rate of rotation. A very important physical law is that of the *conservation of angular momentum:* this states that in any system which is not acted upon by external forces, the total angular momentum remains unchanged. The law holds regardless of what takes place within the system.

One of its consequences is that if a spinning body contracts it must, in the absence of other forces, increase its rate of spin.

ANGULAR VELOCITY. The rate of change of the direction of a moving point as measured from a point at rest. It is usually expressed in radians per second, but can also be put as revolutions per minute, etc. All points on a rotating disc have the same angular velocity about the centre, but their speed of motion increases with increasing distance from the centre.

ANNULAR ECLIPSE. An eclipse of the Sun in which the Moon obscures the central part of the Sun's disc but leaves a thin ring of light showing. (See **Eclipse.**)

ANOMALISTIC YEAR. The time between two successive passages of the Earth through **perihelion.** (See **Year.**)

ANTARCTIC. The South Polar region within the **Antarctic circle.** It contains the large land mass of the Antarctic continent. *Cf.* **Polar Regions.**

ANTARCTIC CIRCLE. The parallel of latitude 66° 33′ South. It is the limit of the area in the southern hemisphere within which the Sun does not set in mid-summer.

ANTENNA. A conductor linked to a radio transmitter for radiating, or to a receiver for intercepting radio waves. The size and shape of an antenna or *aerial* depend on the frequencies for which it is intended, on its directional properties, on the reflectors and wave guides with which it may be linked and on the *gain* required.

DEEP-SPACE ANTENNA resembling a flower with curved petals. It can be housed aboard a space craft in a flat package one yard in diameter. Once outside the atmosphere, it unfurls automatically to provide a large reflective surface for radio waves.

If an antenna could be designed which would radiate energy equally in all directions it would be described as *isotropic* and would have, by definition, unit gain. Similarly, if all the energy could be radiated in one direction in a beam of infinitesimal width the gain would be infinite. In practice something between these two extremes is required; for example a television broadcast transmitter would be required to radiate equally in all directions in a more or less horizontal plane, but is not required to radiate much above the horizon. The highest practicable gain is desirable for a radio telescope, both to provide the highest practicable *resolution* and to ensure that a minimum of extraneous noise from directions off the *boresight* will interfere with the wanted signal.

ANTI-MATTER. Most of the elementary particles (see **Quark**) we know have 'partners' with identical mass and spin, but opposite electric charge. These are the anti-particles. When a particle and its anti-particle collide, they are annihilated and the rest mass of the particle and its anti-particle is turned into energy, mostly in the form of radiation. This process plays an important part in certain theories of Cosmology.

For example, a collision of a proton with an anti-proton leads to their transformation into a meson and an anti-meson, which in turn decay rapidly and emit a neutrino, an anti-neutrino and gamma radiation as well as giving rise to an electron and a positron. An electron and a positron which are themselves in collision also annihilate each other, leaving nothing except gamma radiation.

The anti-particles are just as stable (or unstable) as their corresponding particles and as far as we know they could form a complete anti-world. It has been suggested that some galaxies are composed of anti-matter and, in fact, that the huge energy output necessary in **radio galaxies** and **quasars** is obtained from the annihilation between matter and anti-matter.

APHELION. The point farthest from the Sun of a planet's or comet's **orbit.** Contrary to what might at first be expected, the Earth passes through aphelion during mid-summer (of the northern hemisphere), the difference between the seasons being due much more to the inclination of the Earth's axis than to the changes in its distance from the Sun. *Cf.* **Perihelion.**

APOGEE. The position in the orbit of the Moon or artificial satellite which is farthest from the Earth. Opposite of **Perigee.**

APOLLO. A small asteroid which in 1932 came within 7 million miles of the Earth; track was lost of it as it receded again, and its present position is not accurately known.

APOLLO PROJECT. The U.S. project, initiated in 1961 and executed in 1969–72, for placing two astronauts on the Moon and returning them to Earth by a lunar orbit rendezvous technique. A third man remains in lunar orbit while the other two carry out the actual landing and ascend back into lunar

THE APOLLO 11–17 LANDING SITES. *Apollo 13* was damaged by an explosion, and returned without attempting a Moon Landing.

THE FIRST MOON LANDING TEAM OF APOLLO 11: Neil A. Armstrong (*centre, in command, first to set foot on the Moon*); Colonel Edwin E. Aldrin, U.S.A.F. (*left, Lunar Module Pilot*); and Lt.-Colonel Michael Collins, U.S.A.F. (*right, Command Module Pilot*). All three took part in Gemini flights.

orbit to rendezvous with the third before abandoning the landing vehicle in lunar orbit and returning to Earth. This technique is the most economical in terms of weight required on the Earth–Moon trajectory but has a number of difficult phases, of which the take-off from the Moon and rendezvous in lunar orbit are perhaps the most critical. All the equipment necessary for the mission can be launched by the Saturn 5 launch vehicle, which is capable of placing almost 3,000,000 lbs. in orbit.

The spacecraft consists of three main sections. The *Command Module*, which is the only part to return to Earth, contains the three astronauts for most of the mission together with most of the navigating and computer equipment. The *Service Module* has a 20,000 lb. thrust rocket motor and a large part of the environmental control and life support equipment and power supplies. Finally, the

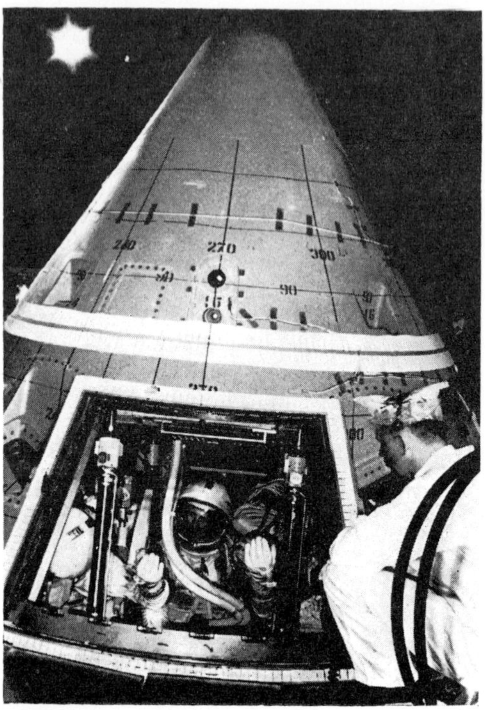

EIGHT-DAY TEST OF APOLLO 008 service and command modules. It was chiefly devoted to checking the heat balance between different parts of the manned system when it was 'soaked' in extremely low temperatures and kept in a vacuum equivalent to a height of 87 miles.

EMERGENCY LAUNCH ESCAPE SYSTEM can be activated before, during or shortly after lift-off, manually or by the Automatic Sequencer. The escape rocket motor pulls the *Apollo* command module clear of the aborting launch vehicle; wing-like canards stabilize the tumbling structure so that the command module is placed blunt end forward; the Tower Jettison Motor fires and pulls the tower clear; the cover of the parachute sub-system is jettisoned, and the 5-ton command module floats downwards.

Lunar Excursion Module (LEM) itself consists of two parts: the *descent stage*, which has the descent motor, landing legs, power supplies and stowage for experimental equipment required on the lunar surface; and the *ascent stage*, which contains the two astronauts, their life support requirements, and all the propulsion and guidance equipment required to enable the astronauts to control and time their lift-off into lunar orbit so that they can rendezvous with the minimum expenditure of propellent with the Command and Service modules. It is in fact the Command module pilot who does most of the manoeuvring to achieve rendezvous, so all that is required is for the lift-off to be timed correctly and for the launch direction to be such that both orbits are in the same plane. The descent stage is used as a launch pad and is left on the lunar surface, and after rendezvous and transfer of the astronauts back to the Command module the ascent stage is also abandoned.

A typical mission begins with lift-off of the Saturn 5 launch vehicle, which requires a short burn of the third stage to place the spacecraft with the third stage still attached in a parking orbit round the Earth. After a pause of one or two revolutions round the Earth which is used to give the spacecraft systems a

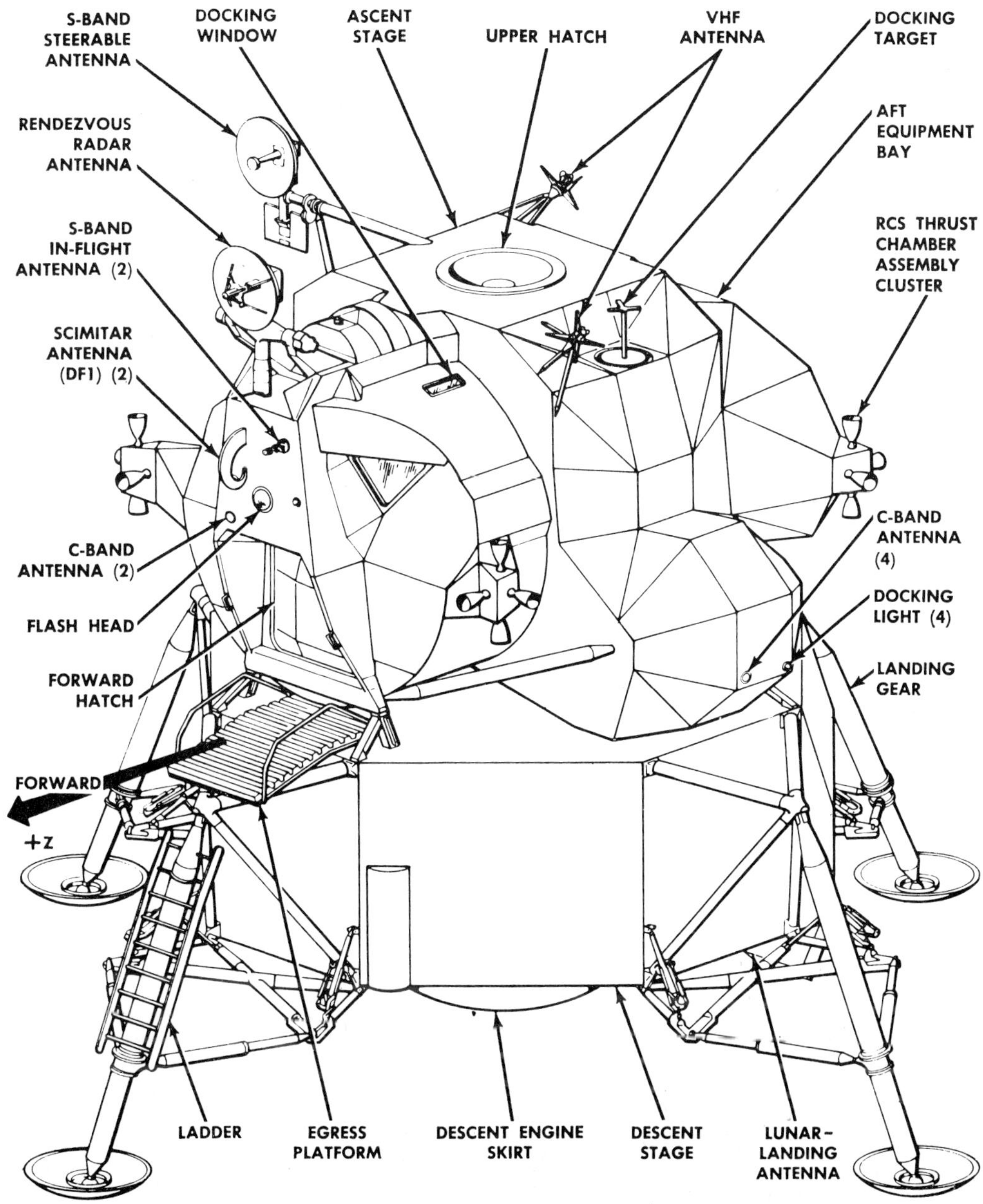

APOLLO Lunar Module

check over and to refine the trajectory, the third stage is fired again to boost the spacecraft to just over escape velocity in the trans-lunar injection manoeuvre which places the spacecraft on the correct course for the Moon. The Command and Service modules are then detached from the third stage, and turn about until the apex of the conical Command module can be inserted into the top of the ascent stage of the LEM to extract the whole of the LEM

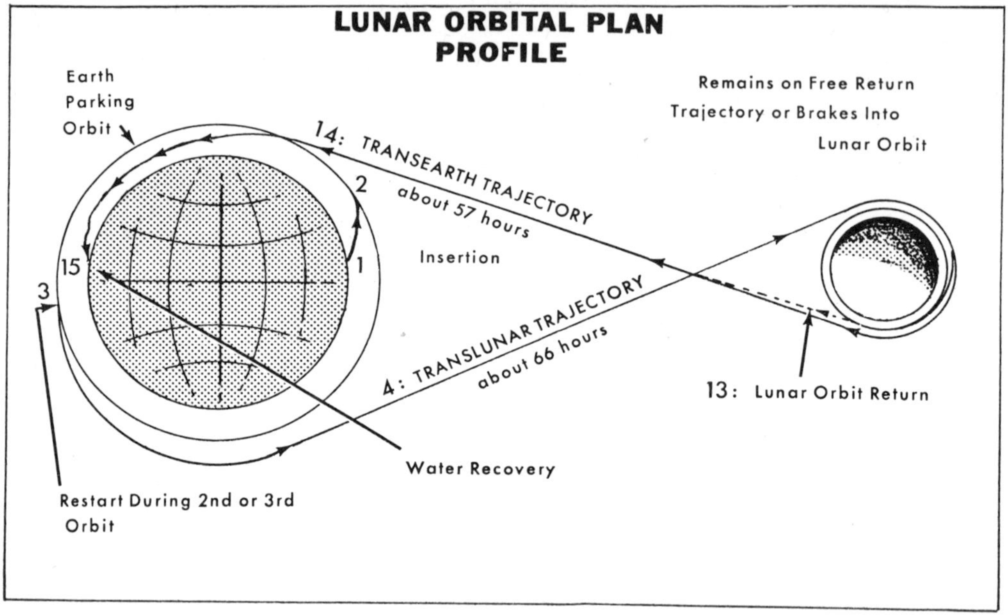

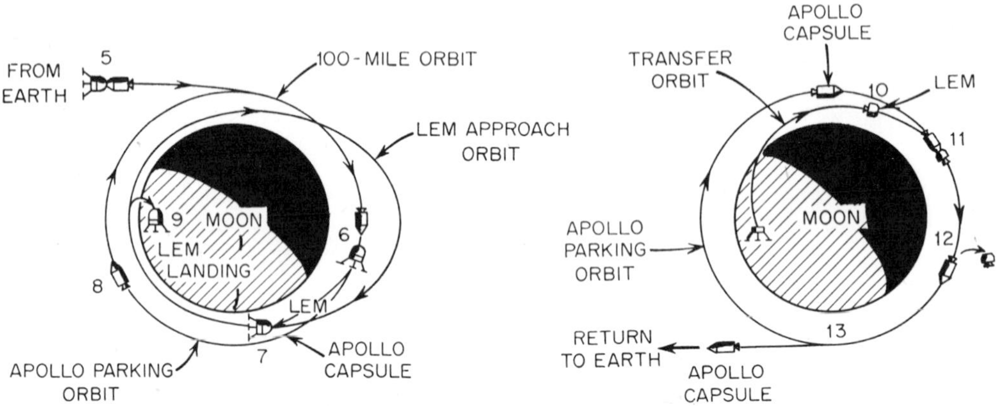

APOLLO MANOEUVRES. *Top:* Service and Command Modules flight plan. *Middle:* Lunar Module descent, ascent and orbital rendezvous at (11). *Bottom:* Turn-round of Apollo prior to extracting Lunar Module, and Lunar Sampling.

from its housing at the top of the third stage. The Apollo spacecraft is then in the configuration in which it approaches the Moon – with the complete LEM attached to the apex of the Command module. When the spacecraft is far enough away from the third stage, the third stage is fired for the last time to place it in a trajectory which ensures that it will be far enough away never to be a collision hazard later in the mission. If trajectory modifications are required on the way to the Moon these are carried out either using the Service module manoeuvring thrusters if the velocity required is small, or using the main Service module propulsion system if larger corrections are necessary. If the original injection has been accurate enough planned trajectory correction manoeuvres may not be required and can be cancelled as they were on the Apollo 8 lunar orbit mission. On the approach to the Moon, the spacecraft systems are given another thorough check, and if all is well preparations are made to fire the Service propulsion system to slow the spacecraft into lunar orbit. A subsequent firing alters the orbit from an ellipse to nearly a circle, and if all is still going well two of the astronauts enter the LEM through the hatches in both vehicles which form the docking tunnel and separate the LEM from the Command module. After satisfying themselves that all is in order they orientate the LEM in the correct direction and fire the descent motor to bring them out of orbit and slow down the LEM. This allows the LEM to fall out of orbit towards the Moon along the planned descent trajectory. Further firing of the descent motor reduces the velocity to zero just above the surface, and enough propellent is carried to permit a short hover to find the smoothest spot in the area for the actual touch-down. Feeler rods projecting from the legs of the LEM control the last few feet of the descent once one of them has touched the ground.

After the stay on the surface, during which the work described under **Apollo Summary** in the **Appendix** is carried out, the ascent stage is prepared for lift-off, and when the Command module is at the correct range and angle above the horizon it is fired. After rendezvous and transfer of lunar samples and any other impedimenta together with the two lunar explorers, the LEM is cast adrift and preparations are made for the Service propulsion firing (which has to take place

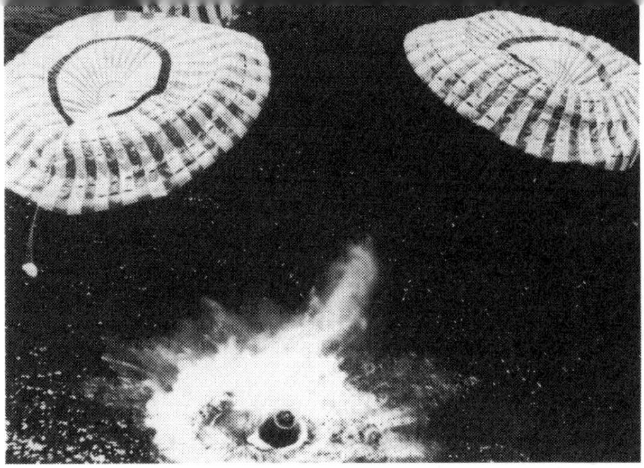

APOLLO 15 SPLASHDOWN. The third of the 88-foot parachutes failed to deploy properly. A small drogue is visible below the parachute on the left. For the first time the returning crew did not have to go into quarantine.

behind the Moon and out of sight of Earth stations) for trans-Earth injection, i.e. for setting the Command and Service modules on course for Earth. This trajectory has to be very exact indeed because the re-entry 'corridor' is only some seven miles wide. Shortly before re-entry the Service module is jettisoned, and the remaining Command module is turned base (heat shield) foremost for re-entry. If the re-entry angle is too steep (or the trajectory misses the corridor on the side nearer the Earth) the heat pulse will be too great for the heat shield to cope, and the drag forces would be so great that the astronauts would probably not survive anyway, and if the angle is too shallow the spacecraft will 'skip' out of the atmosphere and take longer to return than the limited life support systems available without the Service module. In practice this sounds worse than it is, as there are almost three days to ensure that the trajectory is correct, and a number of tracking stations together with observations made on board the spacecraft help to ensure that sufficient data are available to compute the path with quite adequate accuracy. The mission ends with splash-down in an ocean recovery area and lifting of the Command module on board the recovery aircraft carrier. The astronauts may be lifted on to the carrier separately from the module by one of the recovery force of helicopters.

Much of the hardware developed for the initial lunar landing is designed to serve also for later lunar missions and for the Apollo Applications Programme, which is intended to use the Saturn 5 launch vehicle's capabilities in heavy-weight Earth orbital missions. Part of this programme is devoted to the development of an **orbiting astronomical observatory.**

The chief advantage of using a separate lunar module instead of taking the entire spacecraft down to the Moon is that only a relatively light structure has to be braked on descent and lifted back into lunar orbit; all the fuel required for the return to Earth, and much other weight besides, remains in orbit throughout. Each phase of the programme leading up to *Apollo 11* was carefully rehearsed by *Apollo 8* (circum-lunar orbit), *9* (L.M. manoeuvres in Earth orbit), and *10* (L.M. descent to 9 miles from the Moon). For summaries of the six successful landings, please see **Appendix.** (C.W.M.)

δ–AQUARIDS. This **meteor** shower is best seen in the southern hemisphere, and is active for several days, during which the **radiant** is found to be in motion, with a maximum on May 5. The relation with **Halley's Comet** is doubtful, and radio-echo measurements give an orbit of period 11 years.

η–AQUARIDS. Best seen in the southern hemisphere, this **meteor** shower also occurs in historical records. The **radiant** is diffuse and has a daily motion of about 1°. The orbit is surprisingly small but very eccentric.

ARABIA. An ochre-coloured 'desert' region on **Mars.**

ARCTIC CIRCLE. The parallel of latitude 66° 33′ North. It is the limit of the area in the northern hemisphere within which the Sun does not set in mid-summer, and encloses the *Arctic.*

ARGON. A colourless, odourless gas of the **Helium** group which does not react chemically under ordinary conditions. It forms 0·94 % of normal air.

ARTIFICIAL SATELLITE. A man-made space vehicle placed into an **orbit** about one or more members of the solar system. An artificial *Earth satellite* revolves around the Earth as its primary, but it may also embrace the Moon in its orbit and gravitate – subject to many serious perturbations – about the common centre of gravity of the Earth and Moon; this centre lies within the Earth. A *Sun satellite* moves round the Sun and therefore constitutes an *artificial planet.* An artificial satellite may be simply the object of observations from the Earth (*passive*) or it may record or transmit observations made by its instruments (*active*).

This article begins with a discussion of the general theory and uses of artificial satellites, and continues with a description of individual vehicles and a comparative table.

GENERAL THEORY: LAUNCHING PROBLEMS. Let us assume that it is intended to launch a satellite into its orbit round the Earth at 300 miles altitude. It is desired to have perigee, the closest approach to the Earth, not less than 200 miles above the ground. This could be done, for example, with an orbit having a semi-major axis of 4,500 miles. In such a case, if perigee were at 200 miles, then apogee, or the point of farthest recession from the Earth, would be at 800 miles.

ORBITS PRODUCED BY PROJECTING A SATELLITE PRECISELY HORIZONTALLY

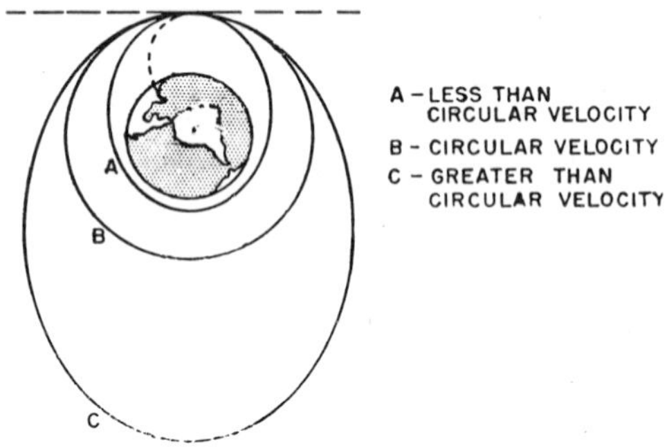

The necessary speed for entry into this orbit is:

$$v_o = 25{,}500 \text{ ft/sec.}$$

The circular velocity corresponding to 300 miles altitude is:

$$v_c = 25{,}000 \text{ ft/sec.}$$

Simply launching the satellite at the given speed will not ensure, however, that the perigee point will be at or above 200 miles. This must be achieved by careful guidance of the satellite so that the initial flight direction does not depart too greatly from the horizontal. If this angle with the horizontal is less

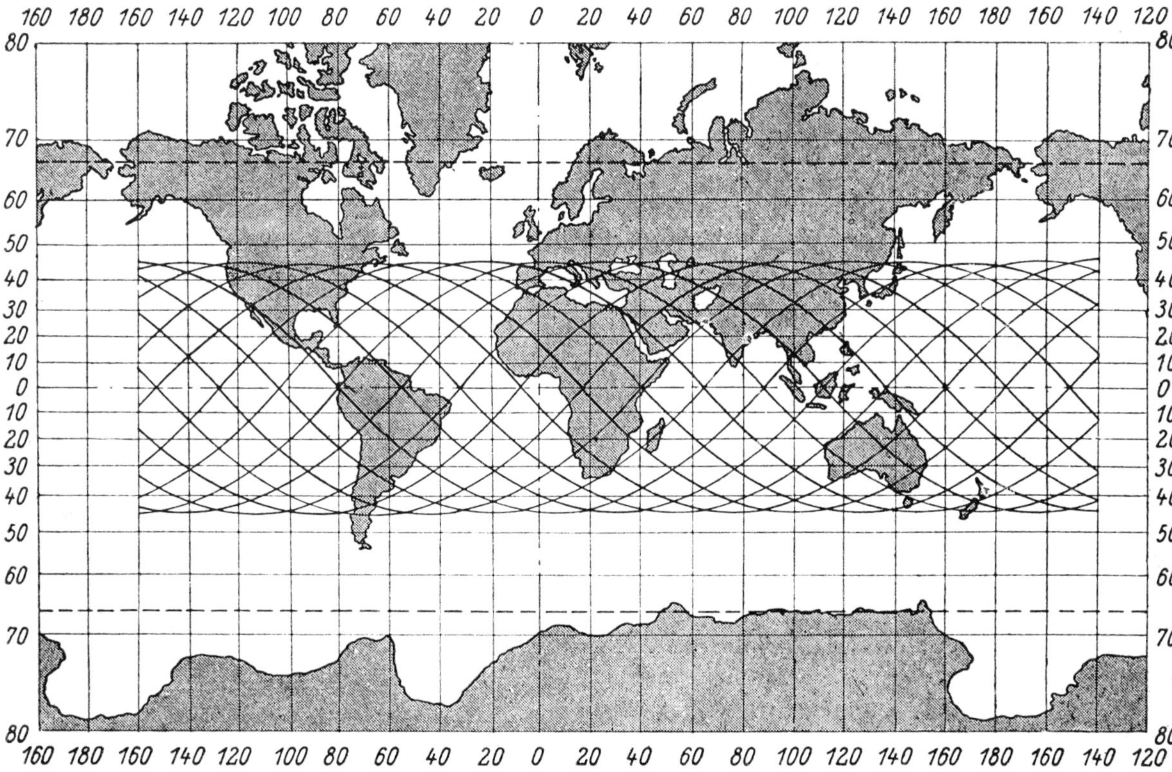

THE PATH OF A SATELLITE in a circular orbit at a height of 346 miles, with the orbit inclined at 45° to the equator. While the Earth turns once round its axis, the satellite completes 15 circuits. At each return its path lies farther to the West because the Earth has turned 24° to the East under it. The limits of visibility are shown by the broken lines in latitudes 67° North and South. At least twice each day the satellite is above the horizon of any point between these limits. The launching point may be anywhere between latitudes 45° South and 45° North.

than 2·9°, then perigee will be higher than 200 miles. If it is zero perigee would, of course, be the launching point itself, hence at a height of 300 miles.

The guidance requirements would be even more stringent if the initial projection were made at a lower speed. For example, projecting at the circular speed of 25,000 feet per second, a perigee height of 200 miles or greater will be obtained only if the entry into the orbit is made at an angle within 1·3° of the horizontal. Similarly the guidance requirements would be eased if the initial projection were made at more than the 25,500 feet per second considered.

If we assume for the moment that the Earth is spherical, then the plane of the satellite's orbit is fixed in space and, once established, does not change. The same is true of the orbit itself. Both are unaffected by the rotation of the Earth. But the track of the satellite over the ground is not necessarily fixed, where by 'track' is meant the path followed by the point in which the line from the centre of the Earth to the satellite intersects the surface of the Earth. Since the orbital plane contains the centre, at any instant of time it intersects the Earth's surface in a great circle. But the eastward rotation of the Earth causes the points on this intersection to slide westward on the ground. Only in the case in which the track is the equator is there no change in the position of the track on the ground. In this equatorial case the principal effect of the Earth's rotation is to alter the apparent period of revolution of the satellite. In the very special case in which the satellite revolves eastward in the equatorial plane at 22,000 miles, it remains constantly above the same spot on the equator, since its period of revolution, being one day, coincides with that of the Earth's rotation.

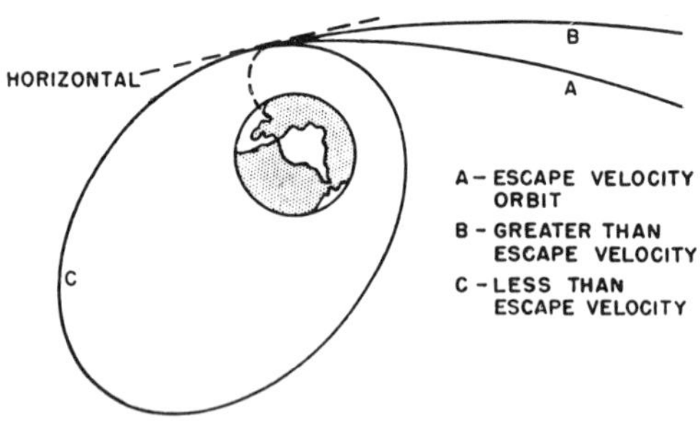

When the satellite's orbital plane is inclined to the equator, the track on the ground weaves back and forth between a maximum northern latitude and an equal maximum southern latitude. In general the track in time pretty thoroughly criss-crosses the orbital belt between the northern and southern latitude extremes, although there are special effects when the period of the satellite bears some simple relation to one day. The reader may be interested, for example, to consider what the track would be if the satellite's period were exactly one day, corresponding to a circular orbit at 22,000 miles altitude; or 1/12 day, corresponding to a circular orbit at about 1,000 miles altitude.

For a polar orbit the satellite will pass over each pole once each revolution. In general its track will criss-cross the entire Earth, although again there are special cases which the reader may wish to consider.

If the satellite were launched at a point of latitude L, then its orbit would have an inclination to the equator of at least L degrees. The inclination would be exactly L degrees if the vehicle were launched either due east or due west. When fired either north or south of east or west, the satellite would follow an orbit inclined at more than L degrees to the equator. Thus an equatorial orbit could be obtained only if the launching were made directly above the equator, and then only by launching due east or due west.

EARTH SATELLITES – THE ACTUAL EARTH. The discussion of the preceding section was based upon the assumption that the Earth is a perfect sphere. Many of the results so obtained are adequate both qualitatively and quantitatively to define the problems involved in launching an artificial satellite and in placing ground stations for observing it. But there are some important effects which are missed in the approximate treatment.

The Earth is *not* a perfect sphere (see **Hayford Spheroid**). The polar radius is some 13 miles shorter than the radius at the equator. This equatorial bulge shows up in a modification of the gravitational field of the Earth.

Qualitatively the effect of the bulge can be thought of as follows. As the satellite approaches the equator in its motion around the orbit, the excess mass in the bulge pulls the satellite out of its orbital plane. If the satellite revolves with an eastward component,

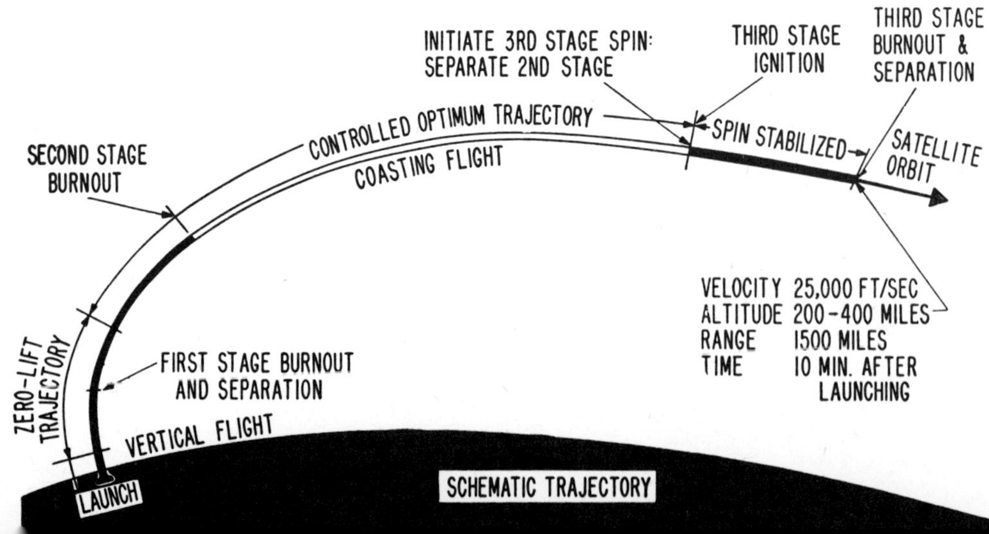

SCHEMATIC TRAJECTORY

THE FIRST ARTIFICIAL SATELLITE SPUTNIK I separating from the nose of its rocket carrier preceded by its conical cover. The four transmitting aerials are partially unfolded. The thin spherical shell is highly polished so as to reflect heat and offer the minimum resistance to the outer atmosphere.

the effect is to cause it to cross the equator to the west of where it would otherwise have crossed. The overall result is that the orbital plane revolves slowly in space, and the equatorial crossings, or nodal points, slide westward around the equator.

For a satellite near the Earth revolving in an orbit with a very small inclination to the equator, the period of this regression of the **nodes,** as it is called, is about 44 days. This motion also causes the track of the satellite to slide westward over the ground, and adds to the much greater westward movement caused by the Earth's rotation. The period of the nodal regression increases with the secant of the orbital inclination to the equator, becoming infinite for a polar orbit.

In the case of a satellite revolving in a westerly direction, the nodal regression carries the equatorial crossing points eastward around the equator. The resulting motion of the satellite's track then subtracts from the westward motion due to the Earth's rotation. Except for this the westward case is equivalent to the eastward one.

ECHO inflatable satellite under test. *Echo 1* and *2* are passive communications satellites which relay by simple reflection radio speech, Morse, television, teleprinted photographs, etc., partly in a co-operative programme with the U.S.S.R.

Period of Revolution of a Satellite in a Circular Orbit about the Earth

Height above the Earth (Miles)	*Approximate Period*
200	90 minutes
1,000	2 hours
22,000	1 day
235,000	1 lunar month

A second effect of the equatorial bulge is to cause the positions of perigee and apogee to advance around the orbit. As the satellite approaches the equator, nearing perigee, the excess mass speeds up the satellite temporarily in its orbit, causing it to overshoot the original perigee point and to arrive at perigee a short while later. This motion of perigee and apogee will cause the height at which the satellite passes over a given region of the Earth to vary with time.

A third, but very much smaller, perturbation affects the motion of small satellites close to the Earth's surface. It arises from irregularities in the distribution of mass in the Earth's crust.

To ignore the atmosphere, as was done in the preceding section, is completely unrealistic. Although the air is exceedingly rarefied at the altitudes under discussion, the speed at which satellites move is so great that there is a noticeable drag effect for orbits with a low or moderate perigee. This causes a steady degeneration of the orbit, the satellite tending to move closer to the lower atmosphere. The drop in mean height represents a loss of potential energy which shows up as increased kinetic energy, i.e. during this regime the satellite actually speeds up. It is therefore not proper to say that the drag 'slows down' the satellite at this stage.

Eventually the orbit degenerates to the point where enough of the satellite's motion is through denser air to cause drag to have a dominating influence, and the vehicle begins to slow down. Energy is now taken up by tremendous aerodynamic heating which, in the absence of special re-entry techniques, burns and vaporizes the whole or part of the vehicle. The satellite disintegrates like an exploding meteor, lighting up the whole sky for a few moments from a height of some 65 miles.

SATELLITE TRACKING. For most applications it is essential to know the orbit of a satellite accurately enough to identify the satellite positively among a constantly changing space population of over a thousand satellites, rocket stages and components. Very high accuracy indeed is needed when studies are being made of the exact extent of perturbations to the orbit – as has been done in recent years in connection with studies of the shape of the Earth. Optical tracking using specially designed tracking telescopes is the

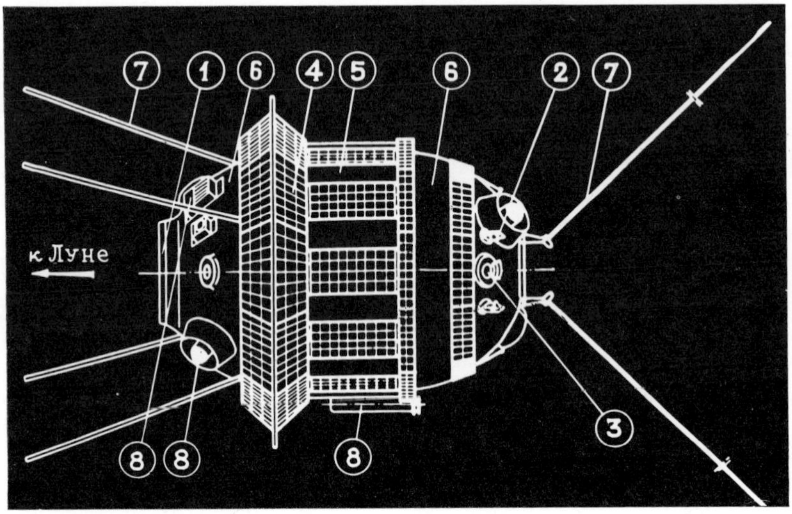

LUNIK III, also called *Cosmos III.* This is the vehicle which carried out a photo-reconnaissance of the Moon. (1) Camera lenses. (2) Motor of orientation system. (3) Solar monitor (sun-seeker). (4) Solar battery panels. (5) Shutters of temperature regulating system. (6) Part of heat screen. (7) Aerials.

It circumnavigated the Moon and photographed 70 % of the averted side. All equipment in this satellite was duplicated. If any part broke down, it could be replaced from the reserve equipment by a radio signal from the Russian ground stations.

On reaching a position 40,000 miles from the Moon on a line between the Moon and the Sun, gyroscopic apparatus stopped the satellite's spin. Then one end was directed towards the Sun by a **sun-seeker**. This roughly aligned the other end, containing the cameras, with the Moon. Another optical unit trained the cameras more precisely until a 'Moon in' signal triggered off the automatic photographic process.

The satellite then set itself spinning again to equalize temperature conditions. Shielded from cosmic rays as far as possible and under conditions of weightlessness, the films were developed, fixed and dried. When the vehicle had circumnavigated the Moon and was again fairly close to the Earth it began to transmit the pictures to the ground stations.

T.S. *Vandenberg*, one of several tracking stations used in the U.S. space programme. It can follow a number of spacecraft at the same time, relay transmissions from them, and issue commands to them in certain cases.

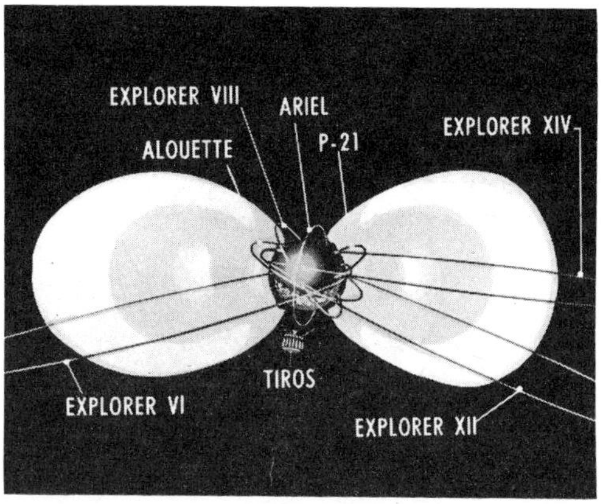

EIGHT SATELLITES TO EXPLORE THE VAN ALLEN BELTS and their orbits. The shaded area is the cross-section through the inner belt. Tiros is primarily a meteorological satellite; Alouette was built in Canada, and the instruments of Ariel were made in the United Kingdom.

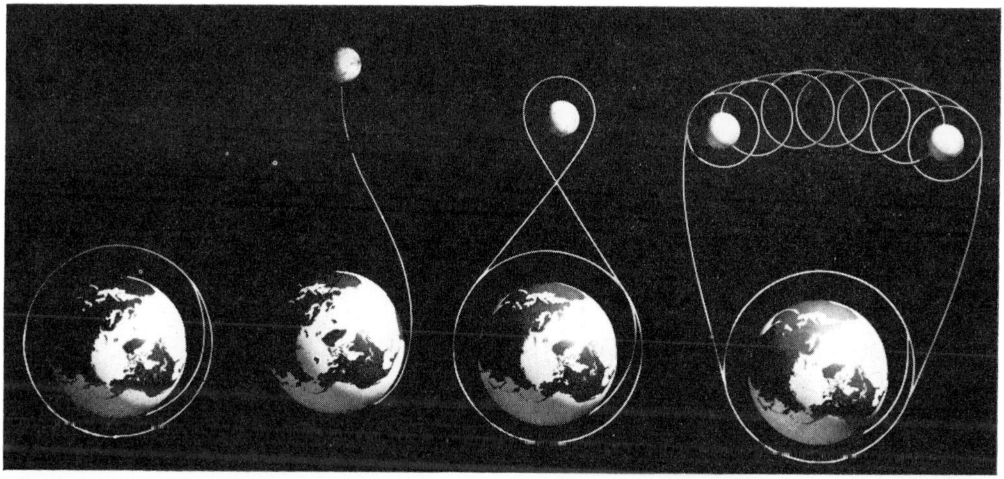

FOUR TYPES OF SATELLITE ORBIT. *The first* is a simple Earth satellite orbit viewed from above the North Pole. As in the case of the others, the launching is towards the East to derive maximum benefit from the eastward rotation of the Earth. *The second* is the path of a vehicle which falls directly on the Moon. The dot break in the line indicates the *neutral point*, where the Earth's gravitational pull is exactly counterbalanced by that of the Moon. From the time of all-burnt up to reaching the neutral point, the rocket loses speed and its orbit curves towards the Earth; beyond the neutral point, it gathers speed and turns towards the Moon. It strikes the surface with a speed not less than the Moon's escape velocity, unless retro-rockets are fired to give a soft landing. Each of the other orbits passes through a neutral point of its own.

The third type allows circumnavigation of the Moon and return to Earth without the rocket firing after injection of the vehicle from Earth orbit into trans-lunar orbit. *The fourth* type is a possible one for a lunar orbiter. Rocket firings are needed for ascent into trans-lunar orbit, injection into circum-lunar orbit and injection into trans-Earth orbit, with possible mid-course corrections. If it began with a 'figure of eight' it would be more economical with power. As the Moon circles the Earth, it carries the orbiter with it.

SYNCOM INSTRUMENTATION. The *Syncom* series of active communications satellites was the first to overcome the great difficulties of achieving a synchronous orbit, i.e. placing a satellite at 22,000 m. height in the plane of the equator so that it remains above a fixed point on the Earth. *Syncom 1* failed to transmit after injection into its orbit; *Syncom 2* and *3* were moved several times after placing in orbit – one is now above the Indian Ocean, the other above the Date Line. The drawing below shows how three Syncom satellites in synchronous orbits can provide virtually global coverage if they are stationed above, for example, the mid-Pacific, the Caribbean and the Indian Ocean.

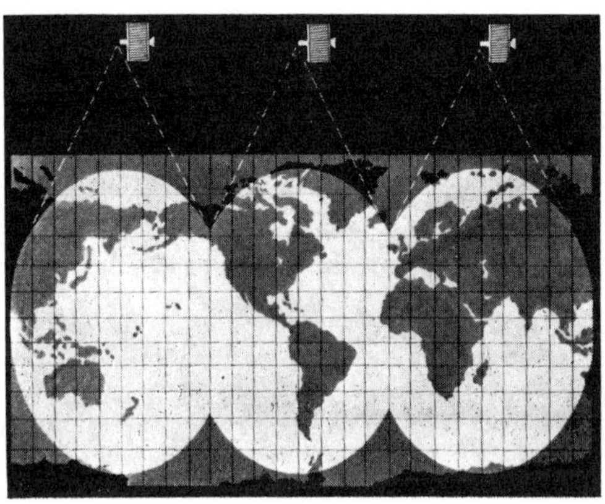

most accurate way of determining the direction of a satellite, and if sufficient observations are made it is also the most accurate way of determining orbits. However, the most accurate tracking telescopes use a photographic recording process which makes rapid results difficult to obtain. Furthermore, optical tracking is only possible for those parts of the orbit where the satellite is in sunlight and the telescope in darkness – or when the telescope is in darkness and the satellite carries an optical beacon.

Radar tracking is often used where the highest accuracy in orbit determination is not essential, and the orbit can be defined at any time the satellite happens to be above the radar site's horizon – a situation which occurs much more frequently than the combination of weather and required position for optical tracking. For satellites in near-Earth orbits either *skin-tracking* (tracking the echo from the surface of a body which does not itself transmit) or *beacon tracking* can be used. Beacon tracking will provide the most accurate results for a given installation cost, and if the spacecraft carries a transmitter of accurately known (and stable) frequency, measurement of the **Doppler** shift permits accurate measurement of the range rate. For satellites in high orbits many thousands of miles above the Earth, the power required for skin-tracking radars becomes enormous and only beacon tracking is normally feasible.

SATELLITE APPLICATIONS. One or two satellite applications have already been hinted at, and communications satellites have been advocated since 1945. Several fields have only become apparent since the technology required to exploit them has become available. Satellites are also used for more fundamental research into the nature of the Earth's environment, and have been responsible for many recent discoveries of far-reaching importance to the understanding of the Sun and solar system.

Applications for which satellites have been used in recent years include (apart from pure research into the space environment and development of hardware to exploit it) communications and navigation, meteorology, military reconnaissance, survey of the Earth's agricultural and mineral resources (a growing field) and geodetic survey.

Satellite communications is now so taken for granted that it is difficult to remember that

PIONEER IV. This satellite was intended as a Moon probe. Like its Russian counterpart, *Lunik I*, it did not quite achieve the planned trajectory and became one of the first artificial planets. It is still orbiting the Sun, as is the final stage of its rocket. The fibreglass outer cone is gold-washed to provide conductivity. The striped pattern is painted on to regulate temperature by controlling reflection and absorption of solar heat.

ATS 2 (*Applications Technology Satellite*) has tested the 'gravity gradient' method of stabilizing satellites. Weights are fixed at the ends of tubular booms 123 feet long. Since the lower weights are fractionally closer to the Earth, gravity acts more strongly on them, and this tends to keep the vehicle 'upright' with its television cameras constantly pointing at the Earth. The stabilizing force is extremely weak, but involves no expenditure of energy.

without it trans-oceanic television is virtually impossible. However, a variety of systems have been shown to be feasible. Systems employing a satellite *repeater*, which receives signals from the ground stations, amplifies them and re-transmits them to the ground on a different frequency, are universally used. Experiments have been carried out in bouncing signals off large balloon-type reflectors and other space 'junk' with a large reflecting surface, but although the satellite has no active parts and is therefore very reliable, it must be in a fairly low orbit to keep the total path length short. If this were not done, excessive transmitter power would be required to produce a measurable signal at the distant ground station. Even in fairly low orbits, the bandwidth or information-carrying capacity of such a system is limited, a large number of satellites are required to give anything approaching continuous availability on any specified transmission path, and the transmitter power necessary for it makes ground stations expensive to instal and to operate. More ground stations would be required (because of the bandwidth limitation) than would be the case if repeater-type satellites were used in the same orbits.

It so happens that most of the heavy communications traffic in the Western world is concentrated into a latitude band between 65° N. and 65° S. This is fortunate, because it means that the whole of the area of interest is potentially visible from a satellite in a sufficiently high orbit in the plane of the Earth's equator. If the period of the orbit is

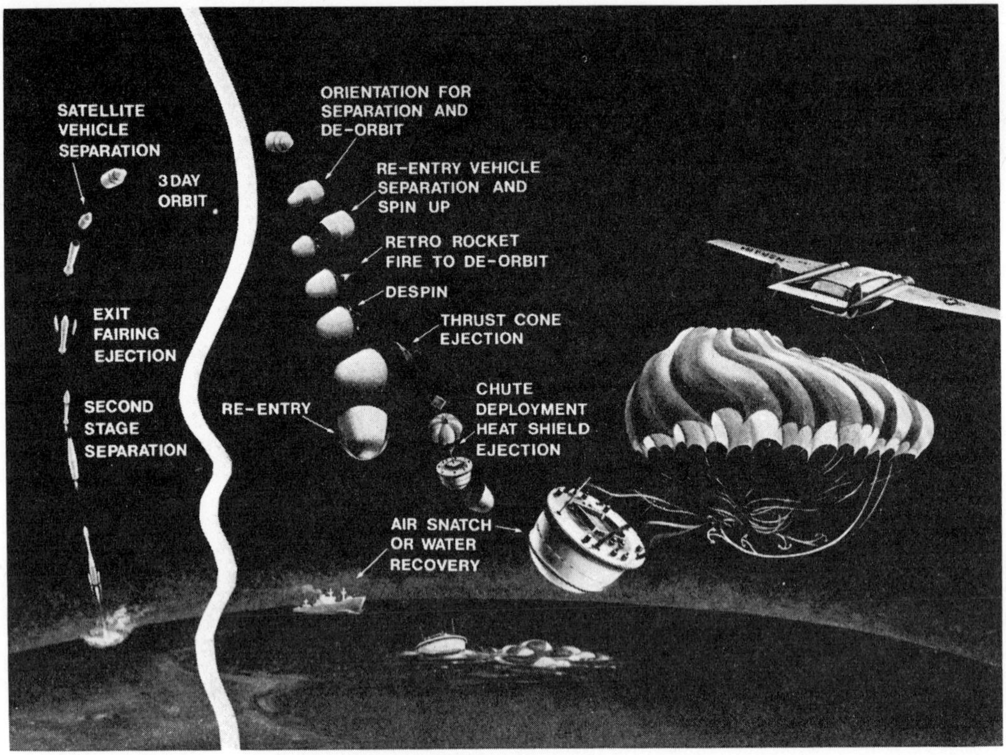

TYPICAL FLIGHT PLAN of an unmanned research satellite whose payload must be recovered intact – such as a biosatellite used in the study of the effects of weightlessness, radiation, acceleration, etc., on living organisms from viruses and algae upwards. The air snatch can be executed with the aid of a long wire with a hook or loop which is trailed by an aeroplane as it flies in an arc past the parachute.

one sidereal day, the satellite will (subject to very minor perturbations) remain fixed above the same point on the equator. This so-called *synchronous* orbit has the added advantage that, once the required satellite positions are decided on, only very small movement of ground station antennas is required to follow the satellite. Expensive high-speed tracking systems capable of handling large and varying Doppler frequency shifts are therefore unnecessary. The feasibility of satellite communications using repeater-type satellites in fairly low orbits was demonstrated as long ago as 1961 in the *Telstar* experiments, but after it was shown with *Syncom* 2 (the first successful synchronous satellite orbited in July 1962) that the transmission delay imposed on telephony by the distance to synchronous altitude (22,000-odd miles) and back would be acceptable, all subsequent commercial American communications satellites have been in synchronous orbits, and developments have been concentrated on more elaborate and more powerful (and heavier) satellite payloads which will permit access by a number of ground stations at the same time.

Navigation satellites were initially developed for military purposes, primarily to enable missile-firing submarines to monitor their inertial navigation systems for long-term accuracy (see **Inertial Guidance**). They are now used for more general navigation purposes. In its simplest form a navigation satellite in an accurately known orbit need only transmit a signal on an accurately controlled frequency. The user listens to the signal and records the Doppler frequency shift due to the motion of the satellite. If the orbital path and the satellite's position along it are known by reference to the satellite's orbital elements, two observations give a calculable position for the observer.

Weather satellites have done much to fill the gaps in the Earth-based network of stations collecting meteorological data for weather forecasting. Until April 1960, when the first Tiros weather satellite was flown, vast areas of the surface of the Earth were unobserved. Cloud cover photographs from space compensated for many of these gaps and permitted the whole life of such phenomena as hurricanes to be observed. Infra-red observations permit indirect measurement of cloud and surface temperatures, but direct measurement of such things as wind speed, rainfall, etc., from space is difficult. However, if automatic recording stations could be sited in remote areas there is no important reason from the purely engineering point of view why these stations should not be interrogated periodically by a passing satellite and the readings transmitted to Earth when the satellite is above the horizon of a suitable ground station.

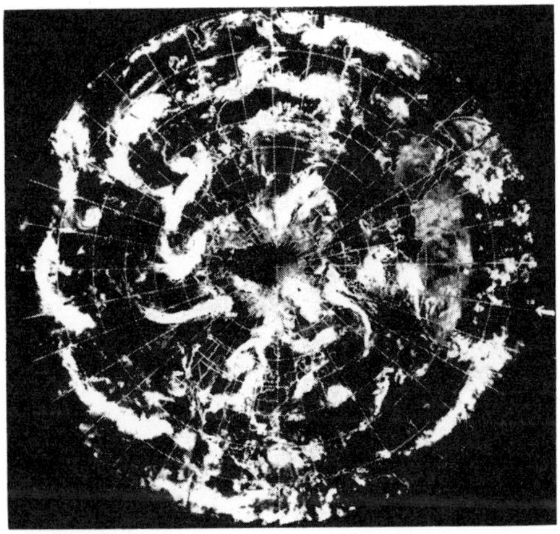

CLOUD COVER in the Northern Hemisphere as recorded by an ESSA weather satellite on successive passes in a near-circumpolar orbit. The geographic North Pole is at the centre and the Greenwich Meridian is indicated by the white arrow.

Reconnaissance satellites became desirable after the celebrated U2 affair, and were probably the first purely military spacecraft. In this category come satellites intended to detect radiation from nuclear explosions and the heat from large rocket launches as well as those with photographic payloads. There is little doubt that many of the techniques developed for military observation of the Earth have been used subsequently for purely peaceful purposes.

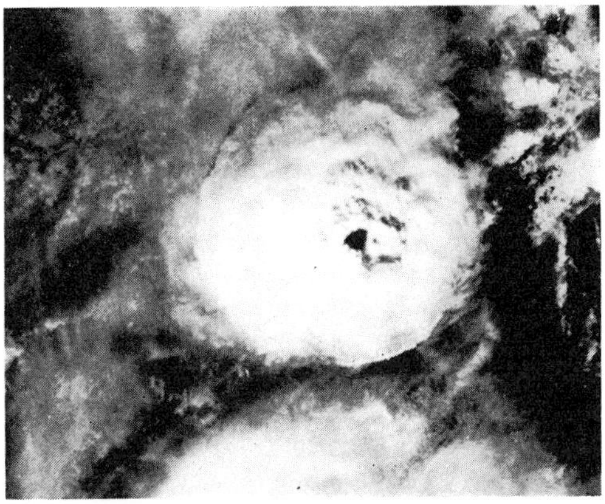

THUNDERHEAD OVER SOUTH AMERICA. A nearly vertical view from Apollo 9 in Earth orbit.

Earth Resources Survey is the area to benefit most from techniques developed for military reconnaissance, and satellites have been orbited which can assess the agricultural and forest potential of various areas on Earth, can detect plant diseases often before they are apparent to farmers on Earth, can spot geological formations likely to contain minerals, and measure temperatures of ocean areas. In theory at least this capability could lead to massive exploitation of the mineral resources of small countries by oil and mining companies with exclusive access to information from space, but NASA has shown no tendency to limit distribution of the information for commercial reasons. Perhaps even more interesting is the fact that the President of the United States and the Kremlin are in a better position to assess the prospects for the Chinese rice crop than Chairman Mao! The photographs under **Earth** and the colour plate facing page 105 illustrate some of these points.

A few random examples will illustrate the diversity of the practical applications of satellites: geologists were able to construct geological maps of the Arabian peninsula within a few hours of receiving satellite photographs; hurricane Beulah was located six days before it reached the Caribbean; the disparity in the shrimp population between the lower and upper Florida Bay (of interest to the fishing industry) was accounted for by the detection of submarine eddy currents in the bay; ocean currents generally and the flow of water supplies from melting snows can be studied in new ways with the aid of satellites.

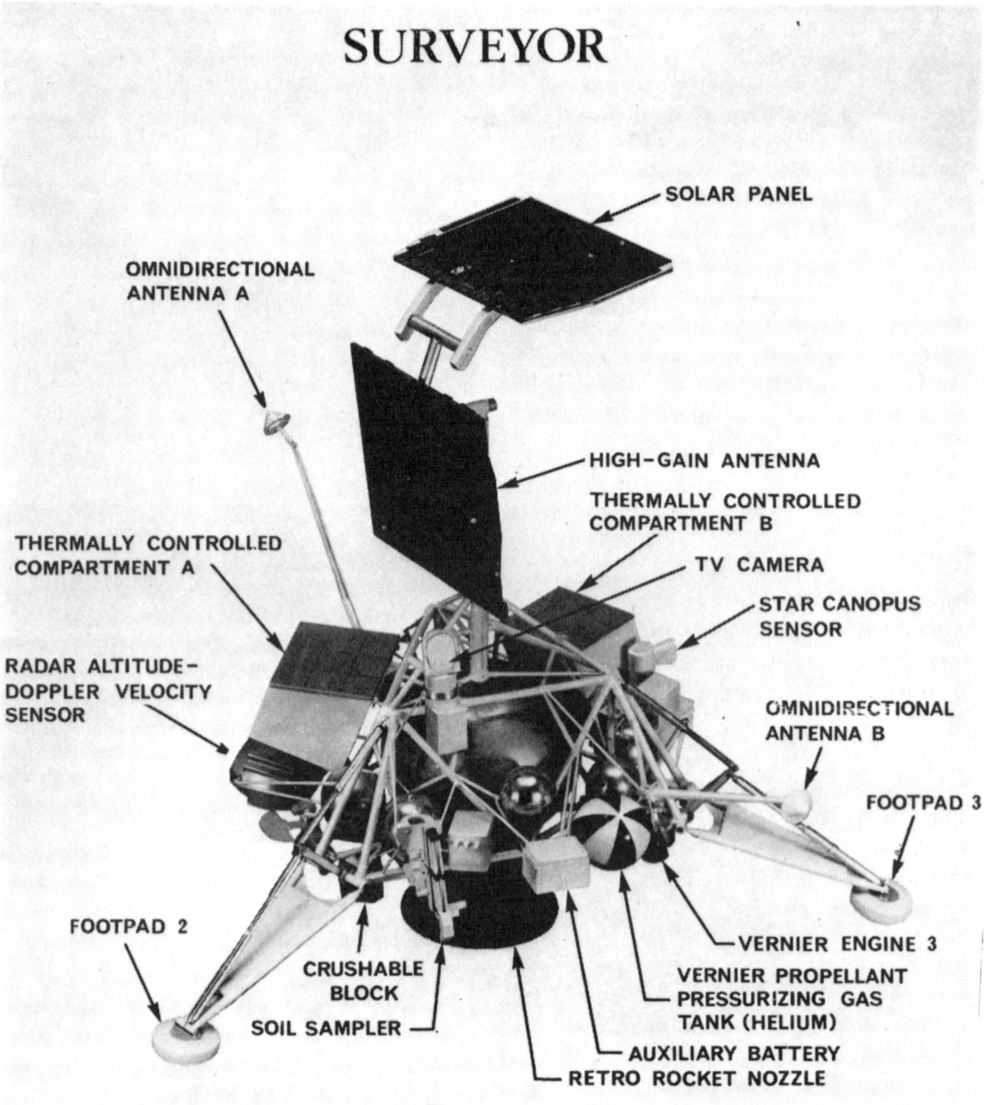

A TYPICAL SURVEYOR DESIGN. Over thirty satellites of the very successful *Surveyor* series have carried out a large variety of research missions. The type illustrated here soft-landed on the Moon.

APPLICATIONS TO PURE SCIENCE. Many scientific enquiries concerning the upper atmosphere and interplanetary space could until recently be pursued only by indirect methods and inference; others could not be carried on from the Earth at all because of the blanketing effect of the atmosphere or, occasionally, because they required the conditions of **free fall** which even in rockets are attained only for a few minutes. Vertical sounding rockets afford no more than a brief glimpse of conditions at any level, while artificial satellites provide a relatively permanent platform for continuous observations over long periods. This opportunity has now been applied to research on the following topics:

DATA	Sputnik I 1957 α U.S.S.R.	Sputnik II 1957 β U.S.S.R.	Explorer I 1958 α U.S.A.	Vanguard I 1958 β U.S.A.	Intelsat 3C 1969–11A U.S.A.
LAUNCH DATE	1957, October 4	1957, November 3	1958, February 1	1958, March 17	1969, February 6
TYPE	Earth satellite	Earth satellite	Earth satellite	Earth satellite	Earth satellite
ORBIT					
initial period (*min.*)	96·2	103·7	114·9	134·2	1,436
inclination	64·9°	65·3°	33·5°	34·25°	0·7°
eccentricity	0·06	0·105	0·14	0·20	0·0002
max. height (*km.*)	939	1,660	2,548	3,968	35,790
min. height (*km.*)	215	212	356	650	35,770
LIFETIME	92 days	5½ months	11 years	300 years	> million years
WEIGHT (*kg.*)					
total	86·3	4,500?	13·9	28	146
instruments		508	4·8	1·5	
SHAPE	Sphere	Cone and cylinder	Long cylinder	Sphere	Cylinder
INSTRUMENTATION	Two transmitters, temperature and pressure meters, magnetometer, radiation counter. batteries	Transmitters, three photomultipliers for spectroscopic studies, cosmic ray counters, biological equipment	Two transmitters, cosmic ray counters, temperature and micrometeoritegauges	Solar- and battery-powered transmitters, temperature meter	
PRIMARY AIMS	Preliminary investigations. (Believed to be tenth launching attempt)	Study of effect of environment on dog passenger	See instruments. Discovered van Allen belts. (Believed second launching attempt)	Radio-tracking and air-density measurements	Commercial communications
LAUNCH VEHICLE	Three-stage (T-4?)	Three-stage	Jupiter-C	Vanguard	Improved Delta

FIVE EARTH SATELLITES

Five Lunar Probes

Data	Lunik II 1959 ξ 1 U.S.S.R.	Lunik III 1959 θ 1 U.S.S.R.	Surveyor 1 1966–45A U.S.A.	Lunar Orbiter 5 1967–75A U.S.A.	Surveyor 6 1967–112A U.S.A.
Launch Date	1959, September 12	1959, October 4	1966, May 30	1967, August 1	1967, November 7
Type	Moon Probe	Moon Probe	Moon Probe	Moon Probe	Moon Probe
Orbit					
initial period		15 days			
inclination	65°	75°			
eccentricity		0·82			
max. distance from Earth (km.)	400,000	483,000	400,000	400,000	400,000
Flight Time	34 h.	6 months	63·6 h.	90 h. to injection	65·4 h.
Weight (*kg.*)	390	278	281	390	300
Size (*m.*)	0·9 (sphere)	1·2 × 1·3	3·1 × 4·2	1·5 × 1·7	3 × 4·3
Launch Vehicle			Centaur	Agena D	Centaur
Mission	Hit Moon in 1°W., 30°N. followed by its rocket. Radiation and magnetic survey	Passed 6,200 km. behind Moon. Decayed in Earth's atmosphere March 1960. Photo reconnaissance	Soft-landed on Moon in 43·2°W., 2·45°S. Rocket passed 17,000 km. behind Moon	Orbited Moon at 196 to 6,050 km. height for several months, then sent crashing into Moon. Rocket in Earth orbit	Soft-landed November 10 in 0·49°N., 1·40°W. Made to hop November 17. Launch rocket in highly eccentric Earth orbit

DATA	Luna 1 1959 μ 1 U.S.S.R.	Mariner 4 1964–77A U.S.A.	Venus 4 1967–58A U.S.S.R.	Pioneer 8 1967–123A U.S.A.	Venus 5 and 6 1969–01A and 02A U.S.S.R.
Launch Date	1959, January 2	1964, November 28	1967, June 12	1967, December 13	1969, January 5 and 10
Type	Moon Probe	Mars Probe	Venus Probe	Space Probe	Venus Probes
Orbit					
period (days)	450	567	283	386	Orbital data similar to Venus 4
inclination to ecliptic	0·01°	2·54°	2°	0·6°	
eccentricity	0·148	0·173	0·22	0·047	
perihelion (A.U.)	0·978	1·109	0·65	0·99	
aphelion (A.U.)	1·318	1·574	1·02	1·087	
Weight (*kg.*)	362	261	383	66	
Size (*m.*)	0·9 (sphere)	1·4 × 0·5	1 (sphere)	0·94 × 0·9	
Launch Vehicle		Agena D		Altair	
Mission	Intended to strike Moon. Passed 6,000 km. ahead of Moon and entered orbit around Sun. First artificial planet	Passed 9,850 km. behind Mars on 1965, July 15. Photographed craters on Mars	Soft-landed on Venus near equator on 1967, October 18. Transmitted detailed information during descent through Venusian atmosphere	Reached aphelion 1968, June 23. Made observations for studies of Sun and interplanetary space	Both probes soft-landed within a day of each other and transmitted data which are being analysed. It is not yet certain whether any of the Venus probes transmitted down to ground level, whether they landed on a mountain, or whether lower Venusian atmosphere interferes with radio transmission

FIVE ARTIFICIAL PLANETS

Five Manned Satellites

DATA	Vostok 1 1961 μ 1 U.S.S.R.	Apollo 8 1968–118A U.S.A.	Soyuz 5 1969–05A U.S.S.R.	Apollo 11 1969–59A U.S.A.	Skylab 1973–27A U.S.A.
Launch Date	1961, April 12	1968, December 21	1969, January 15	1969, July 16	1973, May 14
Type	1-man Earth satellite	3-man lunar orbiter	3-man Earth satellite	manned lunar landing complex	3-man space laboratory
Orbit					
period (min.)	89·3	24,400	88·9	24,400	93·2
inclination	64·95°	30·7°	51·7°	33·3°	50·0°
eccentricity	0·011	0·976	0·002	0·976	0·001
max. height (km.)	315	533,000	233	533,000	439
min. height (km.)	169	174	210	174	427
Flight Time	1·8 h.	6·12 days	3·04 days	8·14 days	15 months
Weight (*kg.*)	4,725		6,000?	28,800	85,000?
Size (*m.*)	4·3 × 2·4	10·4 × 3·9	7·5 × 2·2	11·2 × 3·9	25·6 × 6·6
Launch Vehicle		Saturn 5		Saturn 5	Saturn 5 mod.
Mission	First manned space flight, recovery on land after one orbit. Physiological studies, tests of life-support systems, re-entry and recovery procedures	First manned lunar orbiter. Detailed reconnaissance of possible landing sites. Re-ignition of rocket on far side of Moon for return to Earth. Followed by Apollo 9 and 10, the latter testing lunar module close to Moon	Link-up and exchange of crew members with Soyuz 4. Command module detachable from work and rest compartment. Recovery at sea	First moon landing mission. Two men descend in lunar module and later rejoin command module which waits in selenocentric orbit. Collection of rock samples, reconnaissance of landing site in Sea of Tranquillity, seismic recordings.	Space laboratory manned by crews which arrive and leave in Apollo command modules. Programme incudes solar geophysical, physiological and zero gravity research. Equipped to sustain crews for several months

(1) The intensity of electromagnetic and corpuscular radiation from the Sun at different levels. Many probes have carried instruments (spectrographs, photomultipliers, electrostatic fluxmeters or ionization gauges and counters with varying shielding and filters) to measure these quantities. It has been established that the **ionosphere** is lower than was previously believed, and that its strata are not sharply defined and are subject to local, transient anomalies.

(2) **Cosmic rays.** Particles have been recorded with energies up to a billion billion electron volts, and the flux of heavy nuclei is found to be very low.

(3) Meteoric and interplanetary dust. Explorer I, Lunik I and later probes have carried **micrometeorite** gauges which measure the erosion of the surfaces of thin plates exposed to the impact of the dust by detecting changes in the conductivity of the plates. Piezo-electric pick-ups and microphones have been used to record the kinetic energies of micrometeorites striking their surfaces. It is assumed that Lunik III ceased to transmit upon being punctured by a **meteor.**

(4) The shape and extent of the Earth's magnetic field, and the **van Allen** radiation belts. The existence of these belts was first demonstrated by Pioneer I and Pioneer III; magnetometers flown in the lunar probes have helped to map them to a distance of over 90,000 miles from the Earth.

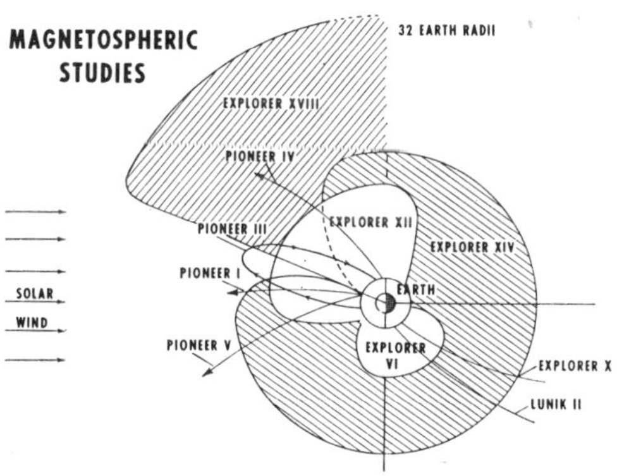

(5) Distribution of mass in the **Earth** and its equatorial bulge; checks on longitude determinations and the **Astronomical Unit.** Irregularities in the conformation of the Earth cause **perturbations** of satellite orbits.

PICK-UP STATION IN CHITA, U.S.S.R. whose chief function is to relay television programmes via the *Molniya I* satellite. A number of such stations are spread throughout Russia. The network can also be adapted for communications with research satellites and for tracking purposes.

(6) Conditions in space near the **Moon;** its magnetic field and the appearance of the 'far' side. *Lunik II* confirmed the belief that the Moon's magnetic field is extremely weak; *Lunik III* photographed the hidden side and relayed the pictures to Earth. *Ranger*, *Lunar Orbiter*, *Surveyor* and later manned *Apollo* space vehicles mapped the entire lunar surface photographically, examined the texture of the surface after soft landings, measured radiation and temperature on and near the Moon and outlined the gravitational anomalies due to heavy mass concentrations (*mascons*) below the surface.

(7) Conditions in interplanetary space generally and near Mars and Venus in particular. These have been studied by means of **planetary probes,** some of which can be regarded as artificial solar satellites, i.e. artificial planets.

(8) Effects of radiation and prolonged free fall in animals, particularly mammals including man.

(9) **Airglow, radiation pressure,** and thermal equilibrium of a vehicle in space.

(10) Time measurements comparing a clock in a satellite with one on the ground to test relativity predictions on **time dilatation.**

(11) Photography of other members of the solar system from space, where there is no atmospheric absorption and perfect **seeing.** The Earth's cloud cover.

(12) Orbital decay, **air resistance.**

This list is by no means exhaustive, and the art is still very young: not very many years have elapsed since Sputnik I was launched. The accompanying tables and illustration give a few representative examples of developments, during this period. At the time of writing, some twelve hundred artificial satellites have been launched, with lifetimes varying from a few days to millions of years; the number of satellites, rockets, space vehicles and pieces of debris in orbit round the Earth, the Earth-Moon system or the Sun is 2 946.

ASTEROID or *Minor Planet*. The many thousands of small bodies, revolving round the Sun mainly between the orbits of Mars and Jupiter, are known collectively as asteroids; alternative names are minor planets and planetoids. All are insignificant, the largest (**Ceres**) being less than 500 miles in diameter.

DISCOVERY. In 1772 Johann Bode drew attention to a curious numerical relationship between the distances of the planets, discovered some years earlier by Titius of Wittenberg (see **Bode's Law**), and concluded that there should be an extra planet revolving between Mars and Jupiter. In 1800 six German astronomers, headed by the famous lunar observer Schröter, determined to make a serious effort to track down the missing body; but before their scheme could be brought into working order, Piazzi, at Palermo, discovered a star-like object which proved to revolve at the correct distance from the Sun. It was named *Ceres.*

Olbers, a member of Schröter's 'celestial police', discovered a second small planet (*Pallas*) in 1802; *Juno* followed in 1804, and *Vesta* in 1807. The fifth member of the group, *Astraea,* was not discovered until 1845, but since then minor planets have been found in great numbers. Over 1,500 have now had their orbits computed, and the total number has been estimated as 44,000, though many of these must be too small and faint to be detected.

ORBITS. Most of the minor planets revolve at distances from the Sun intermediate between those of Mars and Jupiter. The average orbital eccentricity is 0·15, but some particular members exceed this greatly; examples are **Albert** (0·54) and **Hidalgo** (0·65). The inclinations also show a wide

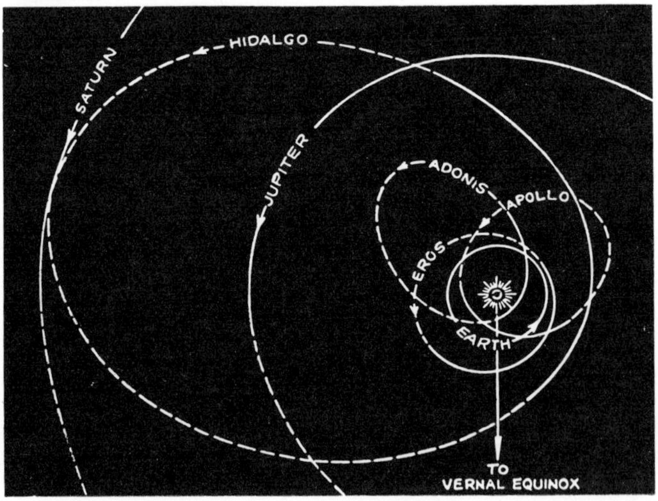

SOME UNUSUAL ASTEROID ORBITS. Most asteroids have nearly circular orbits and move in a belt between Mars and Jupiter, but a number of exceptions are known, and there may be many more. Although the planetary and asteroid paths appear to cross each other when drawn on a flat sheet of paper, collisions are impossible because the orbits are inclined at various angles and pass above or below one another. The broken lines indicate the parts of orbits which lie 'below' the plane in which the Earth moves.

range, from 43° (Hidalgo) down to almost nil. No minor planet with **retrograde motion** has been discovered.

In 1886, D. Kirkwood showed that minor planets are scarce in regions which would involve a period which is a simple fraction of Jupiter's. Jupiter causes, in fact, definite gaps in the main asteroid zone, known generally as the **Kirkwood Gaps.** A similar effect is seen with regard to Saturn's rings, due to the influence of the satellites of that planet (see **Saturn**).

Some exceptional asteroids have orbits which carry them well away from the main swarm. Some, the *Trojans*, revolve in the orbit of Jupiter; others have orbits which carry them even beyond – Hidalgo moves from the orbit of Mars out almost to that of Saturn – and yet others invade the inner regions of the solar system, one asteroid (**Icarus**) having a perihelion distance less than that of Mercury.

NAMES. The first asteroids to be discovered were dignified by mythological names (*Ceres*, *Hygeia*, *Metis*, *Psyche*, *Thalia* and others). Unfortunately the supply of goddesses soon ran out, and some of the more recent names are hardly in keeping. For instance, No. 674, discovered by a student at Drake University, is named 'Ekard' – 'Drake' spelled backwards. Asteroids with exceptional orbits are usually given male names.

DIMENSIONS. Ceres, with a diameter of 429 miles, is by far the largest of the minor

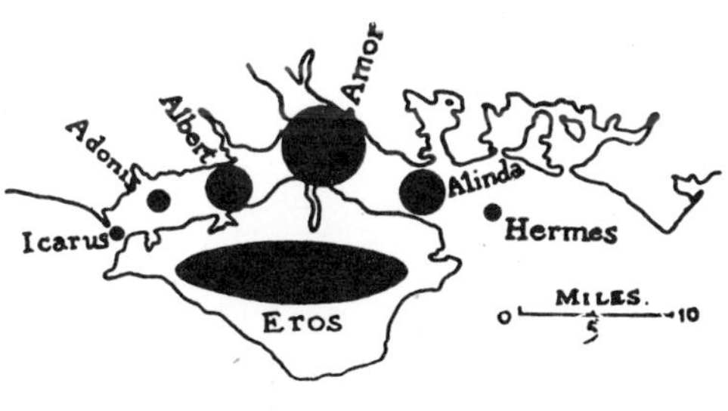

Some asteroids compared with the Isle of Wight. The exact size and shape of Eros are not known, but regular fluctuations in its brilliance make it seem almost certain that it is an irregular, roughly oblong body spinning slowly.
(*From 'Guide to Planets', by P. Moore*)

Asteroids are usually discovered from photographs such as this. During the exposure, while the telescope follows the stars, the asteroid moves against the background and leaves a short trail.

planets. Pallas is 304 miles across; no other exceeds 300 miles. The brightest of the group, by virtue of its high albedo, is Vesta, which can attain the sixth magnitude and can then be glimpsed without optical aid. All the asteroids put together would form a body with a mass less than 1 % of that of the Earth, and it is scarcely necessary to add that even Ceres can retain no trace of atmosphere.

THE TROJANS. No. 719, *Thule*, at a mean distance of almost 400 million miles from the Sun, was long regarded as the outermost minor planet; but in 1908 came the discovery of the first of the Trojans, which move in the same orbit as that of Jupiter. There are two Trojan groups, one 60° in front of Jupiter and the other 60° behind it. Thirteen Trojans are now known. The largest, *Achilles* and *Patroclus*, are over 150 miles in diameter, but their great distance makes them difficult to observe.

EARTH-GRAZERS. The so-called earth-grazing minor planets are those which have orbits which bring them well within the orbit of Mars, and thus close to the Earth. The most famous member of the group is **Eros,** No. 433, which was discovered by Witt in 1898. It is an irregularly-shaped body with a longest diameter of 15 miles, and in 1931 it passed within 17 million miles of the Earth, proving very useful as an aid to determining the value of the **astronomical unit.** The next close approach will be that of 1975.

Other earth-grazers are *Albert* (minimum distance from the Earth 20 million miles, diameter 3 miles); *Amor* (10 million miles, 10 miles); *Apollo* (7 million miles, 2 miles); *Adonis* ($1\frac{1}{2}$ million miles, 1 mile); and *Hermes* (200,000 miles, 1 mile). The closest approach on record is that of Hermes, in 1937, when the minimum distance was 400,000 miles – less than twice the distance of the Moon. Fortunately, the chances of our being hit by a minor planet are extremely small.

Owing to the small masses of the earth-grazers, Eros excepted, their orbits are difficult to compute accurately, and several of them have unfortunately been lost.

HIDALGO. This remarkable object was discovered by Baade in 1920. It has the greatest orbital eccentricity and inclination of any minor planet, and was at one time believed to be an unusual comet, though subsequent investigation has not borne out this view.

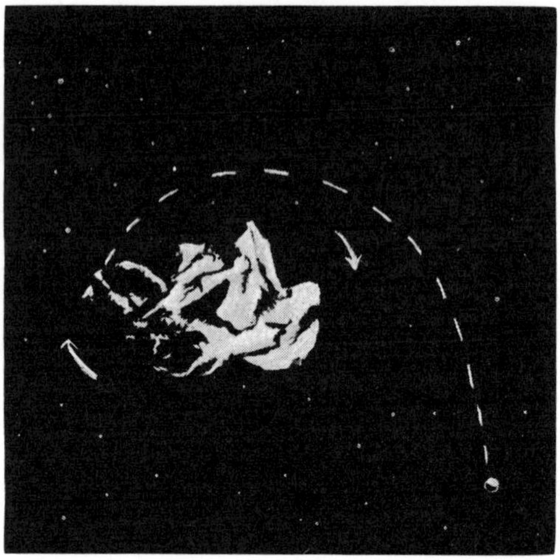

AN 'UNRIDABLE' ASTEROID. The masses of many minor planets are so small that, over most parts of their surfaces, gravitation is too weak to overcome the centrifugal force set up by their rotation. Any object placed loosely on such an asteroid would at once be gently but irrevocably flung into space and would drift away along a spiral path. Surface gravity on the asteroid ranges from slightly positive near its axis of spin to markedly *negative* on its 'equator'.

ICARUS. Icarus, discovered by Baade in 1948, may be regarded as an earth-grazer, since it can approach to within 4 million miles of our world. Its aphelion distance from the Sun is 183 million miles, so that it is then well beyond Mars; but at perihelion it is a mere 19 million miles from the Sun, closer even than Mercury. The sidereal period is 409 days. Icarus experiences great extremes of temperature, and at perihelion it must be red-hot. In 1968 it passed the Earth at 4 million miles.

ORIGIN OF THE MINOR PLANETS. Olbers suggested that the asteroids were caused by the disruption of a former major planet that used to revolve between the orbits of Mars and Jupiter. This remains a possibility, even though there is no clue as to the cause of the disaster. On the other hand, it is equally possible that the asteroids never formed part of a larger planet, and in this case they may be termed the débris of the solar system.

LANDING ON A MINOR PLANET. It is clear that the minor planets are airless, lifeless worldlets. Even if they are reached in future centuries, it is difficult to see how they can be put to much use, and they are among the least interesting bodies of the solar system. Only those with exceptional orbits, such as Icarus and Eros, are likely to engage the attention of either amateur or professional astronomers. (P.M.)

ASTRAEA. An **asteroid** with a diameter of about 60 miles.

ASTROLABE. An instrument for measuring the **altitude** of celestial bodies. In its ancient form, which goes back to the 3rd century B.C., it is a circular disc marked off in degrees along its rim and fitted with a movable arm carrying sights through which the star is viewed. When in use, it is held suspended in a vertical plane. It is now entirely replaced by the sextant, but its name has been adopted for a modern instrument which gives very exact measurements on firm ground.

ASTROLOGY. The ancient practice of divining the course of future events from observations of celestial bodies. The planets, certain stars and the signs of the Zodiac were each

GERMAN ASTROLABE, 16th century. Stars are viewed through the holes in the two end plates of the movable arm.

considered symbolic and capable of exerting influences towards war, anger, love, etc., and their conjunctions, oppositions and other configurations at the moment of birth of a person were believed to be of great importance in casting his *horoscope*.

As a means of prophecy, astrology is entirely worthless and without foundation. But through stimulating detailed observation of celestial motions from prehistoric times it became the parent of **astronomy**.

ASTRONAUTICS is the science of space flight.

ASTRONOMICAL UNIT. The mean distance of the Earth from the Sun. It is used as a unit for measuring and expressing distances within the solar system and among the nearer stars and is equal to about 149·6 million kilometres (93 million miles). Since the distances and motions of satellites, planets and stars are based upon the astronomical unit, it is most desirable that its value should be accurately known. Until recently the best results were obtained from **parallax** measurements made on those **asteroids** that pass within the Earth's orbit round the Sun. Much more accurate results have now been obtained by radar observations of Venus (see **Radar Astronomy** and **Distances**).

ASTRONOMY can claim with justice to be the oldest of the sciences. From the earliest times men have watched the stars and wondered what they were. Living, as they did, in small communities, many of them nomadic, they were much closer to Nature than we are today. There were for them no artificial lights to brighten the skies at night and to dim the stars; the town dwellers of today no longer see the beauty of the heavens, the stars in all their glory.

From very early times men knew the stars. They were their guides at night. They gave to many of them names; they divided them into groupings or constellations, named after animals, mythological deities, or common objects, as they thought they saw some resemblance, or as fancy took them. The names of the constellations in use today were mostly given about 2700 B.C. by people living in the region of Mesopotamia.

They imagined that the celestial bodies influenced their lives. The Sun, in its annual passage amongst the stars along the Zodiac, marked out the year with the alternation of the seasons and was clearly of great importance. The Moon, with the succession of its phases, was useful as the measure of a shorter interval of time; with the evolution of a priestly caste, it became general for religious festivals to be associated with the phases of the Moon. The Moon was believed to affect men in various ways and we have a survival of this belief, for instance, in the words 'lunacy' and 'lunatic'. It was not a great step further to suppose that all the other celestial bodies had some influence on human activities.

Misguided though they were in this belief, it prompted the observation of the celestial bodies, the recording of phenomena such as eclipses of the Sun and Moon, and of the times of New Moon. They found that the celestial bodies were of two types: there were the 'fixed' stars, whose positions relative to one another did not change in the course of their daily motion across the sky, nor from night to night, nor from year to year; there were also the 'moving' stars, whose positions relative to one another and to the fixed stars changed from night to night and from year

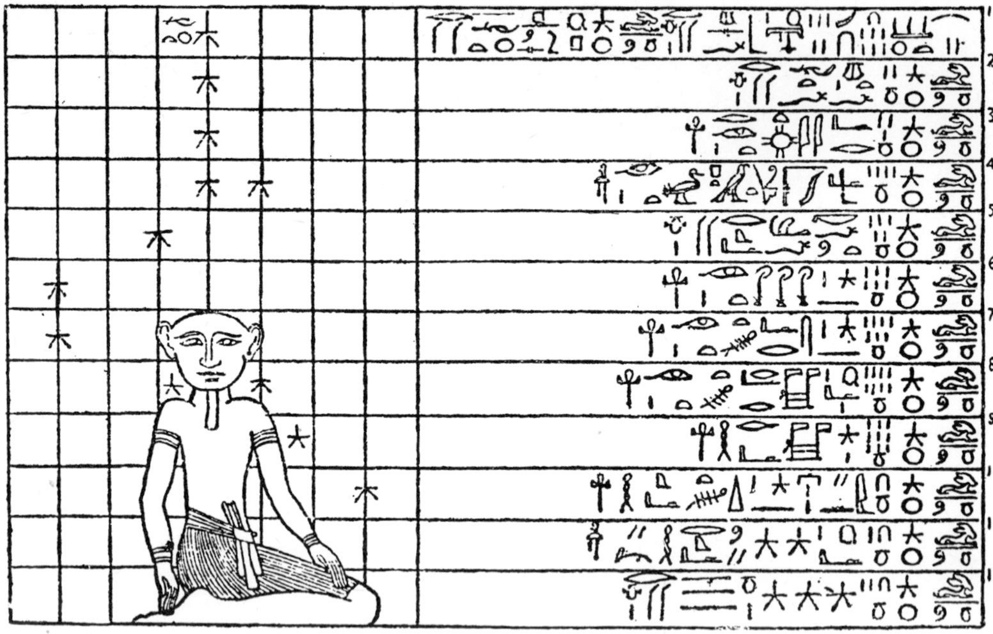

AN ANCIENT EGYPTIAN STAR CHART of about 1800 B.C. It lists the positions of stars as they appear on a certain date relative to the figure of the astronomer's assistant who is seated a few paces north of the astronomer himself.

The chart is for the 16th day of the month of Phaophi, when in the

1st	hour	the Leg of the Giant	(is) above the middle.
2nd	hour	the Heel of the Giant	above the middle.
3rd	hour	the star Arje	above the left eye.
4th	hour	the Head of the Bird	above the left eye.
5th	hour	his Tail	above the middle.
6th	hour	the Thousandfold Star	above the left eye.
		. . . .	. . . .
9th	hour	Orion	above the left elbow.
10th	hour	the star that follows the soul of Isis (*Sirius*)	above the left elbow.

to year. There were seven of these moving stars, Sun, Moon, Mercury, Venus, Mars, Jupiter and Saturn, which from early times were associated with the names of deities and to which were attributed the qualities of these deities. These moving stars or *planets* – a term whose literal meaning is a moving star – were always to be found within the belt circling the sky that was called the Zodiac. But their motions were not easy to explain nor their positions easy to predict. The search for an explanation provided the urge to observe and record. So astronomy and astrology grew up, as it were, hand in hand.

Comets, or 'hairy stars', seemed to be of a different nature. They appeared without warning and at infrequent intervals. No sort of prediction as to when a comet might appear seemed possible. A bright comet, with its flaming tail stretching far across the sky, is a magnificent spectacle; to primitive people it was awe-inspiring and came to be regarded as an ill-omen, a harbinger of disaster, of plague or flood. Possibly some bright comets had appeared at or before such happenings and had consequently evoked such beliefs.

Three periods of time of importance to mankind were marked out by the celestial

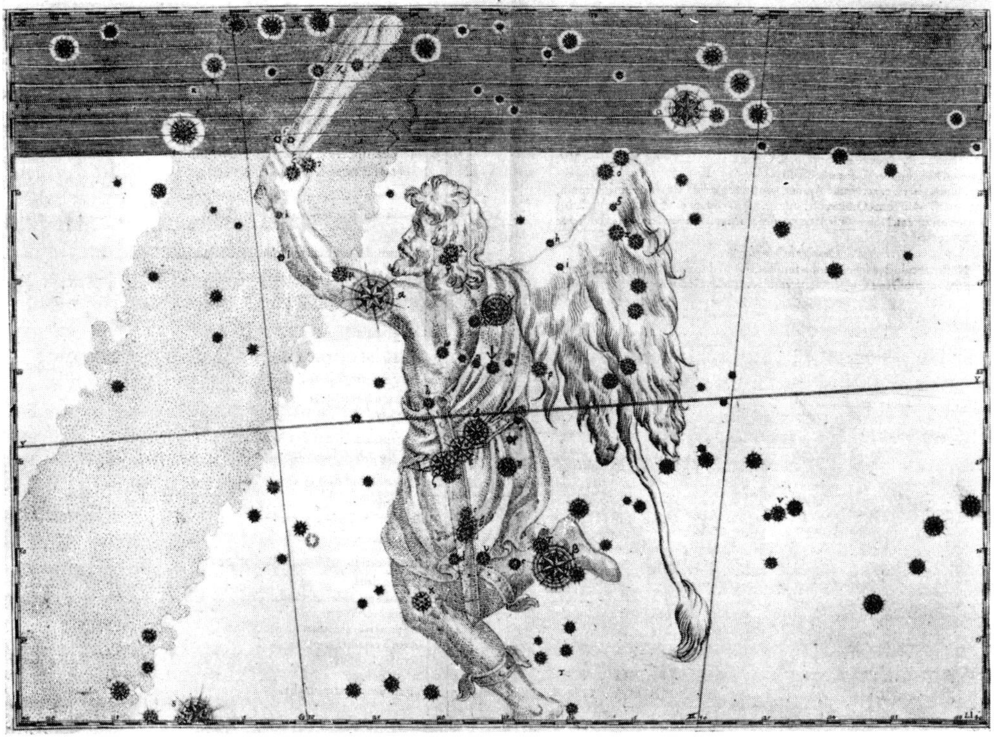

ORION, from Johann Bayer's *Uranometria* (1603). The stars in the constellation are marked with Greek letters in order of apparent brightness. The dotted band on the left is part of the Milky Way. Two groups of three stars form the *Belt of Orion* and the *Sword of Orion*.

bodies: the day, with its alternation of daylight and darkness, in which the vault of heaven appeared to make a complete rotation; the lunar month, in which the Moon went through its sequence of phases; and the year, with the succession of the seasons. Attempts were early made to combine these into a calendar for the regulation of human activities, for fixing the times for the observance of religious festivals, for determining the times for the sowing of seeds and so on. But for a long while neither the length of the lunar month – which, indeed, is variable – nor the length of the year was known. The year might be marked out by the alternation of a dry period and a wet period, by the alternation of summer and winter, or by the annual flooding of a river, as of the Nile. But none of these phenomena was sufficiently regular to serve as a useful guide. The difficulty of forming a calendar lies in the fact that neither the year nor the lunar month contains an exact number of days, nor does the year contain an exact number of lunar months. The day usually began at sunset, a reminder of which we find in the account of the Creation in the Book of Genesis: 'And the evening and the morning were the first day'. The month began with the first appearance of the crescent Moon in the evening sky.

It was soon found that twelve lunar months, each recorded by the first appearance of the crescent Moon, were shorter than the year, for with this reckoning the calendar year and the seasons soon fell out of phase with one another; also that thirteen lunar months were longer than the year. To keep the year in phase with the seasons, some years were given twelve months, and others thirteen; it was usually left to the priestly caste to decide when a thirteenth month should be added. From the accumulation of records, it was eventually found that 19 years and 235 lunar months were very nearly equal in length. This made a more regular arrangement possible in which during a cycle of 19 years, 12 years were each given 12 months and 7 years were each given 13 months, the longer years being given definite fixed positions in the cycle. From the records, it was also found that eclipses of the Sun and Moon recurred after a period of 18 years and 10 or 11 days, which was called the Saros, and used for the prediction of eclipses.

Amongst the Romans, the intercalation of the thirteenth month was left to the Pontiffs, who manipulated the calendar for their own purposes, so that by the time that Julius Caesar became Pontifex Maximus, the seasons were grossly out of phase with the calendar year. The reform of the calendar by Julius Caesar in the year 45 B.C. made the great step forward of cutting the month completely free from the phases of the Moon, giving to each month a fixed number of days (the lengths of the months then assigned being those that are still in use), and to the year a normal length of 365 days, with an extra day every fourth year. The further reform by Pope Gregory XIII in 1582 modified the rule for fixing leap years, so as to bring the average length of the calendar year into closer agreement with the true length of the year; introduced into England in 1752, it is the calendar in universal use today.

Primitive people, looking at the heavens, formed the idea that the number of the stars was immensely great. Their number was compared with the number of grains of sand on the seashore. This view though (as we now know) correct was surprising, for only a few thousand stars can be seen with the naked eye. When the first ideas began to be formed about the Universe, it was thought that the stars were all attached to a crystal sphere, lying just beyond the sphere which carried Saturn, the most distant of the then known planets. The revolutionary view propounded by Copernicus in 1543 – that the stars were at rest and that their apparent diurnal motions were due to the rotation of the Earth on its axis and, furthermore, that the Earth revolved round the Sun and not the Sun around the Earth – made it no longer necessary to assume that the stars were all at the same distance. It was, however, only gradually that the view came to be accepted that the stars were scattered throughout space to a great distance; it was supposed that the differences in brightness of the stars were due mainly to differences in distance and that, with increase in distance, the stars became fainter and fainter until they ceased to be visible. So men came to have a much grander view of the Universe; instead of the small compact Universe pictured by the early Greek astronomers, it was thought that the Universe might even be infinite in its extent.

The Greek astronomers made many attempts to account for the observed motions of the planets. In their view the circle was the perfect curve and their theories of the motions of the planets were all based on the assumption that it must be possible to account for these motions by a combination of circular motions. This representation was not satisfactory; as observations improved in accuracy, more and more of these circular motions had to be added and the theories became more and more artificial. Even so, this view was still held in the time of Copernicus, though he made some simplification by placing the Sun instead of the Earth at the centre of the Universe.

This Gordian knot was eventually cut by Kepler in the 16th century. Using the observations made by the great Danish astronomer, Tycho Brahe, he was able to show that the planets moved in ellipses, with the Sun in one of the foci, and in such a way that the line from the planet to the Sun traced out equal areas in equal times. This was an immense simplification. Kepler's laws of motions of the planets would appear to indicate that there must be some basic underlying reason.

But the idea that the Earth, the home of man, of man made in the image of God, was fixed at the centre of the Universe, had been accepted for so many centuries and was so firmly rooted in men's minds that the very revolutionary ideas of Copernicus were slow in gaining acceptance. The Copernican theory and the laws of Kepler they chose to regard merely as a convenient mathematical representation. It was the invention of the telescope at the beginning of the 17th century and the discoveries made with it by Galileo that slowly but surely caused the old ideas to be

Opposite:

GASEOUS NEBULA IN GEMINI, photographed in red light. (*Mount Wilson – Palomar*)

abandoned. Galileo found that there were spots on the Sun, whose positions changed from day to day in a way that could be explained only by supposing that the Sun was rotating about an axis. If the Sun rotated, why not the Earth also? He discovered the four major satellites of Jupiter, and observed their changes of position night by night, which proved that they were revolving around Jupiter, analogous to the Copernican view that the planets revolved round the Sun. He showed that Venus went through a complete cycle of phases like those of the Moon, which could be explained on the Copernican theory but not on the old Greek theories.

Kepler's famous three laws of planetary motion were deductions from observations. No theory was involved, and they remained uncoordinated until the genius of Newton provided an explanation by his theory of universal gravitation. He supposed that every particle of matter in the Universe attracted every other particle with a force proportional to the product of their masses and varying inversely as the square of their distance apart. By this grand unifying conception, he showed that the same force that causes an apple to fall to the ground also holds the Moon in its orbit round the Earth; the Moon instead of moving outwards into space can be thought of as continuously falling towards the Earth under its gravitational attraction. Newton showed that this same theory could account for the main phenomena of the tides, for Kepler's laws of planetary motion, and for the principal inequalities in the motion of the Moon.

> Nature and Nature's laws lay hid in night;
> God said 'Let Newton be', and all was light.
> (*Pope*)

Halley, the contemporary and friend of Newton, who had defrayed the costs of publication of Newton's immortal work, the *Principia*, felt certain that the comets must also move under the control of the Sun's gravitational attraction. With this in mind, he collected all the observational data that were available of bright comets that had appeared in preceding centuries and worked out their orbits. He hoped that a future comet would be found to be moving along the path of an earlier comet and would therefore prove to be a return of a comet that had previously appeared. Having computed the orbits of 24 comets, he found that the paths of bright comets that had appeared in the years 1531, 1607 and 1682 were so nearly identical that they must relate to one and the same comet. He therefore predicted that the comet would appear again about the end of the year 1758. As that time approached, very great interest was aroused as to whether Halley's prediction would be verified. The comet duly appeared, being first detected on Christmas Day, 1758. This most famous of all comets has become known as Halley's Comet. The success of the prediction was a triumphant vindication of the correctness of Newton's theory, which had not gained general acceptance on the Continent, where Descartes' theory of vortices was preferred. It proved moreover that the comets, like the planets, move according to law and not by chance and that, even at the enormous distances from the Sun to which they recede in the course of their orbital motion, they are still subject to the control of the Sun's gravitation.

The paths of the Moon and the planets are not strictly ellipses because, in consequence of the universal nature of gravitation, there is mutual attraction between all the members of the solar system. The motion of any particular planet is therefore perturbed by the attractions of all the other planets, which are continually varying as their distances change. The accurate prediction of the position of any planet at a particular time is consequently not an easy matter. During the 18th century many investigations were undertaken, particularly by Laplace, who has been called the Newton of France, to account in detail for the observed motions of the planets and the Moon. For this purpose, the mass of each of the bodies involved must be found, because the gravitational attraction is proportional to the mass of the attracting body. The outcome of these and of later investigations was to show that Newton's theory of gravitation could account in detail for all the observed motions, with two exceptions: the motion of the perihelion point of the orbit of Mercury showed a small but definite discordance from theory, and there were anomalies in the observed positions of the Moon that could not be accounted for.

The explanation of these residual discordances has been found only during the present century. Though Newton had formulated his law of universal gravitation, he attempted no explanation of gravitational attraction.

Apollo 9 docked with its Lunar Module.

Apollo 15: Lunar Module 'Falcon' at Hadley Base.

Lunar Module during trials over the Atlantic.

Apollo 8 Recovery.

By the formulation of his theory of general relativity, Einstein provided a new conception of time, space and gravitation. According to this theory, the properties of space are modified by the presence of matter; space becomes curved. The shortest path in a curved space is not a straight line. According to Einstein's theory, a planet moves round the Sun in a curved path because that is the straightest path in a curved space. But the path proves to be slightly different from that computed on the basis of Newton's law of gravitation; the difference accounts exactly for the discordance in the motions of the perihelion of Mercury. No other differences from strict gravitational theory which are large enough to be detected by observation are involved.

The anomalies in the observed positions of the Moon have been proved in recent years to be the consequence of slight variations in the rate of rotation of the Earth. The predicted positions are based, of course, on the assumption that the rotation, which provides our basic standard, is uniform. The variations in the Earth's rotation also produce divergences between observed and computed positions of the planets but, as the planets are at much greater distances than the Moon, the divergences are much smaller for them; it is only the improved accuracy of observation that has enabled them to be detected.

The slightly erratic time-keeping of the Earth has become of importance in recent years; time is now required with very high accuracy by the radio engineer and the physicist, for the control of precision standards of frequency. Comparisons between the observed positions of the Moon and the accurate theory of the Moon's motion now provide the required control over the behaviour of the Earth and enable a uniform time to be derived from the slightly non-uniform time based on the Earth's rotation.

In 1718 Halley found, by comparing current positions of stars with those observed by Hipparchus about 130 B.C., that some of the bright stars had certainly changed their positions. The stars were therefore not fixed, as had always been believed. If some are moving, all are probably moving. It was therefore no longer logical to suppose that the Sun, or any particular star, is at the centre of the Universe.

The first attempt to formulate a structure of the Universe, based on observation, was made by William Herschel about the end of the 18th century. He found it to be a much-flattened, almost disc-like structure, the Sun being near its centre. The broad belt of hazy light, stretching across the sky, called the Milky Way, consists of a vast number of faint distant stars; the Milky Way marks the great extension of the disc-like system. Herschel was unable to give any information about the dimensions of the system, for the first measurements of the distances of stars were not made until some years later.

The measurements of stellar distances, extended by various indirect methods, have provided in recent years a much more detailed and more accurate picture of our stellar system or galaxy. As Herschel showed, it is a highly flattened system extending in its central plane for about 100,000 light years. It has a central massive nucleus, in the direction of the brightest portion of the Milky Way. The Sun lies nearly in the central plane, but far out from the nucleus, at a distance of about 30,000 light years. The whole system is in slow rotation, under the general influence of its gravitation, and therefore not rotating like a solid body; the nearer to the centre, the faster is the rotation. The solar system makes one revolution around the nucleus in about 250 million years, travelling at a speed of about 150 miles a second. Along the central plane of the system there is a great deal of matter, in the form of gas and small solid particles (dust) that have not condensed into stars. This dusty matter dims or obscures the light from the stars beyond it and was responsible for Herschel's erroneous conclusion that the Sun was near the centre of the system. The total mass of the system is about one hundred thousand million times the mass of the Sun.

Observations in the last few decades have shown that our galaxy is but one of many other galaxies, extending outwards to the greatest observable distances (about 3,000 million light years), which show characteristics generally similar to those of our galaxy. They are mostly of a spiral structure, with arms spiralling out from the central nucleus. More recently it has been established that our galaxy has this typical spiral structure and that the solar system lies in one of the spiral arms. It is estimated that there are several hundred million of these systems, their average distance apart being a few million light years.

They are found all to be receding from our

Johannes Hevelius and his wife making observations with their double octant, A.D. 1659.

galaxy, and also from each other, the relative velocity of recession of any two being proportional to their distance apart. We may express this in another form: the whole Universe is in a state of expansion, carrying the galaxies with it. What this means is still a matter for discussion, and is dealt with in the article on **Cosmology.**

The relationship of our Earth to the Universe thus proves to be very different from the early ideas, which placed the Earth at the centre of a small Universe, and believed it to be the most important part of the Universe, around which everything else revolved. It now proves to be merely one of a system of planets, revolving around the Sun which is but one of many thousands of millions of stars comprising our galaxy. This galaxy is in turn one of many millions of galaxies scattered through space to distances beyond limits that can be reached by the most powerful telescopes yet made.

Because astronomy has been able to tell us so much about the nature and vastness of Creation, it can be called the queen of the sciences. It has an inherent appeal to the human mind. We, on our little Earth, endeavouring to plumb the remote depths of space, can only feel humble at the insignificance of our home, the Earth, in the vast scheme of things. Many questions instinctively come to our mind, which remain as yet unanswered. How did the Universe begin and how will it end? Or has it existed through the infinite past and will it continue to exist for an infinite future? Was there an initial act of creation, or is creation a process that is continuously operating, new matter being created, from which stars and galaxies will form? If there is no continuous creation, we should eventually appear to be alone in space, for one by one the galaxies, by the process of expansion, will pass beyond the range of observation. What is man's place in the Universe? Is there life on other worlds and, if so, may it possibly have developed into forms much higher, much more intelligent, much more competent than man? Everywhere in the Universe, we find evidence of law and order: chance plays no part in its evolution. Is this evidence for a Divine Creator and for divine control?

There is much that we should like to know. There is much for the astronomers of the future to investigate. At present we can say, with St. Paul, that we see as through a glass darkly. And as we look, awed and mystified, we cannot but echo the words of the Psalmist 'What is man, that Thou art mindful of him?'
(H.S.J.)

ASTROPHYSICS. The application of the laws and principles of physics to all aspects of stellar astronomy.

Astrophysics embraces so large a part of astronomy that its component topics are inevitably scattered through the pages of this volume. We indicate below the headings under which these topics will be found.

The method by which the physical conditions which obtain near the surface of a star may be deduced from a study of its light is described in the article on **Spectroscopy.** Certain characteristics of this light determine the **Spectral Classification of Stars.** The mass and radius of a star, and sometimes other interesting information, may be found directly in some cases, principally those of the **Binary Stars.** From the knowledge so acquired it is possible to hazard a guess as to the conditions and processes occurring inside the star. The two main sources of energy in stars are the *carbon-nitrogen cycle* and the *proton-proton process*, which are described under **Stellar Energy.**

The scales that are applied to the measurements of a star's brightness are given under **Magnitude** and **Bolometric Magnitude.** The **Astronomical Unit,** the **Parsec** and the **Light Year** are units commonly employed to express distances, and are compared in a table under **Distances, Astronomical**; the article on **Parallax** should also be read in this connection.

Under **Star** the methods are described by which stellar masses, densities, gravitational fields and rotations are found. **Proper Motion** deals with the *apparent* movement of stars on the celestial sphere; their *true* motions are discussed in **Stars, Motions and Distances.**

In recent years many new branches of astrophysics have developed. Some of these are described in the articles on **Interstellar Matter, Cosmic Rays** and **Nucleogenesis.** Also in recent years, there has been a tendency to refer to different branches of astrophysics by the part of the **electromagnetic spectrum** that is being discussed. See, for example, the articles on **Gamma-Ray Astronomy, X-Ray Astronomy** and **Infra-red Astronomy. Neutrino Astronomy** may give direct information about the conditions at the centres of stars.

Having learnt about stars individually, we can attempt to fit them into a pattern and see if any conclusions can be drawn about **Stellar Evolution.** The most promising pattern is the **Hertzsprung–Russell Diagram,** from which the notion of *Main Sequence* is derived.

ASYMPTOTE. A (usually straight) line which approaches nearer and nearer to a curve without ever touching it. (See figure under **Conic Sections.**)

ATMOSPHERE. The gaseous envelope surrounding a star, planet or satellite. Whether such a body can retain an atmosphere permanently depends chiefly on its mass, size and surface temperature. The molecules in a gas are in constant random agitation, and their average speed increases with increasing temperature. At any moment, a proportion of these molecules will be moving away from the surface of the body, and those whose speed is greater than the **escape velocity** for that particular star or planet will leave it altogether and drift into interstellar space. Thus a small, light body will constantly lose some of its atmosphere unless its surface is so cold that virtually no molecules can reach escape velocity.

A more massive body may actually increase its atmosphere by the reverse process of *accretion.* As it travels through space, it sweeps up interstellar matter by its gravitational attraction, and this will be added to its atmosphere.

The atmosphere of the Earth is described in the following article, and that of the various planets under their names. Only one satellite in the solar system – Saturn's Titan – is definitely known to have an atmosphere. It consists mainly of methane, a poisonous gas often used as a rocket propellent. It has been suggested that this methane might be used to refuel rockets, but it is probable that by the time Titan can be reached, such propellents will be obsolete.

For those stars that consist largely or wholly of gases, the word atmosphere can have no precise meaning beyond referring loosely to the surface layers. Hydrogen is by far the most abundant element in these layers; helium is common, and neon, oxygen and nitrogen are usually found in varying amounts and states of ionization. There are, however, some interesting extremes: the atmospheres of **white dwarfs** are probably only a few feet thick, the tremendous gravitational pull of these stars compressing matter below that depth into a state which can no longer be called gaseous. (See **Degenerate Matter** and **Critical Temperature.**) Some very hot O and B type stars have, by their rapid rotation, thrown off gases which form a shell surrounding the star. Such a shell is known as an *extended atmosphere*; it rotates independently and more slowly about the central part of the star.

ATMOSPHERE OF THE EARTH. A thin envelope of gases surrounding the Earth. Its approximate composition by volume is: nitrogen, 78 %; oxygen, 21 %; argon, 0·94 %; carbon dioxide, 0·03 %; hydrogen, 0·01 %; neon, 0·0012 %; helium, 0·0004 % and minute traces of other rare gases. In addition, there are varying amounts of water vapour.

The weight of the atmosphere creates a pressure at sea-level of about 14·7 lbs. per square inch (= 1 Atmosphere or 760 mm. of mercury). At first, the temperature falls with increasing height, but at about 38 miles it rises suddenly to 200° F. At 52 miles it has fallen once more to −150°, and above that it becomes warmer again.

ATMOSPHERIC ABSORPTION. The Earth's atmosphere is transparent to visible light, but it absorbs most other frequencies of the **electromagnetic spectrum,** including the X-rays and most of the ultraviolet and infrared radiation emitted by the Sun. Without this cushioning effect of the atmosphere, the Sun's rays would make the Earth's surface unbearably hot and destroy life on it. But to astronomers, and especially to spectroscopists among them, this partial opaqueness of the atmosphere means that large parts of stellar spectra cannot be studied from within the atmosphere. Photographs taken by cameras in high altitude rockets have already considerably extended our knowledge of the solar spectrum, but the much fainter radiations from stars and the planets require prolonged and carefully controlled exposures; artificial satellites provide opportunities to take these.

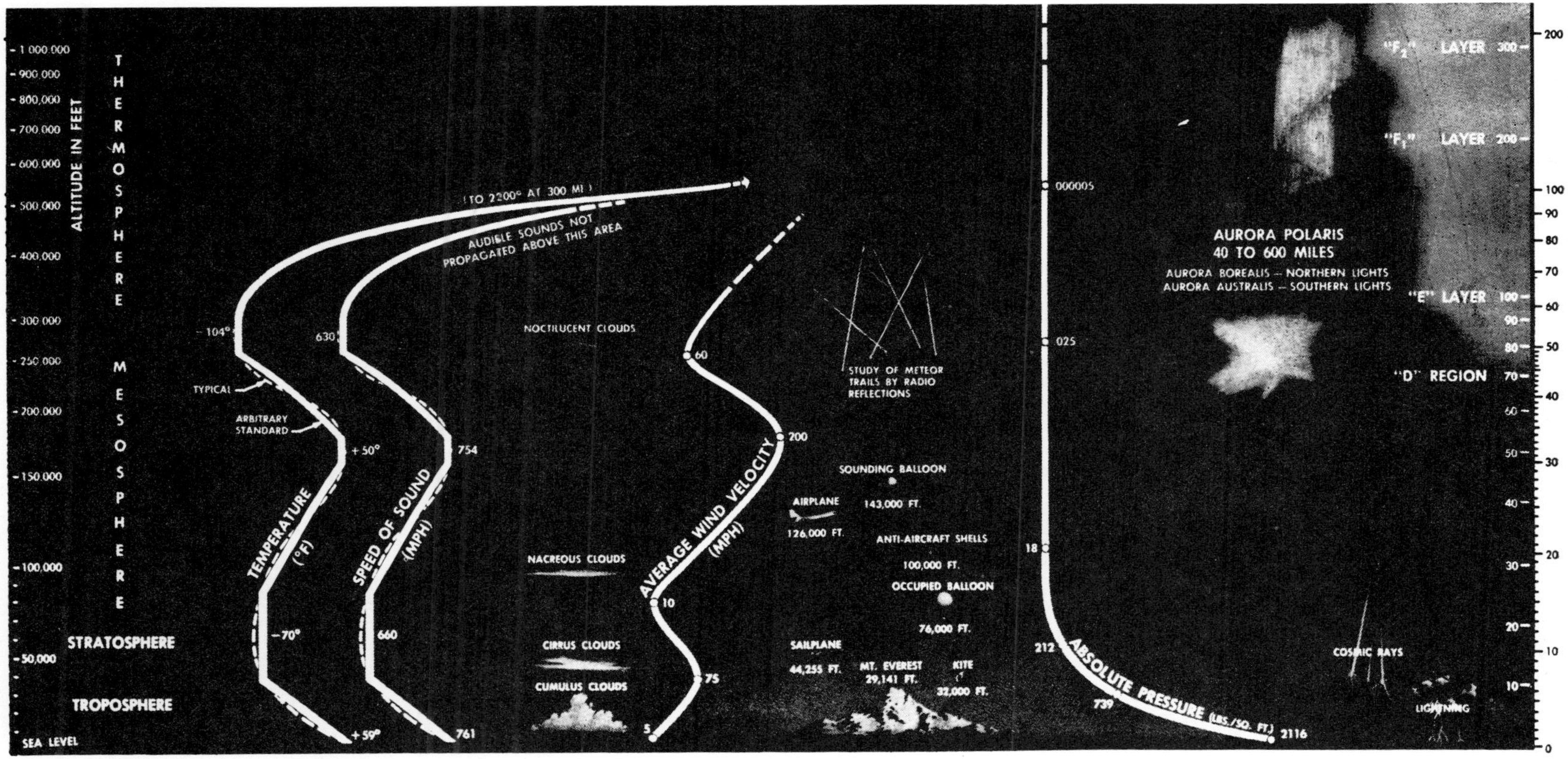

THE EARTH'S ATMOSPHERE up to about 230 miles. In the *thermosphere* (sometimes called *heterosphere*) gases tend to separate into layers of oxygen, helium, and atomic hydrogen. Beyond this lies the *exosphere*, which is virtually equivalent to interplanetary space.

The thermosphere absorbs practically all the hard ultraviolet radiation of the Sun and is thereby raised to high temperatures. These temperatures are subject to three periodic changes superimposed upon each other: a daily cycle due to the Earth's rotation, a 27-day cycle due to the Sun's rotation, and a less regular long-period change linked to activity on the Sun's surface. The fluctuations in turn affect the density, which tends to be higher during the daylight period, as well as the height and extent in depth of the chemical and ionized layers. At night the D-layer practically vanishes, and the E-layer loses about 98 % of its free electrons.

The heat content of the hotter regions of the thermosphere is relatively small because the density is very low. The lower, denser strata of the atmosphere can therefore accept a great deal of this heat by conduction without the fluctuations of temperature making themselves directly felt at ground level.

There are two 'windows' in the atmosphere, the first ranging from the very 'softest' ultraviolet through the visible spectrum into the infra-red (radiated heat); the second covers the shorter radio waves, and it is this window through which **radio astronomy** looks for new knowledge of the stars. (See also **Electromagnetic Spectrum.**)

SCATTERING. The molecules in the air scatter the sunlight in all directions, but the bluer part of the light is affected most. As a result, the Sun appears more yellow to us than it really is, its direct rays having lost a higher proportion of the blue frequencies which have been scattered throughout the sky and give it its colour. At an altitude of 100 miles the air is already too tenuous for this effect to be noticeable, so that there the Sun appears almost white, the sky black, and the stars are clearly visible against the dark background.

The atmosphere is subject to **tides** in the same way as the water layer of the Earth, but the effect is completely overshadowed by the random changes in the weather at ground level.

At about 60 miles the high altitude winds stream around the Earth at speeds of some 130 m.p.h. They are probably caused by the unequal heating of the atmosphere on the day and night sides, and are remarkably steady but contain much turbulence. (See also **Refraction.**)

Observations by artificial satellites have shown the atmosphere to be far more extensive than was previously thought. The lower ionosphere was found to be warmer over the polar regions than elsewhere.

If the extent of an atmosphere were defined as the distance at which its density falls to that of interplanetary matter between, say, Mars and Saturn, then the Earth may be within the atmosphere of the Sun.

ATOM. The smallest unit of an **element** which retains the chemical characteristics of that element. Atoms were originally thought to be hard, indivisible particles, but they are now known to be more like miniature solar systems.

At the centre of the atom is the *nucleus*, compounded mainly of **protons** with a positive electrical charge, and of **neutrons,** which are similar in size but have no charge. For each proton there is an **electron** which circles about the nucleus. The mass of a proton is one thousand, eight hundred and thirty-six times that of an electron, but the negative charge on an electron is equal in strength to that on the proton. Since the numbers of protons and electrons are equal, the total charge of an atom is zero, but if an electron is removed (or added) the resulting *ion* will have excess positive (or negative) charge. Atoms may also combine with each other to form *molecules*, whose chemical and physical characteristics may differ markedly from those of the original atoms.

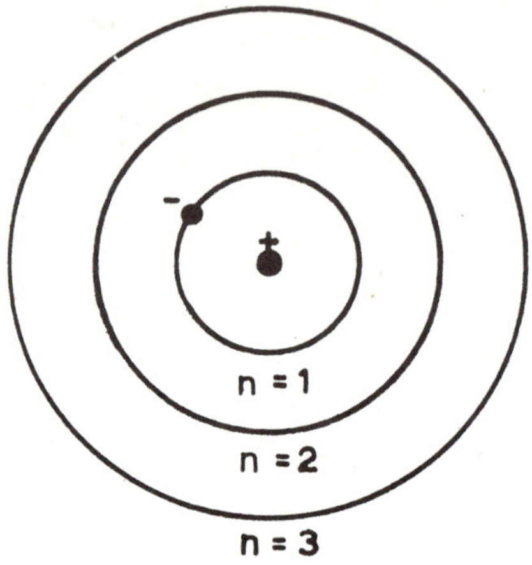

THE HYDROGEN ATOM shown schematically with the proton at the centre and the electron in the orbit corresponding to the ground state.

In a planetary system, gravitational forces are the most important; in an atom, these are completely overshadowed by electrical forces. In 1913 the Danish physicist Neils Bohr showed that the electrons are not free to pursue any orbit round the nucleus of an atom; only certain particular orbits are stable, those in which the electrons are distributed round the nucleus in distinct, spherical shells or layers. Each shell may accommodate a certain number of electrons, and the energies of the electrons moving in these orbits, under the influence of the electric field of the nucleus, have definite fixed values, the smallest orbit corresponding to the level of lowest energy. The chemical character of an atom is determined chiefly by the number of electrons in the outermost shell.

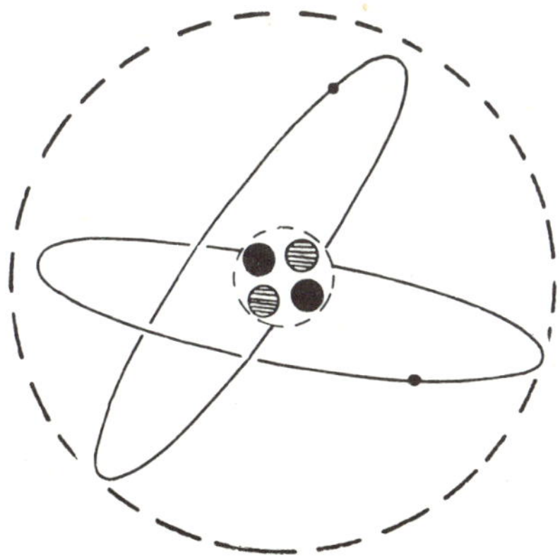

THE HELIUM ATOM: two protons and two neutrons form the nucleus; two electrons travel in orbits around it. The nucleus of the helium atom without the electrons is known as an alpha particle.

An atom may absorb energy, for example in the form of light, and this absorption of energy can cause an electron to jump from one orbit to another of greater radius; the energy is released if the electron returns to a smaller orbit. Each particular change of orbit corresponds to absorption or emission of energy in the form of a **quantum** of radiation of a particular **frequency,** the frequency being proportional to the energy gained or lost by the electron during its transition from one orbit to another. When many atoms of an element undergo such changes, a variety of frequencies will be involved, which will be exhibited in the absorption or emission **spectrum** of the element. Excessive absorption of energy by an atom may lead to the detachment of one or more of its electrons, that is, ionization.

The simplest atom is that of hydrogen, in which a single electron with negative charge revolves around a positively charged proton. For a crude scale model one can think of an orange placed at the centre of the Albert Hall and orbited by a pea at distances comparable with the radius of the concert hall. This perhaps illustrates that atoms should not be imagined as mini-billiard balls but as regions almost entirely void, with local concentrations of energy. The actual hydrogen atom has a diameter of about one Ångström or one hundred-millionth of a centimetre.

The stable electron orbits are labelled in the Bohr theory by the numbers $n = 1, 2, 3, \ldots$ Each of these corresponds to a particular state or **energy level.** There are an infinity of such states, and the smallest orbit ($n = 1$) represents the level of lowest energy or the *ground state.*

The **helium** atom comprises a nucleus of two protons and two neutrons, about which two electrons revolve. The nucleus of a helium atom is known as an alpha particle, and is of great importance in nuclear reactions in stellar interiors (see **Stellar Energy**).

ATOMIC ENERGY. See **Nuclear Reactions.**

ATOMIC NUMBER. The number of protons in the nucleus of an **atom.**

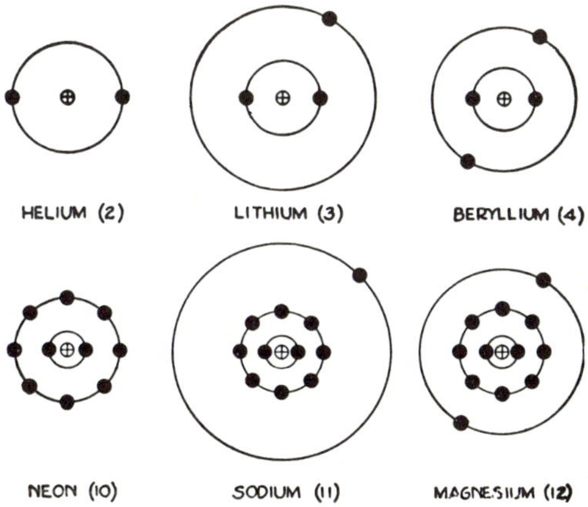

Electron shells of some of the lighter elements, with their atomic numbers. Details of the nucleus are not shown. In the un-ionized atom, the numbers of protons and electrons are the same.

ATOMIC WEIGHT. The number of times that a given **atom** is heavier than one atom of hydrogen is called its atomic weight. It is a ratio, and quite distinct from the actual weight in grams of the atom. (More recently, oxygen = 16 has been substituted as the standard, which makes the atomic weight of hydrogen = 1·008.)

AURORA POLARIS, also called *Aurora Borealis* or Northern Lights, and *Aurora*

Australis or Southern Lights, are glows in the atmosphere at a height of 40–200 miles and higher, often brilliantly coloured. They are caused by streams of electrically charged particles emitted by the Sun striking the upper atmosphere, and since the charged particles are attracted towards the Earth's **magnetic poles,** auroral displays are seen mainly in polar or at least high latitudes. Outbursts of solar flares generally lead to increased auroral activity after a definite time interval, from which the speed of the emitted particles may be calculated. (See also **Van Allen Belts.**)

AZIMUTH. The horizontal direction or *bearing* of a heavenly body calculated from the North or South points of the observer's horizon.

A BRILLIANT AURORAL DISPLAY following a solar flare, photographed in Alaska by Prof. V. P. Hessler. Photographs cannot do justice to the swiftly changing, finely striated patterns of the displays.

B

BACKGROUND RADIATION. Since 1965 observations of the *microwave* radiation with wavelengths between 60 and 0·3 cm. reaching the Earth have been made. The observations are not easy to perform because of the many corrections; notably the effect of terrestrial sources of radiation must be allowed for. After these corrections are made, radiation very much more intense than that expected from known radio sources is found. The radiation is *isotropic* (i.e. the same in all directions) and its spectrum in the observed range is that of **black body** radiation at about 3° K., but until observations have been made at shorter wavelengths it is impossible to know whether it is indeed black body. Big-bang theories of **cosmology** predict the existence of isotropic black body (or relict) radiation and so the observed microwave background can be explained by these theories. The steady-state theory requires a non-cosmological explanation. Some galaxies are strong emitters in the far infra-red and it has been suggested that they may be strong in the microwave region of the **electromagnetic spectrum** as well. The background radiation would then be due to the integrated emission of these objects. (P.R.O.)

BACK-OUT. Procedure for the organized reversal of a **count-down** when the launching of a missile has to be postponed.

BAILLY. The largest crater on the visible side of the **Moon.** It has a diameter of over 170 miles.

BAILY'S BEADS. When the Moon eclipses the Sun, just before the moment of totality the crescent Sun will break up into a string of light patches called *Baily's Beads.* They represent the last direct rays of sunlight to penetrate between the irregularities of the lunar surface, and reveal the exact shape of the mountains at the Moon's edge. The beads disappear almost at once, but are seen again at the west limb after totality. (A photograph appears under **Eclipse.**)

BALLISTICS. The science that deals with the motion of projectiles. *Interior ballistics* deal with the propulsion of the projectile, and *exterior ballistics* with the path it follows when it is not guided.

The exterior ballistician is concerned with the forces acting upon the projectile when it is moving freely by its own momentum, especially the gravitational force, wind gusts, air resistance, rolling and tumbling of the projectile, and the effects of the rotation of the Earth (see **Coriolis Force**). Short range projectiles move in parabolic trajectories modified by air resistance, but for long range projectiles the changing distance from the Earth's centre causes changes in the force of gravity and turns the trajectory into a portion of an ellipse having the centre of the Earth at one of its foci.

By working backwards from the target to the launching site the ballistician can calculate the required launching characteristics such as the muzzle velocity, elevation and deflection in the case of a gun, or the all-burnt velocity, distance from launching pad and direction of motion in the case of a rocket.

A ballistic missile is one which travels through most of its trajectory without guidance or continued propulsion.

BALMER SERIES. A series of lines in the spectrum of the hydrogen atom. Each line corresponds to a transition beginning or ending in the second energy level of the hydrogen atom. (See **Spectroscopy.**)

BETA PARTICLE or *Beta Ray*. A fast electron emitted by certain *radioactive* atoms.

BETELGEUSE, or α Orionis, is the second brightest star in the constellation **Orion.** It is a relatively cool **supergiant,** about 1,200 times as bright as the Sun, 24 million times as voluminous, 15 times as massive, and has a surface temperature about one half the solar value. The density of Betelgeuse is extremely low, and it has been aptly described as a 'red-hot vacuum', in sharp contrast to the **white dwarf** stars which are extremely small and dense. Its brightness and size vary in a somewhat irregular fashion in cycles of a few hundred days.

BINARY STAR, MULTIPLE STAR. A binary star consists of two component stars which revolve about their common **centre of gravity.** A multiple star is a system of more than two components.

As the separation of the components of a binary (or multiple) star is in general very small compared with the distance of the system from the Earth, the components appear very close together. In very few cases is the angular separation of the two stars great enough for them to be seen individually by the naked eye; the eye cannot resolve two points closer together than about three minutes of arc, or one-tenth of the visual diameter of the Moon. Epsilon Lyrae, a fifth-magnitude star near Vega, is however seen as double without optical aid by people with good eyesight, the separation being about 3½ minutes of arc. A telescope reveals that many stars are double. The proportion of double stars with separations greater than ten seconds of arc is quite small; many are less than one second apart.

It is possible for two stars at very different distances to lie in almost the same line of sight, giving what is known as an *optical double.* The celebrated pair in the 'handle' of the Plough, **Mizar** and **Alcor,** is of this nature. Optical binaries are an accident of perspective, and move in straight lines relative to each other; but the vast majority of double stars are true binaries in one physical system.

Binaries are very common indeed amongst the stars, at any rate in our neighbourhood of the galaxy: of the 254 stars known to be within ten parsecs of the Sun (counting each star separately) 127 are members of 61 binary or multiple systems.

Binary stars are of great interest to astronomers because they enable the masses of stars to be calculated; they are the only stars whose masses can be reliably determined observationally. In order to find the mass of each star, we need to know the sizes of the orbits of the two stars, and the period of revolution. Unfortunately it is difficult to measure the positions of both components relative to their common centre of gravity and so get both orbits; usually the brighter component is regarded as fixed and the orbit of the fainter one is measured relatively to it. Such an orbit allows the total mass to be found, but not the individual masses. Further difficulties present themselves: the real size of the orbit in miles, not its apparent size in seconds of arc, must be known, and this involves measuring the **parallax** of the star to find its distance. Also, to achieve accuracy, the binary must be watched over a whole period; but periods often run into hundreds or thousands of years. For the period to be

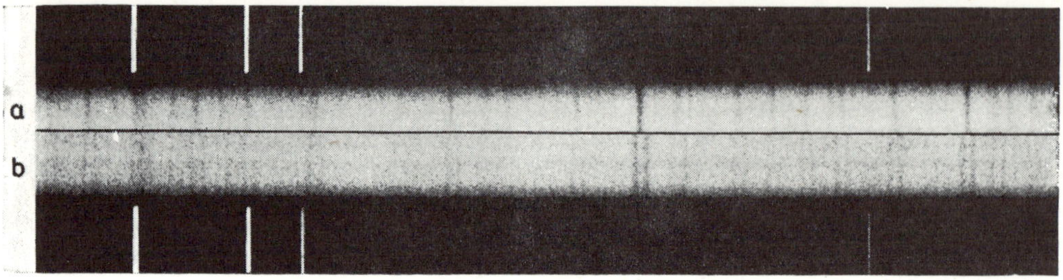

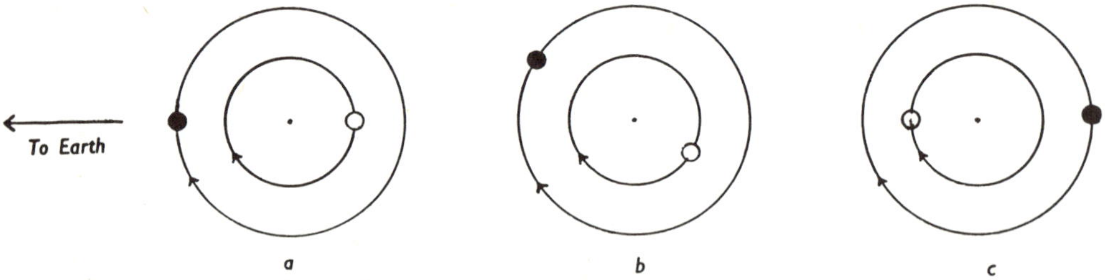

SPECTRUM OF A SPECTROSCOPIC BINARY STAR. The example here is Mizar, which has a period of 20·5 days. (*a*) June 11, 1927. Both components moving at right angles to the line of sight, spectral lines superimposed. (*b*) June 13, 1927. One component is moving towards and the other away from the Earth; their spectral lines therefore undergo Doppler shifts to the left and to the right respectively. When the positions indicated in (*c*) are reached, the spectral lines will again be single. The components revolve about their common centre of gravity.

The top and bottom spectra are comparison spectra obtained in the laboratory, and correspond to situation (*c*). (*Mount Wilson – Palomar*)

short, the stars must be close together, and in order to be resolved by a telescope and to have a measurable parallax they must also be comparatively close to us. Only about 50 binaries have both well-determined orbits and well-determined parallaxes. Even when the masses have been obtained, they may not be representative of star masses as a whole: the masses of close visual binaries near to us are not necessarily typical of all stars everywhere. This is an example of the ever-present problem of *observational selection.*

In *visual binaries* the individual components can be observed directly with the telescope. *Spectroscopic binaries* are recognized from examination of their spectra. *Eclipsing binaries* are detected by their rhythmical fluctuations in brightness.

SPECTROSCOPIC BINARIES. In many binaries the separation of the components is too small to be resolved by even the largest telescope, but the spectroscope may still reveal their double nature. The orbital motion of the components will cause differences in their velocities relative to the Earth, and this will give rise to a difference of the **Doppler shifts** in the light from the two components. If they are about equally bright, the spectroscope will show two spectra whose lines are slightly displaced relatively to each other. If the components are so unequal that the light from one drowns the spectrum of the other, the duplicity may yet be indicated by a periodic oscillation in the positions of the spectral lines, corresponding to the period of revolution of the stars. The spectroscope can, however, tell us nothing if the orbit happens to lie in the plane perpendicular to the line of sight, i.e. if the *inclination* is practically 0°: the line-of-sight or radial velocities of both the stars are then practically constant, and no variation in the Doppler shifts can be detected.

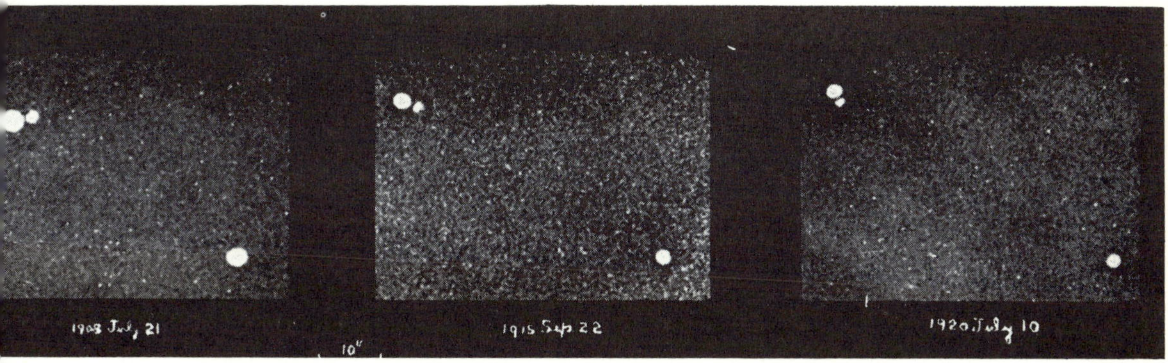

KRÜGER 60, a visual binary star. These three photographs, taken in 1908, 1915 and 1920, show the two components slowly circling each other, obedient to the same laws that govern all motion throughout the Universe. – A third component of the system does not appear on the plates. The star at lower right is not connected with the system. (*Barnard, Yerkes Observatory*)

Spectroscopic binaries often have periods of revolution of days rather than years, although this apparent preference for short periods is in part caused by observational selection. Binaries with short periods have correspondingly high orbital velocities, which are readily detected and measured spectroscopically. Those with periods of several years can only be recognized as binaries after accurate measurements of their radial velocities have been made repeatedly over a time interval comparable with the period. Even today there must remain many undiscovered binaries among the bright stars visible to the unaided eye. Some will yield to continued spectroscopic observation; but some will not – their orbital planes may lie perpendicularly to the line of sight, or, except for quite nearby stars, the separation of the components may be so great that the orbital motion is undetectably small although the apparent separation is still insufficient for the stars to be seen as visual binaries.

Once a binary system has been recognized spectroscopically, the period of revolution is easily found – it is the interval after which the velocity curve repeats itself; and the shape of the orbit can be inferred from a number of determinations of radial velocity at intervals throughout the period. But unless further information is available the inclination of the orbit remains unknown and so therefore does its size; results of some value are nevertheless obtained in such cases by using a statistical mean value for the inclination.

ECLIPSING BINARIES can provide this additional information. About 9 % of all spectroscopic binaries have orbits with sufficiently small inclinations to cause each component to eclipse the other once per revolution. For these stars a whole new range of information may be derived from the manner in which their brightnesses vary. The variation is plotted against time on a graph known as a *light curve*.

The eclipses may be total, annular, or partial; characteristic light curves are shown in the figures. Once in each revolution there is a *primary minimum* and there may or may not be a *secondary minimum* when the star which was eclipsed during the primary minimum passes in front of its companion. When the eclipses are total or annular, or nearly so, a good estimate of the inclination of the orbit may be made. Coupled with spectroscopic evidence, this yields the **eccentricity** and size of the orbit. The radii of the components and their relative brightnesses may be derived from the light curve. In some cases, light from the brighter component is reflected in the fainter: the total brightness of the two rises slightly as the brighter one moves towards us after being eclipsed and shines on more and more of the side of its dull companion which is turned towards us. There is a small secondary minimum as the bright component eclipses the fainter. Obviously the classification of 'eclipsing binaries' depends on perspective, and is not an intrinsic property of the system.

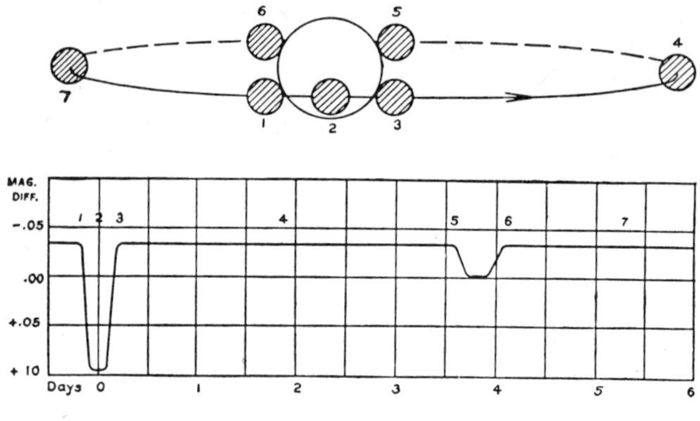

LIGHT CURVE OF AN ECLIPSING BINARY. The numbers on the graph correspond to the positions occupied by the smaller, fainter component relative to its larger companion as seen from the direction of the Earth. Between positions 5 and 6 the smaller star suffers total eclipse. (*After Stebbins*)

In some stars, the edge of the disc or *limb* is darker than the centre, and if such a star suffers an annular eclipse, more light is obscured when the eclipsing body is in front of the centre of the disc, i.e. in the middle of the annular phase, than at the beginning and end of annularity. This gives the minimum a characteristic shape.

Another effect deducible from the light curve is the tidal distortion of close components into ellipsoid shapes; the brightness is then never stationary owing to the constantly changing area of the stars presented to us.

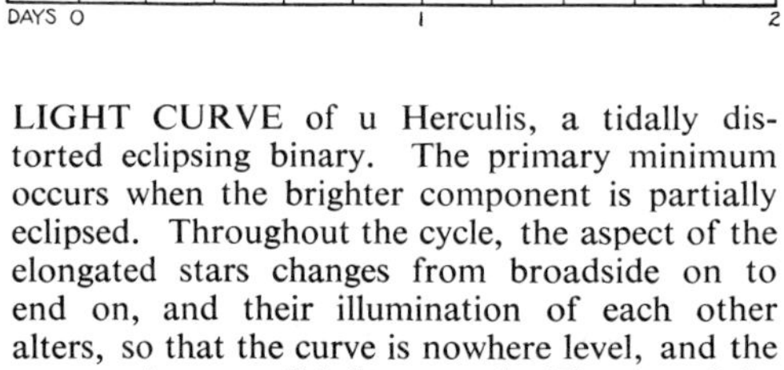

LIGHT CURVE of u Herculis, a tidally distorted eclipsing binary. The primary minimum occurs when the brighter component is partially eclipsed. Throughout the cycle, the aspect of the elongated stars changes from broadside on to end on, and their illumination of each other alters, so that the curve is nowhere level, and the two maxima are slightly unequal. The size of the Sun is indicated for comparison.

(*After R. Baker*)

If a component of an eclipsing binary is spinning about an axis well inclined to the line of sight, one limb will be turning towards the Earth while the other is turning away from it. Just before total eclipse only one limb of the eclipsed star is visible, and if the rotation is rapid enough its period may be estimated by measuring the resulting Doppler shift.

A different aspect of eclipsing binaries is discussed under **variable stars.** (R.G.)

EVOLUTION OF CLOSE BINARY STARS. Close binary stars are those whose separation is of the same order as the sizes of the stars themselves. They can be detected as eclipsing and as spectroscopic binaries, but because of their small dimensions, and consequent rapid periods of orbital rotation, the measurements must be made with large telescopes.

As a star evolves from the early stages in its life of nuclear hydrogen burning to old age, and the development of helium-burning reactions in the interior, the radius of the star increases. This expansion can be as much as a thousand-fold as the initial main sequence star changes to the later evolutionary state of a red giant. (See **Stellar Evolution.**) For this reason the transfer of material from one star to its companion can occur in close binary systems.

Since the *primary*, or more massive star, evolves more rapidly, it may eventually expand to such a radius that its surface is in contact with the inner **Lagrangian point.** Any expansion beyond this, caused by further evolution of the star, will result in the outer layers being sucked off and drawn by gravitational attraction towards the companion star. This causes dramatic changes in the properties of the binary system and of the stars themselves. For the process is found, by computer simulation, to be unstable, and much of the material of the more massive star is rapidly transferred to its companion. So much so, in fact, that the roles of the two stars may be reversed, the originally more massive star shedding so much of its outer layers to the companion star that, when stability is regained, it is the lighter star in the system.

Although this *mass transfer* process is rapid in terms of the normal lifetime of a star, typically lasting for about 50,000 years, this is too slow to cause any observable explosive effects as it proceeds. However, theory predicts that such behaviour must occur, and it is thought that many binary systems, particularly those of the **Algol** type, have undergone this process. Observations of gas streams in several close binaries lend further support to these conclusions. In addition the observations indicate that, in many binary stars, the less massive star has evolved further and appears much older than the more massive companion. This is quite contrary to single stars and the theory of stellar evolution, which requires that the speed with which a star evolves, and ages, increases as the mass of the star increases. This contradiction is resolved if the mass transfer process causes the originally more massive and more rapidly evolving star to become, at a late stage in its evolution, the *secondary*, or less massive component of the binary system.

A number of binary systems are illustrated in the colour plate facing page 64. (G.T.B.)

BIOLOGY OF SPACE, BIOTHERMAL ZONE. See **Life** and **Space Medicine.**

BLACK BODY. An ideal body which absorbs all radiation that falls on it and reflects none. (See **electromagnetic spectrum.**) An object which looks black, e.g. a piece of black paper, absorbs most of the visible radiation, but it may reflect a considerable portion of the ultraviolet and infra-red (heat) rays. Perfectly black bodies in the technical sense do not exist; the nearest approach is an almost completely enclosed cavity in an opaque object. The reason for this is that any ray entering the hole may not be entirely absorbed when it first strikes the wall of the cavity, but that fraction of it which is reflected will strike the wall again elsewhere, will itself be partly absorbed and partly reflected, and so on. At each reflection more of the ray is absorbed until the unabsorbed fraction is infinitesimal. If the opening of the cavity is small enough, the chances of any reflected ray escaping through it before being absorbed are very small.

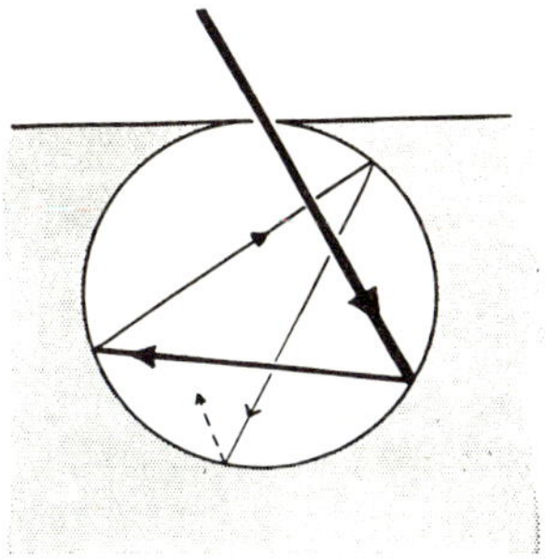

A hole acting as a black body absorber of heat. At each reflection more of the incoming heat ray is absorbed by the walls of the cavity.

If a black body is heated in any way, it will radiate heat better than any less black body, and in strict accordance with **Stefan's Law;** moreover the distribution of the radiated energy over the spectrum will obey another fundamental equation called *Planck's Law.* We can measure the distribution and amount of radiated energy received from the stars. If a star is a perfect radiator, i.e. a black body, or if we know exactly to what extent it differs from a black body, the above laws then enable us to calculate its temperature. (M.T.B.)

BLACK DROP. This is the name given to a curious phenomenon seen during a transit of **Venus.** As Venus draws in front of the Sun, it appears to draw a strip of blackness after it; when this strip disappears, the planet is found to be full on the solar disc. It is therefore impossible to give an accurate estimate of the exact moment of the start of the transit, and the Black Drop reduced the precision of the results obtained for the astronomical unit in 1874 and 1882. See **Transits.**)

BLACK HOLE. Theoretical considerations show that when a star, having expended most of its nuclear energy, collapses under its own gravitational forces through the **white dwarf** stage to become a **neutron star**, the collapse may proceed further: gravity near the immensely dense central mass becomes so great that no light, matter or signal of any kind can escape from its immediate vicinity, which is called a *black hole*. It is correspondingly difficult to design experiments to verify the existence of black holes. One solar mass would yield a black hole 7 km in diameter; its boundary is called the *absolute event horizon*. If normal physical laws hold for the centre, matter there reaches infinite density and is effectively 'crushed out of existence'.

BODE'S LAW. This interesting numerical relationship was first noticed by Titus of Wittenberg, but was brought into prominence by Bode in 1772, and is hence known as *Bode's Law*.

Take the numbers 0, 3, 6, 12, 24, 48, 96, 192 and 384, each of which (apart from the first two) is double its predecessor. Add 4 to each. Taking the Earth's distance from the Sun as 10, this second series gives the distances of the other planets, to scale, with remarkable accuracy, as is shown by the following table:

Planet	*Distance by Bode's Law*	*Actual Distance*
Mercury	4	3·9
Venus	7	7·2
Earth	10	10
Mars	16	15·2
Ceres	28	27·7
Jupiter	52	52·0
Saturn	100	95·4
Uranus	196	191·8
Neptune	—	300·7
Pluto	388	394·6

The Law breaks down for Neptune, and the Bode distance corresponds much more closely to that of Pluto; but when the Law was announced, Uranus, Neptune and Pluto were unknown. Nor was there a known planet corresponding to the distance 28, and the discovery of Ceres, the largest of the asteroids, was regarded as an extra verification of the rule. It is still uncertain whether Bode's Law is fundamental, or is due purely to chance.

BOHR THEORY. The theory, due to Bohr in 1913, explaining the spectrum of the hydrogen atom. (See **Spectroscopy.**)

BOLIDE. A fireball or bright meteor which explodes in the course of its flight through the atmosphere. (See **Meteor.**)

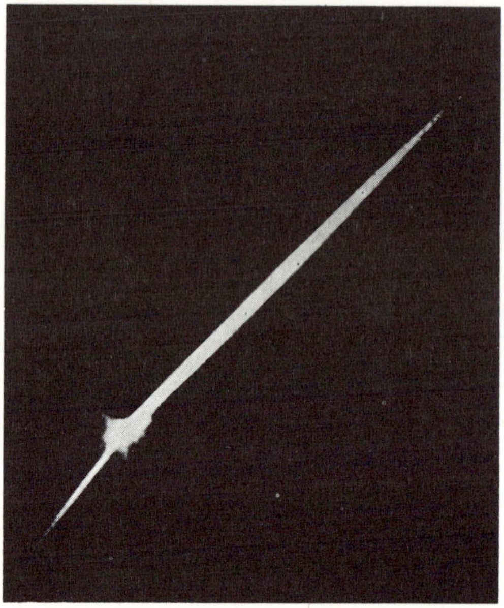

BOLIDE. An exploding Andromeid meteor.

BOLOMETER. A very sensitive kind of thermometer used for measuring weak heat radiation. A thin strip of platinum foil is coated with finely divided 'black' platinum, so that it becomes virtually a **black body.** It is carefully screened from all radiation except that which is being measured, and a weak current is passed through it and through a **galvanometer.** Any radiant heat absorbed by the strip will raise its temperature and thereby lower its electrical resistance, and the increased flow of current can be detected by the galvanometer.

BOLOMETRIC MAGNITUDE. If a **bolometer** is joined to a spectroscope, the distribution of energy radiated in the infra-red region of the spectrum of the source can be measured. From this, and by an application of the **Stefan-Boltzmann Law,** a star's total emission of energy may be calculated. According to this quantity, a star is given a *bolometric magnitude*, and the scale has been arbitrarily adjusted so that for the Sun the visual **magnitude** is equal to the bolometric magnitude. For nearly all other stars, the bolometric magnitude is *less* than the visual magnitude, i.e. a *greater* part of their total radiated energy lies outside the visible range than in the case of the Sun.

BONN DURCHMUSTERUNG. A vast survey and atlas of the sky listing 324,198 stars and their apparent brightnesses. It was compiled for the **epoch** 1855, and various additions were made at later dates. A star is often referred to by its B.D. number.

BOOSTER. A rocket stage used at lift-off of a missile or space vehicle. The term has come to be used colloquially to refer to the whole launch vehicle of a spacecraft, and in this context can have a number of stages. A booster will normally be jettisoned at burnout, but in some cases the booster motors are an integral part of the main structure.

BRAKING ELLIPSES. By following a path known as a series of *braking ellipses*, a rocket returning to Earth or attempting to land on a planetary body could shed much of its speed without using fuel for braking, provided there is an atmosphere.

On its first approach, the missile grazes the outer part of the atmosphere, where some of its speed will be lost through friction and turned into heat. Before the hull is heated to a dangerous extent the missile is carried out of the atmosphere again in an elliptical orbit, cooling off in the very low temperatures of interplanetary space before entering the atmosphere one more. The process is repeated, and on each circuit the rocket loses speed and describes part of a smaller ellipse, until finally its speed is so small that the landing can be made by gliding, parachute, or with a very moderate expenditure of fuel.

If such a missile entered the atmosphere too far or for too long at full speed, it would melt and even be vaporized within seconds by the heat of friction. A small departure from the calculated braking ellipses could therefore have fatal results, and some fuel would still be required to guide the vessel into the right path. Modern **re-entry** methods have made this technique unnecessary, but it remains a theoretical possibility. A space vehicle returning to Earth and striking the atmosphere too obliquely could accidentally go into a braking ellipse. (M.T.B.)

BREAK-UP SYSTEM. A system to ensure the break-up of a rocket vehicle which appears likely to fall outside the safety limits of the launching range. In the event of a guidance or control malfunction it is necessary to destroy the vehicle to limit the possible damage on the ground. Normal Western practice is to appoint a range safety officer who is not normally involved professionally with the vehicle being fired, who monitors its flight and destroys it if it departs more than a predetermined amount from its nominal trajectory. The break-up system consists of a transmitter on the ground and a receiver and associated explosive devices on board the rocket which will ensure rupturing of fuel tanks and immediate cessation of propulsion.

BRITISH ASTRONOMICAL ASSOCIATION. The B.A.A. was founded in 1890 and is open to all persons interested in astronomy. It has as its objects the association of observers, particularly the possessors of small telescopes, for their mutual help and organization in the work of astronomical observation; the circulation of current astronomical information; and the encouragement of a popular interest in astronomy. Monthly meetings are held at the rooms of the **Royal Astronomical Society** in Burlington House, Piccadilly, London, W.1.

BRITISH INTERPLANETARY SOCIETY. Founded in 1933, this society now has about 3,000 members. It holds lecture meetings and publishes a monthly journal devoted to technical aspects of astronautics, as well as a popular magazine entitled *Spaceflight*. *Address:* 12, Bessborough Gardens, London, S.W.1.

C

c. The symbol used to denote the velocity of light *in vacuo*. This velocity is very nearly 3×10^{10} cm. per second (186,000 miles per second). It is a fundamental constant and is the maximum velocity theoretically attainable.

CALLISTO. The fourth satellite of **Jupiter.** It is the largest of the Jovian attendants, with a diameter of about 3,220 miles, but has the very low density of 1·3 times that of water; since the escape velocity is only 0·9 miles per second, there is no possibility of its retaining an **atmosphere.**

CALORIE. The amount of heat required to raise the temperature of one gram of water through one degree Centigrade (from 14·5° to 15·5°).

CANALS, MARTIAN. In 1877, Schiaparelli drew attention to certain narrow, linear features on Mars which he named 'canali'. The canals, as they are popularly known, were studied in detail by Lowell at Flagstaff between 1895 and 1916, and were thought by him to be artificial waterways. This explanation is now discounted, and it seems that the 'canals' are not so narrow or so regular as Lowell believed; they are regarded as natural features. (See **Mars.**)

CARBON. A non-metallic element of atomic weight 12, melting point about 3,500° C. and boiling point 4,200° C. It occurs naturally in three forms, as diamond, graphite or coal.

It is the chief constituent of all forms of living or organic tissue. The carbon atoms can link themselves in long chains and rings to form the highly complicated molecules that are the basis of all living matter. No other atoms except those of silicon possess this property to a comparable degree, and it is reasonable to suppose that carbon is one of the prerequisites of **life.**

Opposite:

BINARY AND TRIPLE STARS. These drawings by David Hardy should be regarded as something between a series of diagrams illustrating points made in the article on **Binary Star** and straightforward visual representation.

Top: ZETA AURIGAE system consisting of a red supergiant and a much smaller blue companion. The latter is about to go into eclipse; it continues to shine through the very tenuous atmosphere and prominences of the supergiant. Absorption lines begin to appear in its spectrum but it will remain visible for several weeks until it is completely obscured by a greater depth of the supergiant. The dip in the light curve of the system is therefore gradual.

Upper Middle: An ALGOL-type system of a semi-detached binary. The orange component is slightly more massive and larger (perhaps 4 times as large as the Sun), but tidally distorted towards the inner Lagrangian Point. Mass transfer is taking place, and the picture indicates streams of gas moving towards the less massive star. From the viewpoint taken for this picture the system is eclipsing, and the light curve displays deep and shallow troughs according to whether the bright component is eclipsing the fainter one, or vice versa.

Lower Middle: ALPHA CENTAURI A and B and PROXIMA CENTAURI, an interesting triple and the solar system's nearest neighbours in space. The three components are, from left to right: α Centauri A, a main sequence star very similar to the Sun in luminosity and spectral type; α Centauri B, a yellow dwarf of absolute magnitude 6·1 and spectral type K5; and Proxima Centauri, a red dwarf of magnitude 14·9 and spectral type M5.

Bottom: View from a hypothetical planet of the red supergiant shown in the top frame. The planet is assumed to be some 500 million miles from the red star and the visual angle is about 35 degrees. Each star casts its own shadow of surface features in which the colour of light from the other star predominates.

The blue star is 3½ times as large as the Sun, but a hundred times as bright.

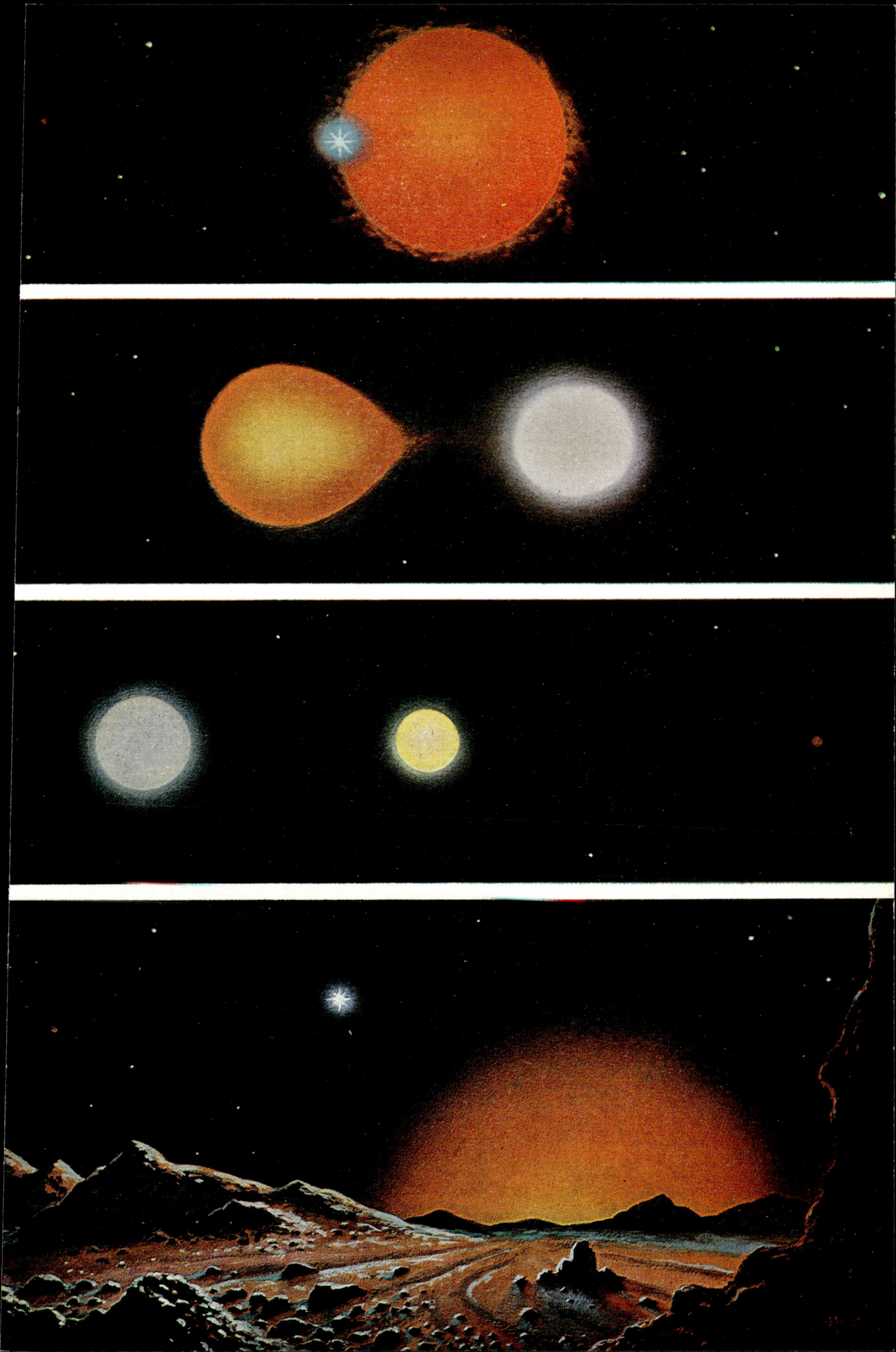

The Control Room in a Cape Kennedy blockhouse during an *Apollo* count-down.

CARBON CYCLE. A nuclear reaction in which the carbon atom promotes the creation of helium from hydrogen. This process is accompanied by a vast release of radiation, and is, together with the proton-proton reaction, the main source of **stellar energy,** where both processes are described.

CARBON DIOXIDE, CO_2, is a colourless gas which forms 0·03 % of normal air. It is produced when carbon is burnt in the presence of oxygen; hence animals, in whose bodies this reaction goes on continuously, inhale oxygen and exhale it again as carbon dioxide. The reverse is true of green plants.

In an enclosed, inhabited chamber such as a submarine or space ship, the carbon dioxide content of the air must be carefully controlled. The act of breathing is stimulated by the presence of CO_2, not by the lack of oxygen. A man breathing pure nitrogen, for example, will notice nothing wrong and will pass out, blue in the face and half suffocated, without a moment's discomfort. On the other hand he will pant violently if the CO_2 content of the air is doubled (perhaps from his own exhalation in a closed, small space) even if there is plenty of oxygen. The use of green algae has been suggested to remove the excess carbon dioxide and turn it back into oxygen, but is too cumbersome. It is far more likely that a space ship would employ nuclear methods similar to those that purify the air in the atomic submarines.

Large quantities of CO_2 have been detected in the upper atmosphere of **Venus,** and smaller amounts on Mars. It would be quite wrong to draw any conclusions from this either for or against the existence of plant life on these planets. (See **Life.**)

CARBON STARS. A fairly rare class of stars of spectral types R, N or S, unusually rich in carbon which exists free in their atmospheres.

CASSEGRAIN TELESCOPE. A reflecting telescope in which the light is reflected from the main mirror on to a secondary mirror and thence back through a hole in the centre of the main mirror into the eyepiece. (See **Telescope.**)

CASSINI DIVISION. The main division in Saturn's ring system, separating Ring A from Ring B. It has a width of 2,500 miles, and can be seen with a small telescope when the rings are well displayed. It is caused, like the other divisions, by the perturbing influence of Saturn's satellites, especially Mimas. (See **Saturn.**)

CASTOR. A multiple star with three visual components; the two brighter ones are themselves spectroscopic binaries, and the faintest is an eclipsing binary, making a total of six observable components. (See **Binary Star.**)

Castor and a nearby star called *Pollux* are together known as The Twins. They are not a perfect pair, as Pollux is almost 0·4 magnitudes brighter than Castor.

CELESTIAL SPHERE. To the naked eye the clear night sky appears domed like part of the inside of a gigantic sphere, and it is convenient to describe the direction in which a star is seen by stating the place it seems to occupy on this imaginary dome, the *celestial sphere*. This sphere is considered to be concentric with the Earth. The points on it which appear to be stationary during its daily revolution are the celestial north and south poles, and are the points in which the Earth's axis produced would cut the celestial sphere.

The plane of the Earth's equator cuts the celestial sphere in the *celestial equator* or *equinoctial*, and the plane in which the Earth's orbit round the Sun lies cuts it in the *ecliptic*. When we look at the Sun we are looking along this plane, and the Sun therefore always appears to occupy a point on the ecliptic, and to move along it.

The celestial equator and ecliptic cut each other in two points. The one through which the Sun passes about the 21st of March is called the *First Point of Aries*, or *Vernal Equinox*, and is often denoted by the sign γ representing the horns of Aries, the ram. The point is named after the nearby constellation Aries, in which it used to lie when early astronomers determined its position, but it has moved gradually away from that constellation, owing to the **Precession of the Equinoxes.** Opposite to it on the celestial sphere is the *First Point of Libra*, or *Autumnal Equinox*, symbolized by the scales, ♎.

Just as on the Earth the position of a place is often given in terms of its latitude and longitude, so the position of a star on the celestial sphere is stated in terms of its

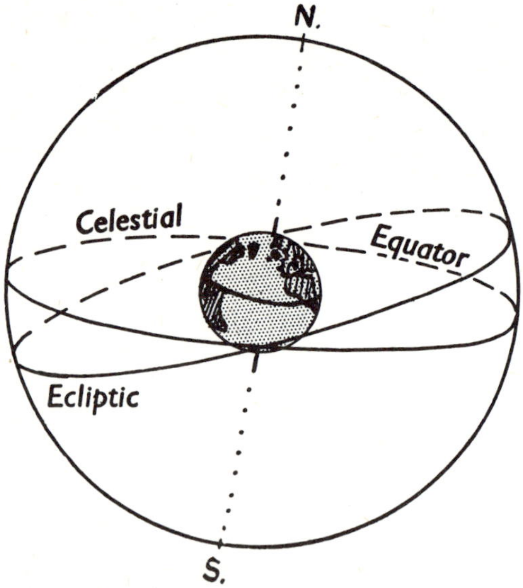

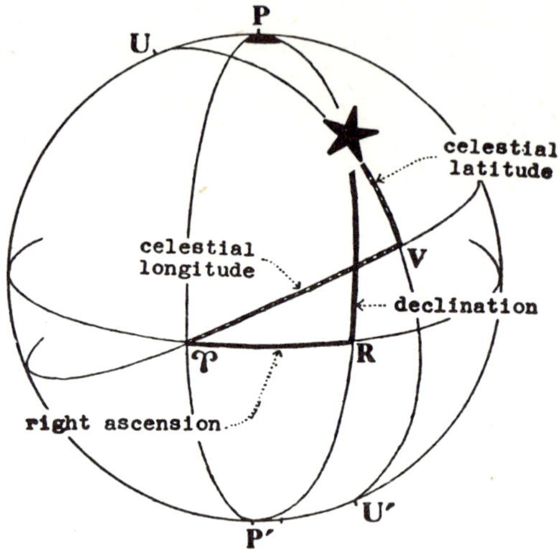

THE CELESTIAL SPHERE: *P* and *P′* are the north and south celestial poles, *U* and *U′* the poles of the ecliptic. Right Ascension is measured eastward along the celestial equator from the intersection of the equator with the ecliptic, i.e. from the First Point of Aries.

(*Admiralty Manual of Navigation*)

Declination and *Right Ascension.* Instead of the Meridian of Greenwich, we draw an imaginary celestial meridian through the celestial pole and the First Point of Aries. The Right Ascension of a star is then the angle between this meridian and the meridian through the star. This corresponds to longitude on the Earth.

The declination of a star is the angle as seen from the centre of the celestial sphere between the star and the point on the celestial equator nearest to it; this corresponds to latitude on the Earth.

(Declination and Right Ascension must not be confused with *Celestial Latitude* and *Longitude*, which are similar in principle but are seldom used. The difference is made clear by the diagram.)

If a line is drawn from a star to the centre of the celestial sphere, which is also the centre of the Earth, this line will cut the Earth's surface in a particular point. This point is then at that moment the *Geographical Position* of the star, and to an observer at that place the star will appear directly overhead, i.e. at his *zenith.* This concept is fundamental to marine navigation.

Globes and a variety of map projections are used to represent the surface of the Earth. The same principles may be used to prepare **star globes** or star charts and atlases, but it is necessary to remember one important difference; places close to each other on a terrestrial map will in fact be close, but stars may appear to be almost next to each other in the sky (and therefore on a star chart) and yet be more removed from each other than others that seem farther apart, i.e. which have a larger angular separation, as one star may be very far *beyond* its apparent neighbour. A star chart therefore gives no information about the true distances between celestial bodies. (M.T.B.)

CENTRE OF GRAVITY. The centre of gravity of any body is the point through which its entire mass may be held to act for most purposes.

Every body with clearly defined limits has such a centre of gravity, whether it be a solid, irregularly-shaped lump of matter or a vast sphere of gas like the majority of stars. The centre of gravity of a sphere is at its centre; that of a uniform rod is in the middle of the rod, half-way along its length; but sometimes it lies outside the body itself, as for instance in the case of a horseshoe, whose centre of gravity is in the middle of the space within the curve of the shoe.

When celestial bodies revolve about each other as they do in binary and planetary systems, they revolve strictly speaking about their common centre of gravity. The common centre of gravity of any two bodies always lies on the straight line joining their separate centres of gravity, and divides that line in the inverse ratio of their masses, so that it is always nearer the heavier body.

CEPHEID VARIABLE. A **giant star** which undergoes periodic changes in luminosity and size. The longer the period, the greater the absolute magnitude of this type of star, and since both the length of the period and the apparent magnitude are readily observable, the distance of a Cepheid variable can always be found. (See **Variable Stars** and **Magnitude.**)

CERENKOV DETECTOR. A detector of fast-moving sub-atomic particles which makes use of the *Cerenkov effect*. This effect is produced when a particle enters a substance such as glass at a velocity greater than that of light in the glass. (The theory of **relativity** shows that no object can travel faster than light in free space.) A 'shock wave' is produced, the particle being rapidly slowed down with the emission of light along the surface of a cone. The angle of the cone depends upon the velocity of the particle in the same way as the angle of a boat's wake depends on the boat's velocity through the water. The Cerenkov detector is constructed to receive only light emitted by this effect, and by positioning the detector to receive light from a Cerenkov cone of particular angle only particles within a given velocity range are detected.

CERES. The largest of the minor planets or **asteroids,** though not the brightest, and the first to be discovered (by Piazzi, in 1801). It proved to have the distance required by **Bode's Law** for the 'missing' planet. It has an estimated diameter of 429 miles, and a sidereal period of 4·6 years. At opposition it can attain magnitude 7·4.

CHARACTERISTIC VELOCITY. The sum of all the velocities that have to be attained, or overcome for purposes of braking, by a rocket intended for a particular journey.

The following example, worked out for a journey from the Earth to the Moon and back, will illustrate the concept. The rocket would first have to reach a speed somewhat greater than the **escape velocity** for the Earth. It would then require no more fuel until it had to brake its fall on to the Moon, and in the absence of an atmosphere this could only be done by firing its rocket motors in the direction opposite to the line of fall. Where there is no friction, exactly as much fuel is needed for deceleration as for a corresponding **acceleration.** The Moon's escape velocity is 2·34 km./sec., and fuel to develop at least this speed in a forward and backward direction will be needed, backward to cancel the speed of fall, and forward for the subsequent take-off. Allowing for **gravitational loss,** and assuming that the landing on the return to Earth can be effected without power by following a path of **braking ellipses,** the characteristic velocity may be calculated from this addition:

Escape from Earth	12·5
Landing on Moon	3·0
Take-off from Moon	3·0
Navigational corrections . .	0·5
Characteristic velocity	19·0 km./sec.

Neglecting slight differences in gravitational loss, this means that the fuel needed for such a journey, if fired all in one burst, would have to be enough to develop this speed. In other words, the characteristic velocity indicates how much fuel (or, more correctly, propellent) is required.

CHROMATIC ABERRATION. A defect inherent in optical **lenses,** due to the fact that they bend rays of light of different colours

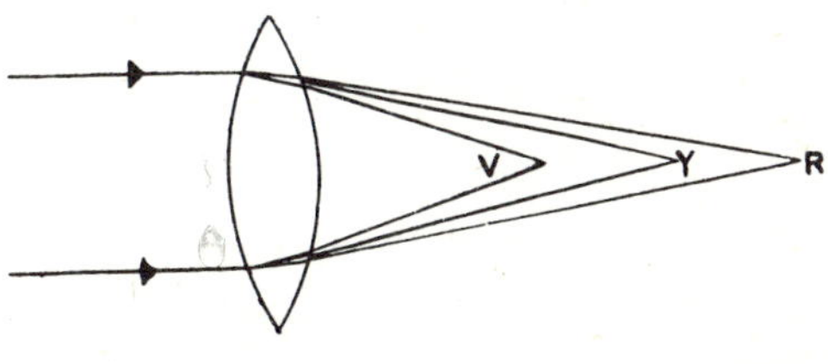

A simple lens bends light of different colours unequally. The three foci shown are those of violet, yellow and red light.

unequally. Violet light is refracted most, and red light least. Though the difference is slight, it does mean that an uncorrected lens will not bring all the colours of an image into focus in the same place, so that the image will be blurred. The coloured fringes seen near the edge of the field of view of an ordinary magnifying glass are an example of chromatic aberration.

The remedy is to use compound lenses containing three or even as many as six components all made of different glasses or other optical materials, such as quartz. Some components are concave and others convex lenses, and they are so arranged that, through partial cancellation of errors, the whole system brings all colours to the same focus, or at least all those colours to which the eye (or the camera) is most sensitive.

What is a nuisance in lenses is fundamental to the action of a **prism** in producing a spectrum, and without chromatic aberration spectroscopy would lack its chief tool.

CHROMOSPHERE (literally 'sphere of colour') is one of the inner layers of the Sun's atmosphere. Its name arises from the pink colour of its light, which is very evident during a total eclipse. (See **Sun.**)

CIRCLE. The path described by a point which moves in a plane, maintaining a fixed distance (the *radius*) from a given point (the *centre*). It is a **conic section** of zero eccentricity.

CIRCULAR VELOCITY. The velocity with which a satellite must move to describe a circular orbit about its primary. (See **Orbit.**)

CLOCK. See **Time Measurement.**

CLOCK STARS. A number of bright stars whose **right ascension** is accurately known from observations made over long periods, used in **time measurement.**

CLOUD CHAMBER. See **Wilson Cloud Chamber.**

CLUSTER. See under **Galactic Cluster, Globular Cluster.**

CLUSTERS OF GALAXIES. These are aggregations of entire galaxies, and are not to be confused with galactic clusters, which occur in and are part of our own galaxy and consist of stars. Clusters of Galaxies are discussed under **Galaxies.**

COAL SACK. An area of the Milky Way in the southern sky where stars are greatly obscured by a cloud of dark material (see **Galactic Nebula**). It appears devoid of stars to the naked eye, although a telescope shows them to be shining dimly through it.

COLLIMATOR. An optical arrangement for collecting light from a source into a parallel beam. It is often simply a converging **lens,** with the source at its focus. It forms an important part of spectroscopes. (See **Spectroscopy.**)

COLOUR INDEX. The human eye differs from the photographic emulsion in its sensitivity to light. It is in fact blind to many parts of the **electromagnetic spectrum** that are registered by the camera. Further, in order to define exactly what is meant by the visible part of the spectrum, a camera with a filter transparent only to certain wavelengths (i.e. the visible range) is used.

The system most commonly used to measure **magnitudes** in different parts of the electromagnetic spectrum is the U, B, V system; where U is the apparent magnitude in the ultraviolet part of the spectrum, B the apparent magnitude in the blue part of the spectrum and V the apparent magnitude in the visual part of the spectrum. These magnitudes are all obtained using filters with the camera and the wavelength range of each filter is specified. The colour index is defined as the difference between any two of these magnitudes, for example:

$$\text{Colour Index} = U - B.$$

The region of the spectrum in which the radiation of a star is at a maximum depends largely on its temperature. The colour index ($B - V$), in fact, is the most convenient measure of its temperature in most cases.

COLOUR-MAGNITUDE DIAGRAM is a variation of the **Hertzsprung–Russell diagram,** with **colour index** substituted for spectral type.

COMA. The vaguely defined area of light at the head of a **comet,** in which the nucleus is embedded.

The word is also used in astronomy for the blurred haze sometimes surrounding the images of stars in the outer parts of a photograph, due to optical imperfections.

COMET. A nebulous body which revolves about the Sun in an elongated ellipse. The name is derived from a fanciful resemblance of the tail to a tress of hair streaming in the wind (Greek *kometes*, long-haired), and this may have led to the erroneous opinion that comets rush across the sky (like meteors) with their tails streaming behind them. The fact that tails of comets always point away from the Sun seems to have been noticed by the Chinese, and the first known observations appear in the annals of the dynasty of Thong about the comet of A.D. 837. In the Western world it was first remarked upon by Peter Apian in 1531, while Tycho Brahe showed that comets are farther away than the Moon.

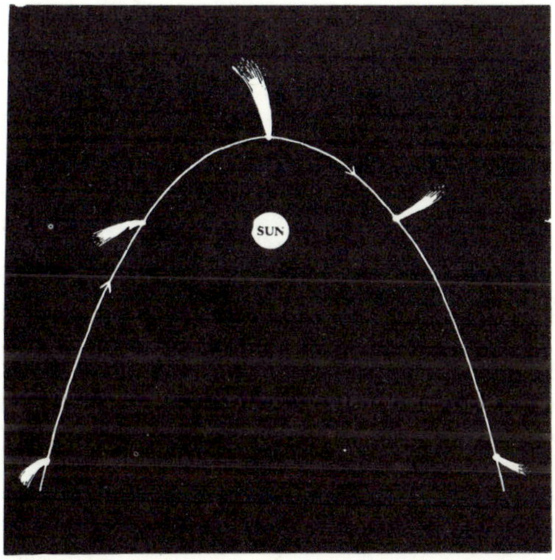

Since their motion is the same as that of the planets, it follows that they have similar apparent movements, remaining in sight for long periods, while moving slowly among the stars. The fainter comets may be seen only for a few days, depending on the circumstances of their discovery and the direction of their motion, whether towards or away from the Sun.

Comets are always brightest when near the Sun, and are then seen in the western sky at sunset, or in the East at sunrise, their tails pointing upwards from the horizon. As a comet approaches the Sun, its tail grows longer and larger, and the whole object brightens. As it swings round the Sun, the tail, still streaming away from the Sun, swings round also, so that it *precedes* the comet in its outward journey. Not all comets are seen like this. Most of them are disappointingly small nebulous objects, vague in form, and without any kind of central condensation. This may develop later, taking the form of a stellar *nucleus*, bright and starlike in appearance. The nebulosity surrounding the nucleus is called the *coma*, the two together forming the *head* of the comet. At a later stage, generally at a distance of about $1\frac{1}{2}$ units from the Sun, the coma becomes unsymmetrical, showing that a tail is beginning to develop; in some cases a definite tail is seen, and may, in exceptional cases, reach imposing proportions (see e.g. **Halley's Comet**). The brightening of the whole object as it approaches the Sun is accompanied by changes in the structure; *jets* (or *beards*) of luminosity are seen, extending in a direction *towards* the Sun, and these are often swept back to form envelopes which are clearly parabolic in shape and which, like the coma, merge imperceptibly into the tail. The coma itself, rather surprisingly, contracts as the comet nears the Sun, and expands again after perihelion passage. Changes in the structure of the tail are frequent, and a comet may develop many separate and clearly distinguished tails.

In the past, discoveries of comets were made entirely by chance, and before the use of the telescope only the brightest comets were seen. With optical aid, it became possible to search for comets, and many discoveries were made by comet-hunters such as Pons and Messier. There is still room for this class of work. In modern times the Japanese comet-hunters, using small telescopes, have been highly successful, and Honda has now 12 independent discoveries to his credit, but is closely followed by Kaoru Ikeya and Tsutomu Seki. In Europe, G. E. D. Alcock has so far made 5 discoveries using powerful 25 × 105 binoculars.

Only in the case of the periodic comets, for which predicted places are published in advance, is any systematic search made with large instruments, and this only in a very few centres. Few observatories can afford to devote the time of their instruments to routine work on comets, and this branch of astronomy is greatly neglected, especially in the southern hemisphere. When large instruments can be employed, they are capable of detecting comets at a much fainter stage than was ever possible in the past, particularly in cases where photography is used with telescopes of short focal length: no less than ten comets have been found in the last five years in the course of the sky survey undertaken with the 48-inch **Schmidt camera** of the Mount Palomar observatory. Similarly, the Czechoslovak astronomers have made something of a record with their discoveries of comets in the course of their routine sweeps of the sky. The number of comets found varies from year to year:

Year	*New*	*Periodic*
1964	3	5
1965	5	5
1966	4	2
1967	4	10

Since in any one year some of the previous year's discoveries will still be under observation, the number actually visible during the year is larger than shown above; in 1951, no less than 22 comets were under observation.

TAILS. The tail of a comet is always its most impressive feature, and even when quite small may show remarkable changes of form. The tails of the great comets may extend half-way across the sky, exciting alarm and superstitious fear. Yet the tail is extremely tenuous, being quite transparent, and having no obvious effect on the light of the stars seen through it. Such a tail may be 50 to 150 million miles in length, and perhaps 5 to 10 million miles wide at the end farthest from the Sun, yet in spite of this great size, it can possess very little mass. There is often evidence of curvature in a long tail, and this effect is greatly enhanced by perspective, the curvature being exaggerated when the observer sees the tail foreshortened.

It is now generally agreed that there are two kinds of comet tails: dust tails and gas tails; they may also be defined as either curved or straight.

The curved tail is a variety which is generally seen in the great spectacular naked-eye comets. Spectroscopic examination of these curved tails reveals that they shine by reflected light and, therefore, must consist of extremely finely divided particles driven out from the region of the nucleus by solar radiation pressure which can for certain size particles exceed the force of gravity and so separate any dust particles from a comet and scatter them behind to form a curved path. Maximum repulsion is experienced on particles of the order of one-third of the wavelength of light which falls on them. This effect is a real one and has been conclusively proved by laboratory experiment. However, the least spectacular straight tails, which are often observed as a separate tail component with the bright comets, are generally found with the fainter telescopic comets of short period during the perihelion side of their orbits. They are much narrower and straighter than the dust tails, and knots of material have been observed moving along them with speeds of several hundred kilometres per second. In the case of Morehouse's comet in 1908, the repulsion force in the gas tail was found to exceed the force of gravity by a factor of 800.

Gas tails are usually straight and narrow and their direction follows closely that of the Sun's radius vector. The study of gas tails has attracted much attention recently, due to the fact that the physical processes taking place in them hold an important key to the development of plasma physics generally. The main constituent of a gas tail, or plasma tail, as it might now be called, is always the single ionized carbon monoxide molecule.

It is generally thought that new comets show predominantly dust tails while short period comets, Encke's, for example, have predominantly gas. The recent bright comets of 1957 had both gas and dust, but the analogous sunward spike of Arend-Roland 1956 *h*, very reminiscent of the bright comet of 1851, was a true dust example, unique in that it was so tenuous that it was observable only in the exact line of the orbital plane when looking through its entire depth. Many comets must possess similar dust tails, but they must often go unobserved since it is extremely rare for the Earth to be so ideally positioned.

COMETARY NUCLEUS AND HEAD. In considerations of the nature and structure of comets, the component called the nucleus is one which attracts most attention. Among the different ideas and theories regarding the physical structure of comets, one of the most acceptable ones is that of Whipple (which is derived from an earlier but less sophisticated model) since it accounts for many of the observed physical peculiarities and yet provides for differences between comets. Colloquially it is often referred to as 'the dirty snowball model'. The idea visualizes the nucleus as a discrete mixture of ice-conglomerate consisting of a highly porous mass of solidified gases – or ices – of water, ammonia, methane and possibly carbon dioxide and dicyanogen, including occasionally solidified particles. Sometimes this model is also visualized as a yeasty raisin-bread.

COMET MOREHOUSE on November 19, 1908, showing the formation of multiple tails. (*Royal Greenwich Observatory*)

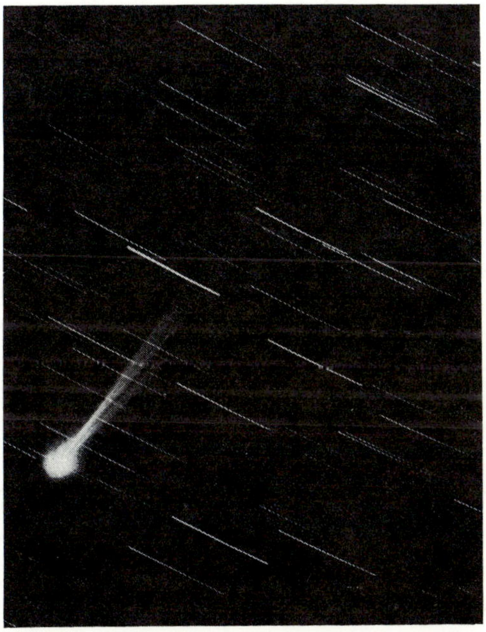

COMET FINSLER, 1937. During the five-hour exposure, the comet moved relative to the background stars. As the camera was kept guided on the comet, the stars appear as trails on the photograph. (*Norman Lockyer Observatory*)

This picture is broadly in agreement with the spectral changes observed in the coma. But all bright comets near the sun also show prominent sodium lines, and spectrograms of the nucleus of the Ikeya-Seki sun-grazer comet, when within a few million miles of the Sun, showed bright lines of neutral iron, nickel, chromium, silicon, manganese and neutral calcium, also the H and K lines of ionized calcium.

An alternative model is put forward by Lyttleton which is similar to some 19th-century ideas. It is known as 'the sandbank model', and envisages the nucleus and the coma of a comet as a continuous whole – in the shape of a gigantic particle cloud of widely scattered gas and dust particles, whose mean distance of separation is large – which is concentrated toward the centre, giving rise to the appearance of a pseudo-discrete nucleus. The much discussed Tungus 'meteorite' which fell in Siberia in 1908, of which no cosmic material has ever been recovered, but which caused tremendous destruction by the effects of blast, has often been attributed to the result of a cometary impact with the Earth.

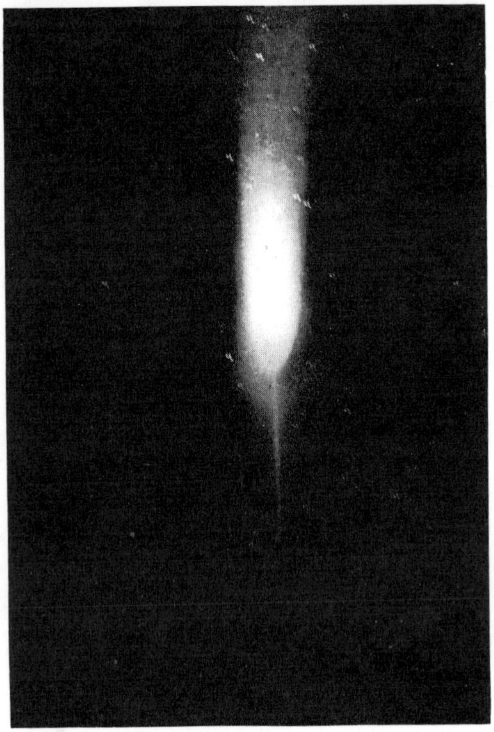

THE AREND-ROLAND COMET, 1956 *h*. Discovered in 1956, this exceptional comet was the brightest to be seen in northern latitudes since the Daylight Comet of 1910. An unusual feature was the long spike projecting from the head *towards* the Sun. The picture shows just over two-thirds of the 7,000,000-mile length of the comet.
(*University of Michigan Observatory*)

Lyttleton's theory would fit in with this explanation, and also with another attractive idea (the micro-particle sandbank model) that the cometary head is akin to a concentrated aggregation of very small and friable micrometeorites, many of which have a fluffy, loosely packed non-crystalline structure (see **Micrometeorites**). If an aggregation of such matter encountered the Earth's atmosphere at cosmic velocity, it would disintegrate into tiny dust particles and no fragments recognizable to the naked eye would reach the Earth's surface intact.

At the present time, however, the discrete nucleus theory has more adherents than the particle-cloud idea. There are authorities who support the idea of the monolith model of a single large block, or blocks of the order of a kilometre upwards, analogous to asteroidal bodies. Earth-based telescope observation unfortunately cannot settle any of these issues, although the work of Elizabeth Roemer, at Flagstaff, showed that, photographically, some of the observed so-called nuclei appeared to be no larger than one or two kilometres. However, her work was unable to give indications of minimum dimensions. Some authorities consider that sizeable nuclei are an absolute necessity if only in order to explain the survival after perihelion passage of the sun-grazing comets.

In general terms then the most acceptable model would consist of a cloud of particles and possibly chunks of matter – most certainly dust and meteoric particles and likely ices in various forms. The picture of a continuous spectrum which cometary nuclei exhibit lends support to this idea.

In contrast to the nucleus (with some notable exceptions, however), the spectrum of the cometary head or coma is not due to reflected sunlight. It is represented by emission bands due to direct radiation by gas molecules, those clearly identified being C_2, CH, CN, OH, NH, NH_2, CH_2, and the ions N_2^+, CO^+, CH^+ and OH^+, all of which are chemically unstable and owe their existence in comas and tails of comets to the extremely low density and freedom from collisions.

LUMINOSITY. It is customary in comet work to quote the brightness of a comet in terms of ordinary stellar magnitudes, but the results are not entirely satisfactory. While it may be possible to estimate the brightness of the stellar nucleus, an estimate of the total magnitude is much more difficult. Attempts have been made to standardize procedure by comparing the total light of the comet with that of a standard star out of focus; but such estimates leave much to the judgment of the observer, and the results are frequently discordant. In spite of this, the measurements are of importance in revealing the remarkable changes that take place in the luminosity of certain comets.

The brightness of a planet or comet must depend on the intensity of illumination it receives from the Sun, and on its distance from the Earth, being proportional to the inverse square of each of these quantities (see **Light**). But few comets follow such a simple law, and although the arbitrary introduction of inverse fourth and even sixth powers has served to predict the brightness of future returns of periodic comets, it is rare for the values to be followed with any accuracy. The rapid variations in the form and brightness of the tail of Comet Morehouse (1908) led to the suggestion that there was some connection between these changes and the presence of sunspots on the Sun, and similar ideas were put forward in the case of certain other comets.

Unfortunately, the period of visibility of a comet is generally too short to allow of an extensive series of measurements of this kind. The only comet for which numerous observations have been published is Schwassmann-Wachmann (1), which is almost continuously visible. Normally a faint object, it undergoes remarkable outbursts, such as that of January 1946, when it brightened from magnitude 16 to 10·2 on January 25 and to 9·4 on January 26; a week later it had dropped back again to magnitude 15. There was at the time a giant sunspot, turned towards the comet, and further outbreaks of the same nature seem to have a definite correlation with solar activity. It has been pointed out, however, that the amount of energy reaching the comet from solar outbursts is too small to do more than initiate some change within the comet itself. These great outbursts must apparently have an origin within the nucleus. This comet has shown much activity again in the middle 1960s, and outbursts are now almost regular annual events – but on these recent occasions there have been no positive correlations between solar activity and the comet's eruptions.

MASSES OF COMETS. Although the nucleus of a comet may be quite large – diameters up to several thousands of kilometres have been reported – the mass of a comet is always small, in comparison with that of the planets. It may be estimated by the absence of perturbative effects, as in the case of Lexell's comet of 1770, which passed within 1½ million miles of the Earth without causing any change in the length of the year. The same comet in 1779 passed so close to Jupiter that it crossed the orbits of the inner satellites, yet although the orbit of the comet itself was completely altered (it has never since been seen) the satellites were in no way affected. It is concluded that the mass of the comet must have been less than one-millionth of the mass of the Earth. But even this is by no means negligible, and a collision with a mass of this order, moving with planetary speed, would be catastrophic.

COMET BROOKS photographed during its return in 1911. Star trails can be seen through its tail and even through part of the head.

Because of their small mass and large size, the mean density of comets is extremely low, about half an ounce per cubic mile. (A cubic mile of air at sea-level weighs 5 million tons.) It seems likely that the nucleus consists of separate blocks of material, or of small particles widely separated. The comet of 1862 and Halley's Comet of 1910 both passed between the Earth and the Sun, but nothing was seen of them as they moved across the face of the Sun.

NOMENCLATURE. Comets are usually named after their discoverers, as many as three names being appended in some cases. For some important periodic comets, the names are those of astronomers who have worked on the perturbations of the orbits; examples are Comets Halley, Encke and Crommelin. At the time of its discovery the comet is designated by the year of discovery, followed by a Roman letter. When the orbits have been determined this numbering is changed to give the order of perihelion passage, the year being followed by Roman numerals. Thus comet 1951 *h* Comas Solá became 1952 VII, while 1951 *l* Schaumasse is known as 1952 III.

ORBITS. Since a comet moves under the gravitational force of the Sun, its orbit must be a **conic section.** Such an orbit can be calculated from three suitably spaced observations, but since only a small part of the path is covered in a short time-interval, it is convenient to assume that the path is a parabola. This simplifies the calculation, and is often sufficiently accurate, since the arc over which the comet is observed is but a small fraction of the whole orbit, and its curvature is almost indistinguishable from that of a parabola. When more observations have been made, it is possible to compute a more accurate orbit, and under modern conditions few parabolic orbits are published. This increase in accuracy enables us to be certain that all comets are members of the solar system, moving in elliptical orbits, which may, however, be so large and eccentric as to be almost parabolic in form, so that **perturbations** may be sufficient to convert them into hyperbolas, in which cases the comets would not return.

Present catalogues of cometary orbits give particulars of 761 orbits, representing 549 individual comets, of which 205 travelled in ellipses, 291 in parabolic and 53 in hyperbolic orbits. The number of periodic comets now known makes it imperative to make a continuous review of their perturbations, in order to make predictions for future returns. The prediction takes the form of an **ephemeris,** giving the position of the comet at stated intervals, and is based on the calculation of perturbations over at least one revolution. This is of little avail if the original orbit is not a good one; hence the work consists of (a) using all available observations to correct the orbit, (b) computing perturbations, and (c) computing the ephemeris. The perturbations in some cases are severe, and entail much labour in their computation; in less extreme cases, approximate methods are sufficient, and these are amenable to treatment by electronic computing machines.

STATISTICS. In any study of the orbits listed in the comet catalogues, it is at once obvious that we are dealing with a very restricted selection of orbits. For instance, the values of perihelion distance are always small, and mainly in the neighbourhood of one **astronomical unit;** this is a natural outcome of the fact that comets are visible to us only when they come near to the Earth. The extreme values for perihelion distance are 5·523 for Comet 1925 I, and 0·00549 for Comet 1880 I.

The orbit of a periodic comet is an ellipse with the Sun in one of its foci. For a given perihelion distance, there must therefore be a definite relationship between the length of the major axis of the ellipse and its eccentricity (see **Conic Section**). It is, in fact, one of the noticeable features of a comet catalogue that the comets of longest period (i.e. those with the largest values for the major axis) have also the largest eccentricities. All the comets with periods of more than 200 years have eccentricities greater than 0·96. The comets of shorter period have a much wider range of eccentricity, two having *e* less than 0·15, while the others lie between 0·3 and 0·993.

The inclinations of cometary orbits to the ecliptic are often stated to be distributed over the whole range from 0° to 180°, and while this is true, the distribution is far from being a random one. There is a definite tendency for the orbits to crowd towards the ecliptic, so that small inclinations near 0° and 180° predominate. This is particularly true of the short period comets, all of which, with only two exceptions, travel in direct orbits with small inclinations. There is some evidence of the effect of Jupiter on these short period comets, since their perihelia show a tendency to crowd towards the position of Jupiter's perihelion. Many of these comets have aphelia which lie near the orbit of Jupiter, and the effect is due to perturbations.

COMET GROUPS. It is always possible to find similarities between the elements of a newly discovered comet and those of an older one. In some cases the resemblances are so striking as to lead to the idea of comet groups, a very large number of which are known.

The most important of these population groups is the one named the **sun-grazing comets** – so called for their very characteristic close approach at perihelion passage to the Sun's surface – often actually passing through the tenuous, hot outer atmosphere. When close to the Sun, they can be observed in broad daylight and often attain a magnitude of −9. Interest in this particular group was revived by the sudden appearance of the spectacular southern comet Ikeya-Seki 1965 VIII. Recent orbital work has shown that a number of the sun-grazers may have originated in a common parent comet. The daylight sun-grazer 1882 II and the Ikeya-Seki comet probably separated at a previous perihelion passage that might well have been represented by the brilliant comet of A.D. 1106.

The sun-grazers have been observed in three distinct clusters: one in the 17th century, the second in the 19th century and the third in progress at the present time. The two strongest concentrations are with the comets 1880 I, 1882 I, the eclipse comets of 1882 and 1887, and then the recent comets 1945 VII, 1963 V, and 1965 VIII (Ikeya-Seki). They form two distinct orbital sub-groups which tend to support the idea of a break-up of the primordial comet at some early perihelion passage – only the main fragments consistently surviving between subsequent perihelion passages. Increased differential motions have resulted from the subsequent accumulation of planetary perturbations. It is the existence of the sun-grazing group which lends considerable support to the Lyttleton concept of the accretion stream idea for the origin of comets. All the sun-grazers appear to return to perihelion from a point in space which approximates near the bright star Sirius almost, in fact, opposite the Sun's way in space. The reason that such comets do not actually plunge into the Sun at perihelion passage is that the centre of gravitational mass of the solar system lies just outside the solar mass. All comets could certainly originate by way of the sun-grazing group. Long-term accumulative planetary perturbations – especially by the major planets – could easily account for the now considerable diversity of orbital characteristics given by modern observations.

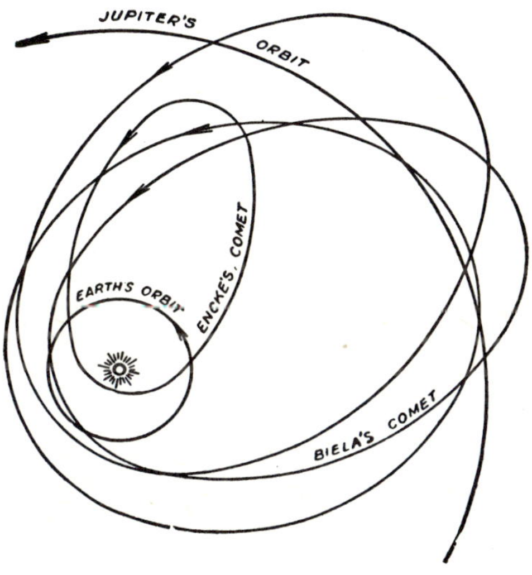

JUPITER'S FAMILY OF COMETS. All the above orbits have been strongly influenced by the great planet, and their aphelia now lie close to its course. Encke's orbit is also the smallest known, and the comet returns every 3·3 years.

SHORT PERIOD COMETS. Those comets which have been seen at more than one return to the Sun are listed in the Table of Comets, in order of period. It was the practice in the past to relate such comets to one or other of the major planets. Thus there is a large group of comets with periods of about 6 years, having aphelia which lie near Jupiter's orbit, at a distance of about 5 units; similar groups have aphelion distances comparable with the distances of Saturn, Uranus or Neptune. This idea of comet families, however, must be regarded as fortuitous, since in most cases the inclinations of the orbits will prevent the comets from passing near enough to the planet to come under its gravitational influence. Thus Halley's comet, which is frequently referred to as belonging to the Neptune family, has the smallest inclination of any comet associated with this planet; yet it actually passes at a distance of 8 units from the orbit of Neptune, although approaching much more closely to Jupiter. In general it may be said that Jupiter is almost entirely responsible for any of these arrangements.

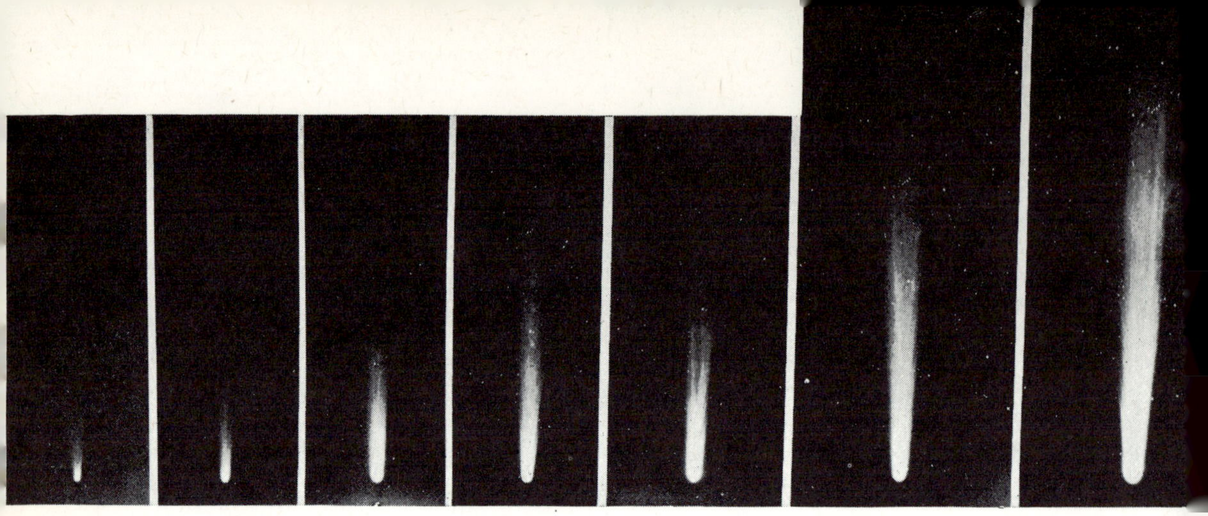

HALLEY'S COMET IN 1910. The alterations in the appearance of the comet between successive photographs are due to changes in distance and foreshortening as well as to actual changes of its shape. (*Mount Wilson – Palomar*)

The shape and size of the short period cometary orbits is clearly a result of prolonged perturbations by Jupiter, and the statistics of these orbits show several interesting correlations with that of Jupiter. Since all of the 6-year comets travel in direct orbits, they will be moving slowly at aphelion in the same direction as Jupiter, and may come under the influence of that planet for lengthy periods. In extreme cases there may be a close approach, when the orbit of the comet will be greatly altered. As an example, Brooks (2) 1889 V had its period changed from 29 years to 7 years by a close approach to Jupiter in 1886; at another close approach in 1921 the elements of the orbit were again altered: the longitude of perihelion has changed very little, but the whole orbit has altered its inclination about the line of **nodes** by about 12°. A more common case is that in which a comet suffers perturbations by Jupiter at each alternate revolution. Comet Pons-Winnecke shows this behaviour clearly, since its period is about half that of Jupiter. Its orbit is steadily becoming more circular and inclining at an increasing angle to the ecliptic. This comet gave rise to a meteor shower in 1916, when the distance between the comet's orbit and that of the Earth was a minimum. Another comet which has given rise to a great shower of meteors during the present century is Comet Giacobini-Zinner, first discovered by Giacobini in 1900, and recovered by Zinner in 1913. (See **Meteor.**)

Perhaps the most interesting of all the short period comets is **Halley's Comet.** Soon after Newton's work on the laws of gravitation, Halley collected observations of 24 comets, and deduced the elements of their orbits. Struck by the similarity between the orbits of the comets of 1531, 1607 and 1682, he concluded that they were three appearances of the same comet, and rightly attributed the difference in the intervals to perturbations by Jupiter. Although no certain method of computing the perturbations was then known, Halley ventured to predict the return of the comet in 1758. Its return on Christmas Day of that year, sixteen years after Halley's death, was the first of many such applications of Newton's Laws. In the following century many new methods for computing perturbations were invented, but most of these are too laborious for use in the case of a comet of such a lengthy period. In 1910 Cowell and Crommelin used a new method to predict the 1910 return, and this was so accurate that it won the *Astronomische Gesellschaft* prize for a successful prediction of this apparition. Subsequently these two computers carried the investigation of Halley's comet back to 240 B.C., and were able to check their figures against many ancient records. The most interesting of these is the appearance of Halley's comet in 1066, an event which is

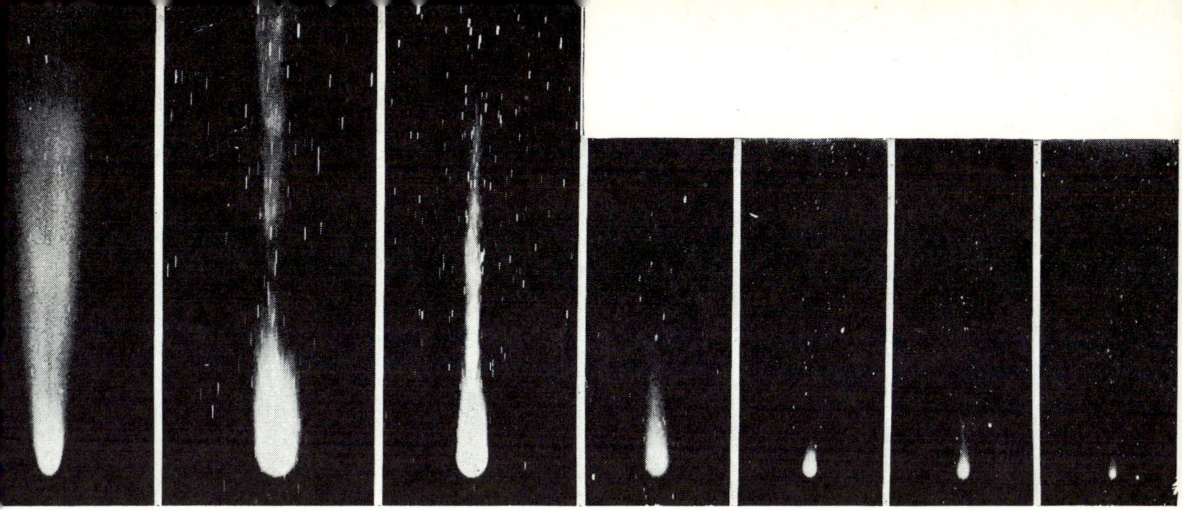

15 23 28 June 3 6 9 11

recorded on the Bayeux tapestry. The next return of this comet will take place in 1986.

It is interesting to note that, although Halley's comet travels in a retrograde orbit, perturbations are sufficiently great to cause the period to vary from 76·0 to 79·6 years. For more than three-quarters of this time the comet is beyond the orbit of Neptune, its speed at aphelion being only 0·91 km./sec. as compared with its perihelion speed of 54·56 km./sec. This is characteristic of all comets, but is most noticeable in the case of the more eccentric orbits, since the perihelion and aphelion speeds are in the ratio of $1 + e$ to $1 - e$.

Encke's Comet, which has the shortest period known, furnished the second instance of the return of a comet. Encke was a pupil of Gauss, who had devised a new method for computing elliptical orbits, and in 1818 Encke computed the orbit of a new comet discovered by Pons. Not only did it prove to have the short period of 3·3 years, but it was also found to be identical with Comets Méchain 1786, Caroline Herschel 1795 and Pons 1805. Encke rounded off this notable piece of work by predicting the return of the comet in 1822, when it was duly recovered. This comet appears to be associated with the **Taurid** meteor shower.

Although most comets travel in eccentric orbits, there are two exceptions whose paths are certainly no more eccentric than those of the minor planets. Comet Schwassmann-Wachmann (1) has the smallest eccentricity of all comets (0·136) and its orbit lies wholly between those of Jupiter and Saturn. It has been seen each year since its discovery in 1925, and its remarkable outbursts of light have already been referred to above.

Comet Oterma which used to be visible at each opposition has an orbit between Mars and Jupiter similar to that of an asteroid of the Hilda group. Between July 1962 and January 1964, it came under the sphere of Jupiter's influence which for a short time caused its motion to correspond to an ellipse round the planet as would a distant satellite. Since, however, it has direct motion instead of retrograde, the comet evaded capture, but moved off in a larger, more eccentric, orbit and so became invisible from the distance of the Earth and suffered a fate similar to that of Lexell's comet. Some of the short period comets have become lost from perturbations changing their orbits so that they leave the solar system for ever in hyperbolic paths or stay at great distances from the Sun in practically circular orbits. Some authorities consider that many lost comets have, in fact, disintegrated completely.

LONG PERIOD COMETS are known with times of revolution up to many thousands of years. The larger values must be regarded with caution; thus the period of 3,910,000 years quoted for Comet 1910 I merely means that the orbit was almost a parabola. The astronomer is more interested in the dimension *a* than in the period, but *a* is strongly correlated with *e* (see **Conic Section**). Such a large value of *a* therefore implies a large value of *e*, and the orbit may well be changed from an ellipse to a hyperbola by planetary perturbations. Nevertheless, long periods must be accepted for the majority of comets, and if this is so, their numbers must be very great. Thus Crommelin estimated that comets might have an average period of 40,000 years, and if 300 new comets are seen in each century there must clearly be at least 120,000 of them revolving about the Sun.

All of the hyperbolic orbits must belong to this group. In all such cases the eccentricity is only slightly greater than unity, and in every case where sufficiently accurate data are available, it has been proved that the hyperbolic nature of the path is due to planetary perturbations, the comet having previously revolved in an elongated ellipse. The facility with which such changes may take place is indicated by the fact that Halley's Comet, which has a speed of 54·56 km./sec. at perihelion, would have its orbit converted into a hyperbola if this speed were increased to 55·1 km./sec. The changes necessary to alter a comet of really long period in this way would be very much less.

One of the most interesting long period comets is *Ikeya-Seki 1965 VIII*, which two Japanese amateurs discovered independently and within a few minutes of each other on September 17, 1965. Preliminary orbits showed that this was a sun-grazing comet similar to that of 1882. The day before perihelion passage, it could be seen in broad daylight shining at about magnitude −9. Two weeks later, when it presented a view in profile, its tail reached a length of over 40°.

At the time of perihelion some observers reported the disintegration of the comet's head into three distinct segments, one of which was brighter than the others. Farther away from the Sun, only a double nucleus was visible which gradually gave way to the form of a single one.

The appearance of this comet was the first favourable opportunity for making physical observations of a really bright comet using modern equipment, especially with high-resolution spectrographs utilizing a broad range of wavelengths. Photographs were secured for the first time showing emission lines of neutral iron, nickel, chromium, silicon and manganese. No radio emissions were detected. The period was calculated at between 875 and 1,040 years.

Another interesting comet was Humason (1961*e*) discovered in September 1961. It was found to have a period of approximately 2,900 years and at perihelion came no nearer to the Sun than 198 million miles (approximately the orbit of Mars). Even at the distance of Jupiter's orbit it was plainly visible telescopically and developed a prominent and very active tail consisting of both gas and dust. This extremely large comet may be representative of a group of comets which revolve round the Sun at great distances, but which are generally too faint to be observed from the distance of the Earth. (A colour photograph of Humason faces page 240.)

DISRUPTION OF COMETS. The actual disruption of a comet has been witnessed on various occasions, but there is no evidence to connect such an event with perturbations or any other external phenomenon. The sun-grazing comets such as that of 1843 must have undergone severe heating and other disruptive effects, since they actually passed through the solar corona, yet their rapid passage round the Sun seems to have caused little change. In the case of the comet of 1882 the nucleus was seen to have broken into separate parts, which were referred to as a 'string of pearls'. The most interesting case of disruption is that of Biela's comet, which was discovered in 1826, and shown to be the same comet as those of 1772 and 1805. It was recovered in 1832 and again in 1846, but in this year the comet divided into two parts. The two comets travelled side by side and were seen again at the 1852 return, rather more widely separated. They have not been seen since, but in 1872 there occurred a spectacular display of meteors from a **radiant** which was shown to have a position in agreement with the elements of the comet. A recent event of this kind occurred in connection with Comet 1955 *g* Honda which was seen at Lick to have twin nuclei 5″ apart.

ANOMALOUS COMETARY MOTION. The so-called non-gravitational secular accelerations and decelerations which occur in specific short period comets (e.g. *Encke*) are most perplexing and unexplained. Recent exhaustive work with very fast computers has shown quite definitely that some non-gravitational force is at work. Attempts to account for these irregularities by encounters with

TABLE OF COMETS *

COMET		No. of Appearances	Period (Years)	Perihelion Distance (in A.U.)	Eccentricity	Inclination
1953 *f*	Encke	48	3·30	0·338	0·847	12°.4
1952 IV	Grigg-Skjellerup	9	4·90	0·856	0·704	17°.6
1954 *a*	Honda-Mrkos-Pajdus-áková	3	5·21	0·556	0·815	13°.2
1951 VIII	Tempel (2)	13	5·27	1·391	0·543	12°.4
1927 I	Neujmin (2)	2	5·43	1·338	0·567	10°.6
1879 I	Brorsen (1) (*lost*)	5	5·46	0·590	0·810	29°.4
1951 IV	Tuttle-Giacobini-Kresák	3	5·49	1·117	0·641	13°.8
1908 II	Tempel-Swift	4	5·68	1·153	0·638	5°.4
1894 IV	De Vico-Swift	3	5·86	1·392	0·572	3°.0
1879 III	Tempel (1) (*lost*)	3	5·98	1·771	0·463	9°.8
1951 VI	Pons-Winnecke	15	6·12	1·159	0·654	21°.7
1951 VII	Kopff	8	6·32	1·495	0·556	7°.2
1948 VIII	Forbes	3	6·42	1·545	0·553	4°.6
1955 *i*	Perrine-Mrkos	4	6·46	1·154	0·667	15°.9
1954 *g*	Schwassmann-Wachmann (2)	6	6·53	2·150	0·385	3°.7
1953 *d*	Reinmuth (2)	3	6·71	1·867	0·469	7°.1
1946 V	Giacobini-Zinner	7	6·42	0·996	0·717	30°.7
1852 III	Biela (*lost*)	6	6·62	0·861	0·756	12°.6
1950 V	Daniel	4	6·66	1·465	0·586	19°.7
1954 *j*	Wirtanen	3	6·67	1·625	0·542	13°.4
1950 II	d'Arrest	10	6·70	1·378	0·612	18°.1
1953 *i*	Finlay	7	6·90	1·049	0·708	3°.4
1953 *b*	Brooks (2)	10	6·72	1·866	0·487	5°.6
1954 *b*	Borrelly	7	7·01	1·448	0·604	31°.1
1906 III	Holmes (*believed lost, recov.* 1964)	4	7·35	2·347	0·379	19°.5
1954 *e*	Faye	14	7·41	1·652	0·565	10°.6
1955 *d*	Whipple	4	7·42	2·450	0·356	10°.2
1955 *c*	Ashbrook-Jackson	2	7·51	2·324	0·394	12°.5
1950 IV	Reinmuth (1)	4	7·65	2·037	0·477	8°.4
1950 III	Oterma (3)	—	7·89	3·390	0·144	4°.0
1952 III	Schaumasse	6	8·17	1·194	0·706	12°.0
1950 VI	Wolf (1)	10	8·42	2·498	0·396	27°.3
1952 VII	Comas Solá	5	8·59	1·766	0·578	13°.5
1949 V	Väisälä (1)	3	10·46	1·752	0·635	11°.3
1951 V	Neujmin (3)	2	10·95	2·032	0·588	3°.8
1938 I	Gale	2	10·99	1·183	0·761	11°.7
1939 X	Tuttle	8	13·61	1·022	0·821	54°.7
1941 VI	Schwassmann-Wachmann (1)	3	16·10	5·523	0·136	9°.5
1948 XIII	Neujmin (1)	3	17·97	1·547	0·774	15°.0
1928 III	Crommelin	6	27·87	0·745	0·919	28°.9
1942 IX	Stephan-Oterma	2	38·96	1·596	0·861	17°.9
1913 VI	Westphal	2	61·73	1·254	0·920	40°.9
1919 III	Brorsen-Metcalf	2	69·06	0·485	0·971	19°.2
1956 *a*	Olbers	3	69·57	1·179	0·930	44°.6
1953 *c*	Pons-Brooks	3	70·88	0·774	0·955	74°.2
1910 II	Halley	29	76·03	0·587	0·967	162°.2
1939 VI	Herschel-Rigollet	2	156·00	0·748	0·974	64°.2
1907 II	Grigg-Mellish	2	164·30	0·923	0·969	109°.8

* For 1973 *f*, Comet Kohoutek, please see **Appendix.**

COMET PERRINE, 1902.
(*Royal Greenwich Observatory*)

dense meteor streams, by spin of the comet head induced by asymmetrical evaporation of ice particles, or (more plausibly) by the expulsion of matter from the head all leave a great deal to be desired.

ORIGIN OF COMETS. Since man first observed comets many ideas have been put forward to explain their continuous presence in the solar system. Today there are three main approaches. The first one is Lyttleton's idea that the Sun during its passage through an interstellar homogeneous cloud of both gas and dust produces condensations within it and impresses on these condensations periodic orbital forms. Oort and Van Woerken consider that the birth of comets took place between one million and one billion years ago through the disintegration of a planet (Olbers' hypothetical planet); subsequent catastrophic perturbations by Jupiter in particular have driven the majority of the comets to form a wide circum-solar belt well beyond the orbit of Pluto. Perturbations induced by nearby stars occasionally force them in towards the Sun, and this would give rise to something like the observed pattern of orbits. They argue that comets and minor planets have a common mode of origin.

The third idea, put forward by Lagrange in 1814, received considerable support in the 19th century. It states that eruptions or volcanic-like explosions have expelled comet material from the planets Jupiter, Saturn, Uranus and Neptune. The great objection comes when the necessary escape velocities are considered: 67 km. per sec. for Jupiter and 42 km. per sec. for Saturn. From a study of known orbits the evidence is overwhelmingly in favour of capture rather than creation by Jupiter. All short periodic comets we see have almost certainly been captured from a long period orbit when the comet passed close to Jupiter at some previous apparition. No orbit has yet been investigated which indicates that the comet could have been ejected from a planetary surface or satellite.

COMET SPACE PROBES. It is doubtful whether the true physical nature of comets will be determined by Earth-bound observations. Only a space probe dispatched through the head of an active comet – with possibly a landing on the nucleus, if such a discrete object exists – offers hope of settling the physical problems. Such fly-bys are now possible and have been planned in detail by various authorities, but at the present time these probes hold low priorities. Only three comets offer reasonable opportunities between 1970 and 1986; Halley's comet when it returns to perihelion in 1986 will present great difficulties owing to its retrograde motion round the Sun. However, a bright new comet with the right orbital parameters could appear unexpectedly at any time. (J.G.P., P.L.B.)

COMPUTER. A machine that performs laborious calculations. Computing machines now play a major part in all branches of science, and in none more than in astronomy and space research. They can do nothing that a skilled human operator cannot do, given enough time, but they work incomparably faster.

There are three main categories: analogue, digital and punched-card computers, and many systems use combinations of these. An *analogue computer* makes use of models whose physical or electrical dimensions vary in a way that is analogous to the mathematical functions of the problems it is designed to solve. They may be regarded as highly elaborated, automatic slide-rules; they are relatively simple and cheap, but cannot normally deal with more than one type of work.

A *punched-card machine* is well adapted for statistical work. Information is entered on a card by making holes in it in certain positions. Thousands of such cards, all punched according to the same system for any given problem, can be fed through the machine in a few minutes. Electric contacts 'feel' the holes, and activate mechanisms which sort and stack the cards and perform other simple operations on them according to a programme which may itself take the form of a punched card.

By far the most interesting and flexible computers are the *electronic digital* ones. They can handle virtually *any* problem provided that: (*a*) there is a system (which may involve arbitrary rating procedures by a human operator) whereby all qualitative terms such as 'worth-while', 'undesirable', etc., are unambiguously translated into numbers; (*b*) the operator can, in a code or 'language' which the machine can accept, state the problem clearly, define the method of solution and the character of the information required, and lay down such criteria and restraints as he considers appropriate. It is also necessary for the operator to realize that he will get an answer only to the precise question he asked, and that he cannot blame the computer if he then draws unwarranted conclusions from the answers concerning, for instance, tactical, strategic, economic or business decisions. A computer can answer as truthfully and deceptively as the Delphic Oracle.

Digital computers may receive their 'input' and their programmes from typewriter keyboards, punched cards or tape, magnetic tape, analogue measuring devices and in many other forms. Everything is then translated into numbers, usually in the *binary* scale, because this lends itself particularly well to further translation into a series of extremely short electric pulses. Writing + for a pulse and o for a pause of definite length between pulses, the first few numbers can be signalled like this:

1	=	+	5	=	+ o +
2	=	+ o	6	=	+ + o
3	=	+ +	7	=	+ + +
4	=	+ o o	8	=	+ o o o, etc.

These patterns of sharp electrical 'kicks' and pauses are the form in which the machine handles all numbers, not only the substantive ones, but also those that embody the instructions for its various components and the routing of partial results. The most complex problems are usually broken down into gigantic sequences of simple additions, subtractions, multiplications and divisions, each one of which may require only fractions of a thousandth of a second. Results and data can be stored in the *memory*, where patterns of pulses are recorded by changing magnetic configurations in cores, on tape or discs, on fluorescent screens and in many other ways. These stored data can be 'read out' from the memory when required with or without erasure, and in some sophisticated machines they may be deliberately 'forgotten' if no use is made of them for a stated period of operation to avoid overcrowding of the memory storage. Final results may be printed out, displayed graphically, or transmitted to another mechanism such as a machine tool or a rocket in flight whose guidance depends on the results of the calculations.

A computer can check its own work for mistakes, indicate its degree of accuracy, learn from experience, translate from one language into another, play chess or poker, find new proofs (including some of great elegance) for geometrical theorems, generate random numbers, and even solve certain intractable problems by efficient trial-and-error (*heuristic*) procedures. It can write bad poetry, conduct dull conversations and simulate human emotions, and could in time do all these things extremely well. It cannot generate prime numbers, give a general theoretical (as opposed to a practical) solution to the **three-body problem,** or protect itself and us from unimaginative and stultifying misapplication of its capabilities. (M.T.B.)

CONIC SECTIONS are curves which may all be obtained as the result of cutting a cone in various ways. If the cut is made parallel to the base of the cone, the shape of the resulting cross-section will be that of a *circle*; if the cut is somewhat inclined, it will be an *ellipse*; if the cut is made parallel to the slope of the side of the cone, it will be a *parabola*; and if the cut is inclined at a still greater angle, it will be a *hyperbola*. Finally, a cut through the very tip of the cone gives a *point*, and one along its edge, a straight *line*. The last two are extreme cases of a very small circle and of a very narrow parabola respectively.

The importance of the conic sections in astronomy is that any body that moves in an unperturbed **orbit** will follow a path that is in fact one of the conic sections. We can define and classify these curves according to their *eccentricity*, as follows:

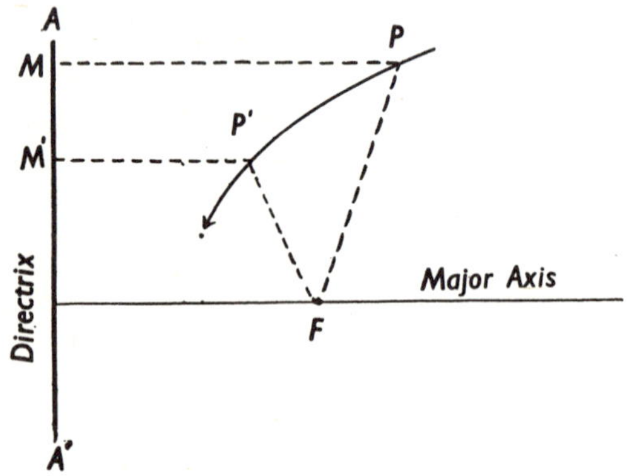

The curved line is part of a conic section. $PF:PM=P'F:P'M'$ for any two points on the curve. In the above example, this ratio is 3 : 4, i.e. less than 1, and the curve is therefore part of an ellipse.

Let there be any fixed straight line, AA' (usually called the *directrix*), and a fixed point F, not on the line itself. Let us now consider any other point P on the same sheet of paper (i.e. in the same plane). Its distance from the fixed point F is PF, and its distance from the fixed line is PM, where PM is perpendicular to AA'. We call the ratio of $PF:PM$ the *eccentricity*; its precise value will depend on whatever position we have chosen for P. If this point P is now allowed to move, but in such a way that the ratio $PF:PM$ remains constant, it will describe one of the conic sections. If the eccentricity $e = 1$, i.e. if PF always equals the distance of P from the straight line, the curve will be a parabola; for a smaller eccentricity, it will be an ellipse, a very elongated one if the eccentricity is almost 1, and an almost circular one if e is nearly zero. (If e is exactly zero, the ellipse becomes a circle; clearly e can only be zero if the directrix is infinitely far away.)

Thus by stating the eccentricity of an orbit, one states its exact shape, though not its size. The point F is called a focus, and the Sun is always at a focus of the orbits of its planets and the comets.

Summing up:

$e > 1$	HYPERBOLA
$e = 1$	PARABOLA
$e < 1$	ELLIPSE
$e = 0$	CIRCLE

THE ELLIPSE. This is the shape of all closed, unperturbed orbits. *Perihelion* and *aphelion* are at opposite ends of the *major axis*.

THE PARABOLA. It is doubtful whether any celestial body ever moves in a truly parabolic orbit, since the slightest perturbation would make e a fraction greater or smaller than unity and so change the character of the orbit, but the difference may be negligible for practical purposes. All projectiles and ballistic missiles move approximately in parabolas once the propelling force has ceased to act.

An important property of a parabolic surface is that it will bring parallel rays striking it to a focus, and will reflect rays from the focus into a parallel beam. Reflecting **telescopes,** car headlights, electric heaters and radar reflectors are some of the many cases in which this principle is applied.

THE HYPERBOLA. This curve has, strictly speaking, two *limbs*. Examination of any point on either part of the curve shows that the condition that its distance from the focus should bear the definite ratio e to its distance from the directrix is still fulfilled.

A remarkable property of the hyperbola is illustrated in the diagram below. For any hyperbola there is a pair of straight lines intersecting mid-way between the two limbs

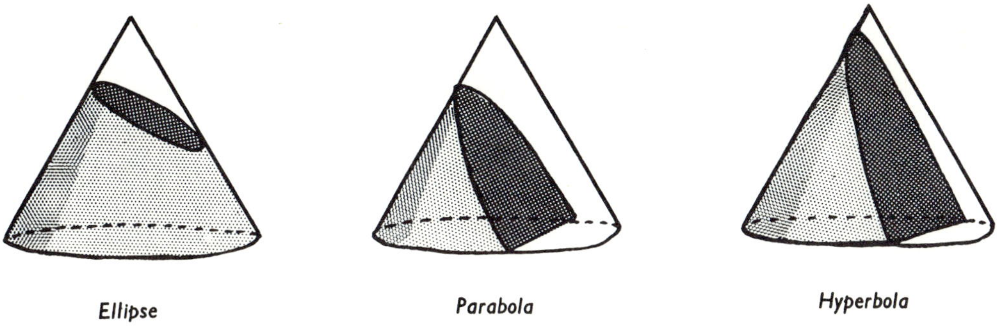

Above: the ellipse, parabola and hyperbola as oblique cross-sections of a cone. *Lower left:* how to draw an ellipse. A loop of cotton is placed over two pins stuck in the foci and is kept taut by the pencil point as the curve is drawn. *Lower right:* the ellipse as a planetary orbit.

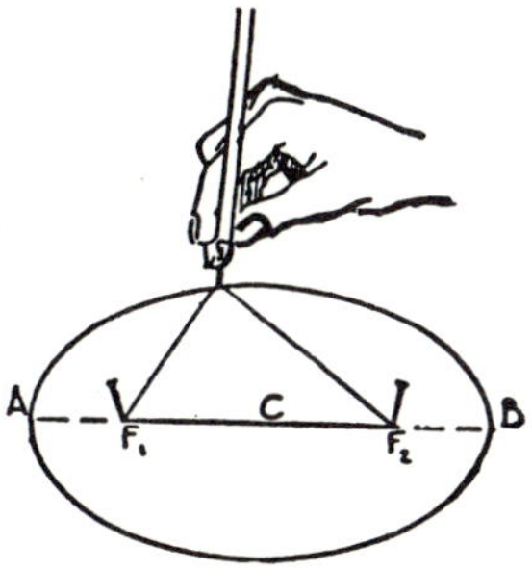

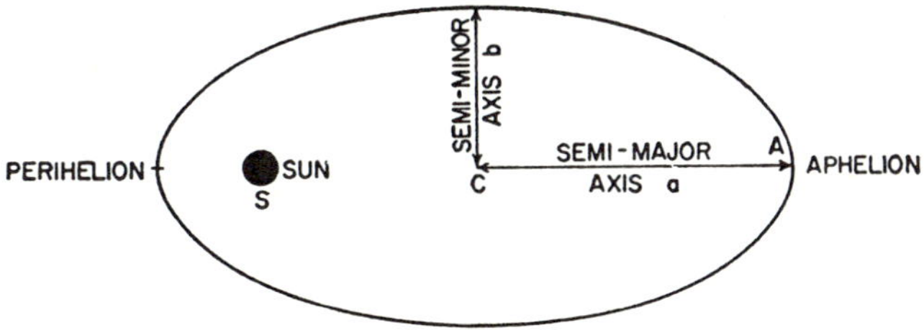

which approach closer and closer to the curve in either direction without ever touching it. These lines are the *asymptotes.*

Any five points in a plane *define* a particular conic. This means that, except in some special cases, there is one and only one curve of the conic section kind that will pass through all these points.

(For a given ellipse or hyperbola, there are actually two possible foci, each with its corresponding directrix, which could take the place of F in the discussion above and yield the same conic, but this is mainly of mathematical interest. For an understanding of orbits, only one focus and one limb of the hyperbola need be considered.) (M.T.B.)

CONJUNCTION. When a planet appears in the same **Right Ascension** as the Sun and is in fact between the Sun and the Earth, it is said to be in *inferior conjunction.* If it is beyond the Sun, it is in *superior conjunction.*

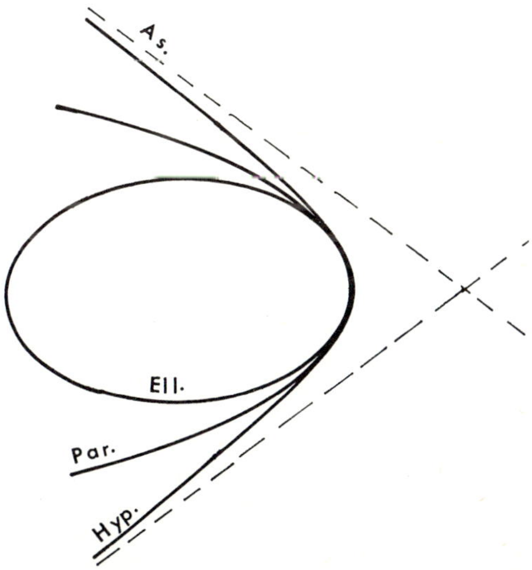

An Ellipse, a Parabola, and a Hyperbola with its Asymptotes. Only one limb of the hyperbola is shown.

If it appears in the Right Ascension opposite to that of the Sun, it is in *opposition.* No inner planet can be in opposition, and no outer planet can be in inferior conjunction.

These distinctions pay no regard to the extent to which a planet may be 'above' or 'below' the Sun.

The diagram also indicates the positions of *quadrature* and *elongation.*

Planets in inferior conjunction may be too close for visual observation, not only because of the brightness of the nearby Sun, but also because their dark side will be facing the Earth, with at most a narrow crescent showing. On rare occasions there is a **transit** of Venus or Mercury at inferior conjunction.

Planets may also be in conjunction with each other or with stars. This simply implies that they lie in the same Right Ascension as viewed from the Earth.

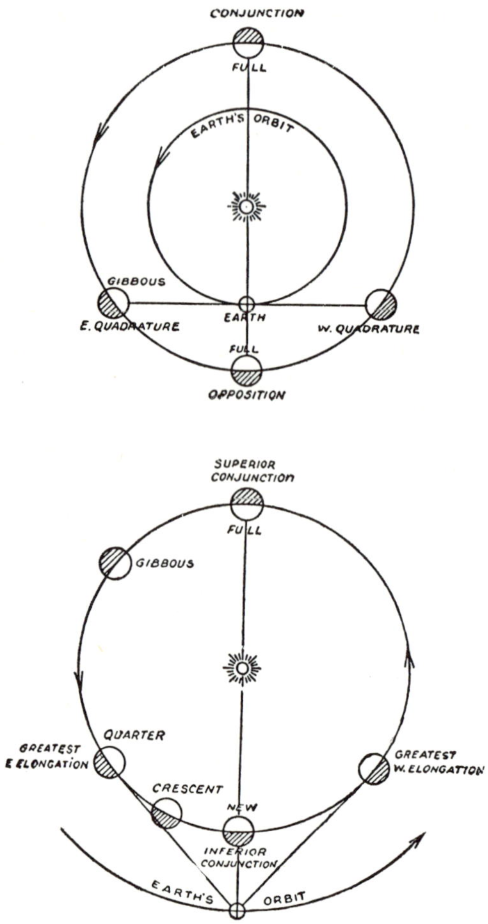

Relative positions of Earth and an inferior planet (*lower diagram*) and a superior planet (*upper diagram*). Only an inferior planet appears to go through the same cycle of phases as the Moon.

CONSTELLATIONS. The fanciful and arbitrary groups into which the stars are divided for easy reference and identification. In most constellations, there is no real connection between member stars, which seem close together merely as a result of perspective. The constellations were named by the Ancients after various gods, animals, etc., but it is vain to look for any resemblance in the patterns except in a few cases.

Stars in constellations are distinguished by small Greek letters prefixed to the name of the constellation, the letters being chosen from the Greek alphabet in the approximate order of brightness of the stars. Thus α Tauri is the brightest star in the constellation Taurus, but like many others it also has a proper name of its own, in this example *Aldebaran.*

(*See charts on the next page opening.*)

CONVECTION. The transfer of heat by movements of an unevenly heated fluid.

If a gas or a liquid is heated from below, the lower layers expand and so become less dense than the layers above. Consequently a circulation is set up, the colder, heavier layers of the fluid sinking down and driving the warmer, lighter layers to the surface. If the newly-risen fluid can cool at the surface it becomes denser again and presently sinks in its turn. Such flows are called *convection currents.*

The granules seen on the surface of the **Sun** are probably due to convection.

COPERNICAN SYSTEM. The theory of the solar system propounded by the monk Nicolas Copernicus in 1543, in which he declared that the planets (and the Earth with them) revolve about the Sun and not about the Earth, and that the Earth rotates about an axis from West to East. The accepted theory at the time was the **Ptolemaic System,** which held the Earth to be at the centre of the planetary orbits. The ideas of the Copernican System were not wholly new, and had been put forward by Aristarchus and other Greek philosophers in the 3rd century B.C., but did not find acceptance until many years later when Galileo, using a telescope, saw the crescent Venus go to full phase, which was not possible in the Ptolemaic System.

Copernicus' main treatise was dedicated to Pope Paul III, and he had received great encouragement and support from the Church. After his death his ideas caused so much uninformed controversy among the laity and disturbed the faith of so many who were unable to reconcile the theory with Scripture, that it was declared false by the Church and became prohibited reading.

COPERNICUS. A 56-mile ray crater on the **Moon,** named after the sixteenth-century astronomer Nicolas Copernicus, the originator of the **Copernican System.**

CORIOLIS FORCE. The 'force' that deflects a projectile during its flight across the surface of the Earth, to the right in the northern hemisphere and to the left in the southern. It is not a true force at all, but an effect due to the rotation of the Earth, and applies generally to all particles travelling across a rotating surface.

Consider a projectile fired in a direction due North from a gun on the equator. In addition to its muzzle velocity it will retain an eastward velocity which it shares with the gun and all things on the ground nearby, because all these things are rotating to the East with the Earth's surface. In the absence of external forces the projectile must keep all its eastward velocity, but as it moves North the rotational speed of the ground under it lessens, and the projectile will draw ahead of the ground in its eastward motion. By the time it falls it may have deviated to the East by several miles, as if a force had acted on it sideways, i.e. the Coriolis force. If the projectile is fired *towards* the equator, it will fly over ground which is moving progressively *faster* to the East, and the projectile will fall to the West of the line along which it was fired. Allowance has to be made for this effect in all gunnery, and in rocket trajectory calculations.

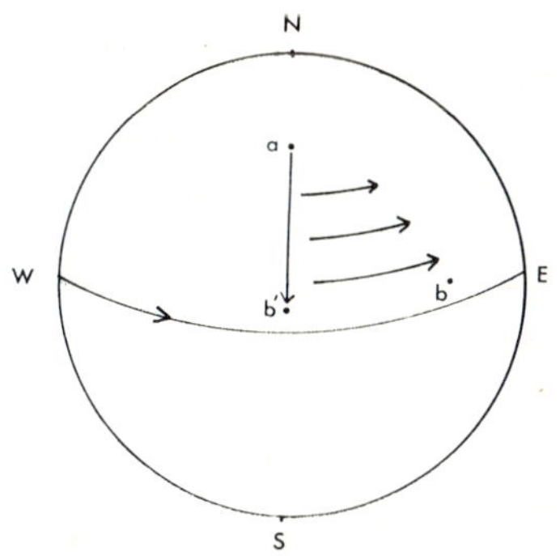

A projectile is fired due South from a point *a* at a point *b*; while it travels, the Earth rotates under it, and the projectile lands at *b′* to the West of *b*.

CORONA, SOLAR. The outer envelope of the Sun. It has up to thirty times the diameter of the Sun itself, and contains highly ionized gases, including gaseous iron, nickel and calcium, at a temperature of some 1,000,000° C. (See **Sun.**)

CORONAGRAPH. A telescope in which a circular disc is used to create artificial eclipses of the **Sun** by blocking the light from the photosphere. It enables solar prominences and the corona to be photographed at any time. Before its invention in 1930 this could be done only during **eclipses.**

COSMIC RAYS. Extremely fast particles continually entering the upper atmosphere from interstellar space.

Ions are being generated all the time in the air to ground level. For ions to be formed, a considerable amount of energy must be given to the parent atoms, and at atmospheric temperatures such energy can only come from some sort of 'ray': this may be either **electromagnetic radiation** of very short wavelength, or a stream of particles travelling at high velocity.

The amount of ionization of the air decreases downwards, e.g. in a mine shaft, and increases upwards. The ionizing rays are therefore most effective high above the Earth, and their origin is outside the Earth: hence the name '*cosmic* rays'. The observed ionization also increases very slightly from the magnetic equator to the magnetic poles. This shows that cosmic rays are affected by the

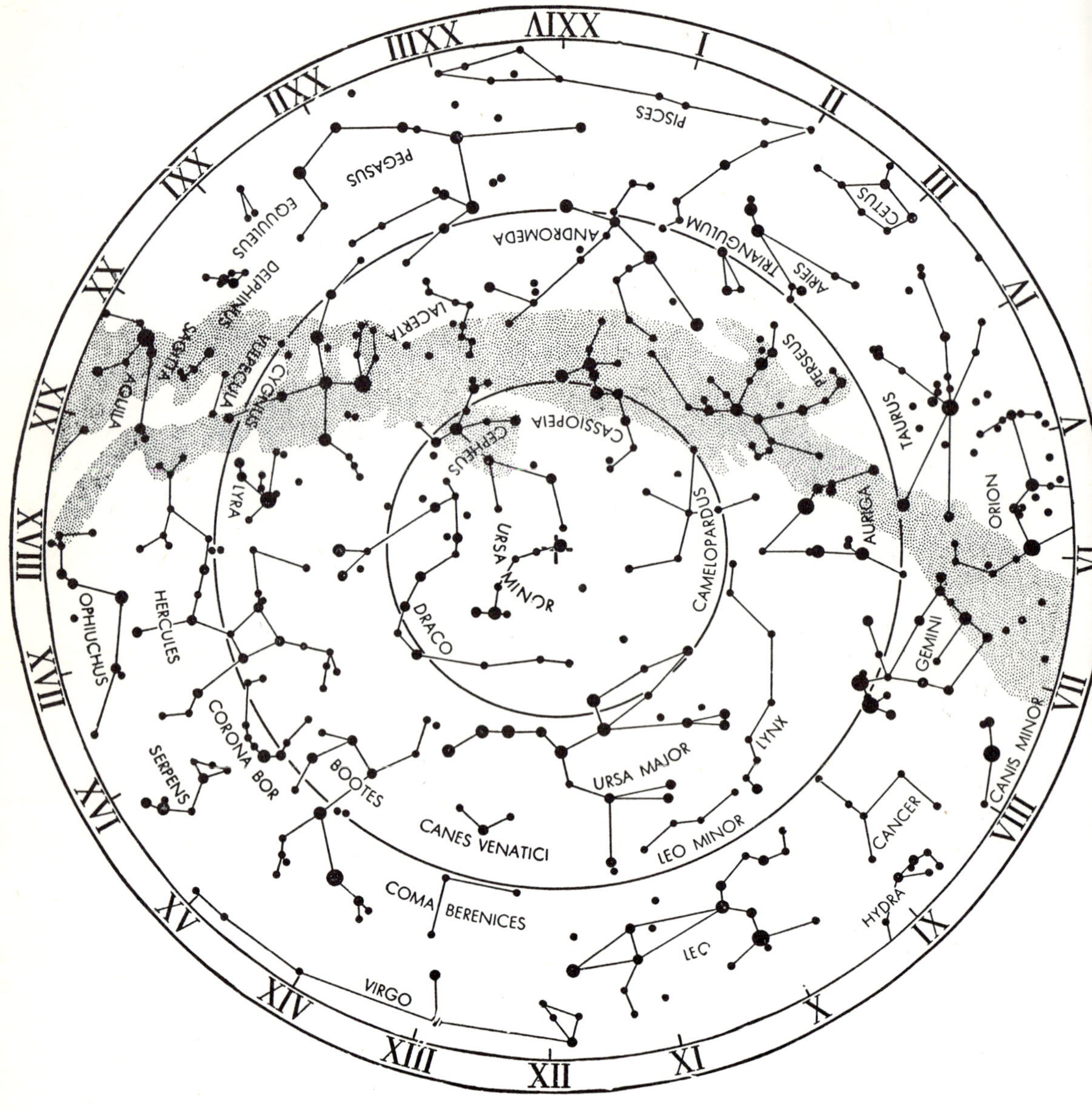

THE NORTHERN HEMISPHERE OF THE SKY. Its centre lies above the Earth's North Pole, and very close to it is the Pole Star. The Milky Way stretches as an irregular band across the upper half of the chart. The brightness of the stars is indicated roughly by the sizes of the dots.

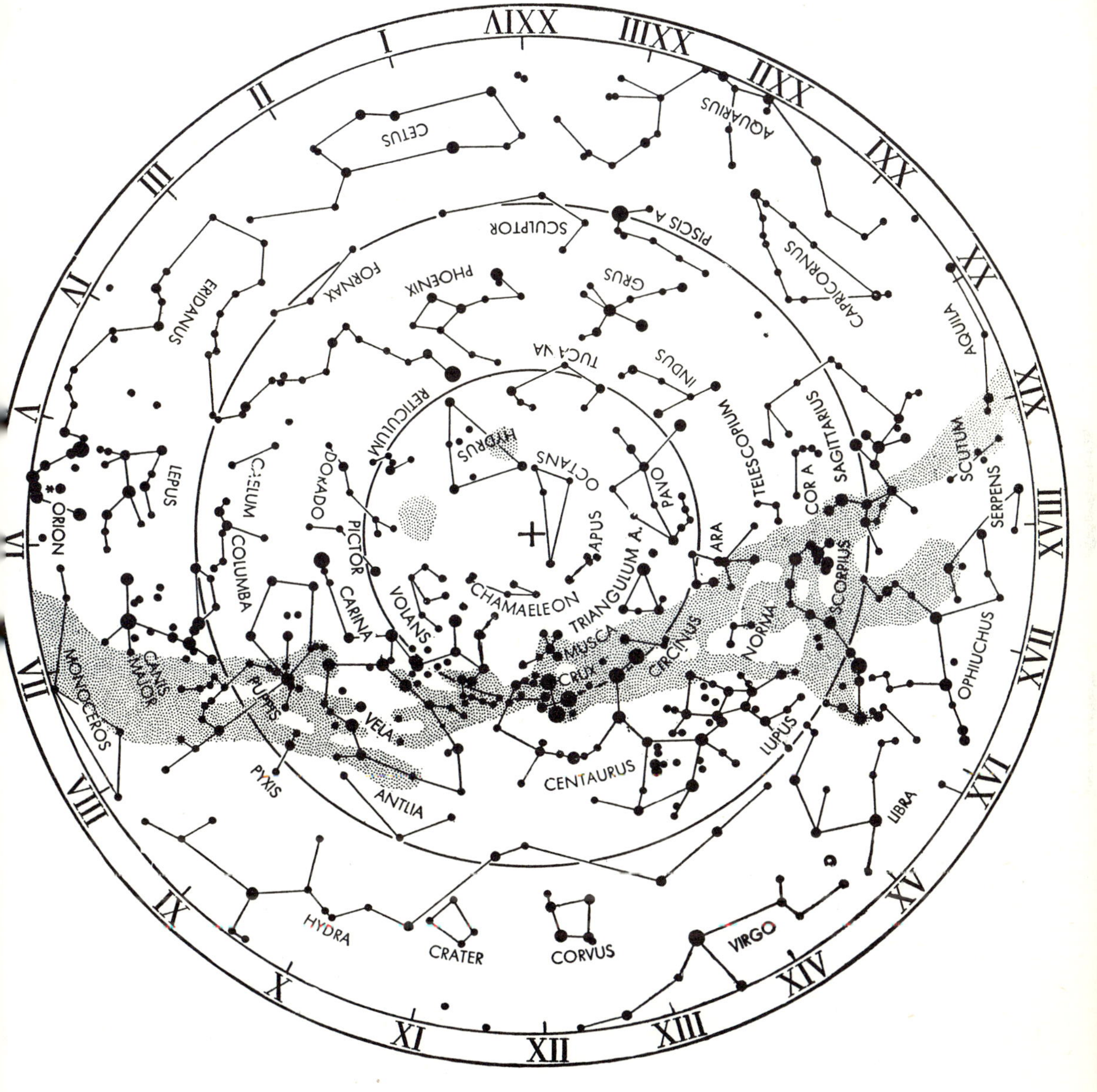

THE SOUTHERN HEMISPHERE OF THE SKY. The constellations near the edge of this chart can be seen from northern temperate latitudes at one time or another during the year. The Roman numerals around the rim give the Right Ascension of stars; for instance, the Southern Cross (Crux) and the gap in the Milky Way close to it known as the Coal Sack have a Right Ascension of just over 12 hours. (A photograph of this group is given in the article on Galactic Nebulae.)

magnetic field and that they must therefore be electrically charged particles, since neutral particles and electromagnetic radiation would not be so affected.

The origin of the rays is still in doubt. Most of them probably come from far outside the solar system, but a proportion of low-energy ones clearly come from the Sun, and energetic solar activity may considerably raise the amount of cosmic radiation for a short time. A solar flare on February 23, 1956 caused a great increase in cosmic radiation all over the world. As the Sun can produce the rays it is reasonable to suppose that many other stars can do so, and the rays originate all over the galaxy. It does not seem possible for the observed energies of some cosmic rays to be even remotely approached by emission from the stars, but a plausible theory has been formulated which attributes the acceleration of the particles to the slight magnetic field which is believed to exist in our galaxy.

The particles which enter our atmosphere are called the *primary* rays. None of them reach ground level themselves, but in colliding with atoms of oxygen or nitrogen at a height of ten miles or more they impart their energies to the fragments resulting from the collisions, and these fragments or *secondary* rays are what we observe at lower levels.

Primary cosmic rays have been registered by rocket-borne cloud chambers, **Geiger** and other tube counters, and in photographic emulsions. Trouble sometimes arises through the fact that a primary ray on striking the metal of the rocket releases a shower of 'spurious' secondary rays, and the equipment has to be arranged in such a way that real and spurious rays can be distinguished. Another danger arises from ionization by cosmic rays of the air inside the instrument section of the rocket, leading to sparking between uncovered high-voltage terminals.

It has been found that primary cosmic rays are mainly atomic nuclei having very large energies because of their enormous velocities. The nuclei are mainly those of light elements, especially hydrogen and helium, but recently, using a photographic emulsion flown from a balloon, several particles with charges greater than ninety times the charge on a hydrogen atom have been found. The relative abundances of the various nuclei in cosmic rays are similar to the relative abundances found in stars; the discrepancies which do exist can be resolved by assuming that the primary rays themselves have been processed by collisions with other atoms.

The biological hazards of cosmic rays in space vehicles are unlikely to be serious except for manned satellites that spend prolonged periods at moderate distances from the Earth and near the polar regions, where the radiation is densest.

COSMOLOGY. The general science of the Universe in all its parts, laws and operations, so far as they can be known by observation and scientific enquiry.

For more than two thousand years the theory of the Universe that was generally accepted was that formulated by Aristotle in the fourth century B.C. He made the first and, for many ages, the only attempt to systematize the whole amount of knowledge of nature that was accessible to mankind; for that reason, perhaps, his views were long regarded as authoritative and were held in great veneration.

In common with most of the Greek philosophers, he believed that the Earth was fixed immovably at the centre of the Universe. At a time when ideas about mechanics were extremely crude, it seemed absurd to suppose that the Earth could be in motion. He also held that the Universe was finite.

According to Aristotle motion in space was of three kinds: motion in a straight line, motion in a circle and motions which are a combination of these. The Universe being finite, motion in a straight line could not continue for ever. The circular motion is the only motion which has neither beginning nor end. The fixed stars and planets must consequently have circular motions.

The world we inhabit was held to be made of the four simple elements, earth, air, fire and water; they possessed the tangible qualities, hot or cold, which were active, and dry or moist, which were passive. Because simple bodies have simple motions, the four elements tend to move in straight lines; thus earth tends downwards, fire upwards. There must be another element, said Aristotle, to which circular motion is natural. As circular motion admits of neither up nor down, this element can be neither heavy nor light; as it is without beginning or end, it must be incapable of increase or change. This superior

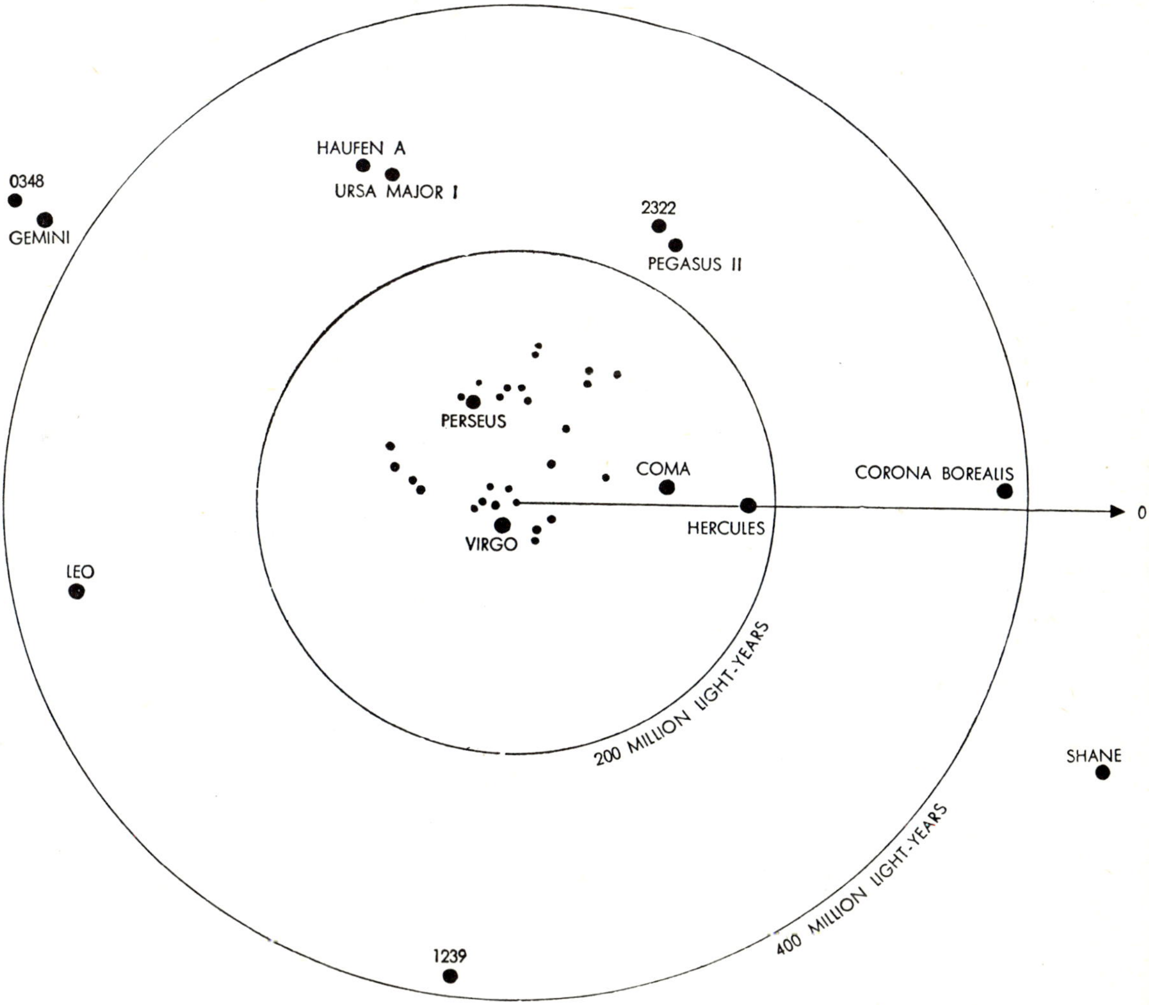

THE DISTRIBUTION OF GALAXIES IN SPACE. Each of the smaller dots represents associations of 50 galaxies or fewer. The larger dots stand for clusters containing more than 50 galaxies. On this scale, the entire local group of some twenty galaxies, including Andromeda and our Milky Way Spiral, is contained in the small central dot. The line from the centre indicates 0° of galactic longitude. Some of the clusters are above, and others below the plane of our own galaxy. The true diameter of the region covered by this diagram is approximately 8,400,000,000,000,000,000,000 miles.

(*After J. Neyman and Elizabeth Scott, from 'Scientific American'*) element was called the aether, and must be more divine than the other four elements, being eternal and changeless. The stars, being spherical and eternal, were supposed to be made of the aether.

The fixed stars were supposed to be attached to a crystal sphere, which made one rotation each day. The seven moving stars or planets (the Sun, Moon, Mercury, Venus, Mars, Jupiter and Saturn) were each attached to separate spheres. The motion of each of these spheres was transmitted to the next inner sphere by a system of intermediate reacting spheres, there being no void or empty spaces between the spheres. The whole space from the fixed stars to the Moon was filled with the various spheres. The aether occupied the whole of this region, which was eternal and unchanging. Below the Moon was the terrestrial region, the home of the four elements, subject to ceaseless change through the strife of the elements and their continual mutual transformation. In this region of change and strife the shooting stars, meteors and comets appeared. Shooting stars and meteors were attributed to exhalations, one of a vaporous nature arising from the water on the Earth, the other dry and smoke-like rising from the Earth, which took fire when they were caught in the rotation of the inner sphere. Comets were explained as exhalations rising from below and catching fire.

The Aristotelian conception of the Universe has been briefly described because it dominated men's minds over many centuries, for far longer than any other cosmological theory has done. It was not without its difficulties, for its ability to explain the actual motions of the planets was very limited. The great changes in the brightness of the planets, and particularly of Venus and Mars, could not be explained if the distance of each planet from the Earth remained the same. To attempt to account for these changes in brightness and also to provide a better representation of the motions of the planets as observations of their positions accumulated, various ingenious mathematical theories were developed, based upon a combination of circular motions. These were looked upon merely as geometrical representations; there was no insistence on their physical truth.

The reaction against the cosmology of Aristotle was commenced by Copernicus, in the 16th century, who advanced the theory that the Sun instead of the Earth is at the centre of the Universe and that the Earth is merely one of the planets revolving round it as a satellite. The diurnal motion of the stars was attributed to the rotation of the Earth on its axis, the sphere of the fixed stars being brought to rest. The heliocentric doctrine of Copernicus was not a mere hypothesis, as the cosmology of Aristotle had been, but a theory worked out in detail to account for the motions of the planets. Two important consequences followed from it. In the first place, it necessitated a considerable enlargement in the size of the Universe; relative changes in the positions of the stars, as the Earth moved round the Sun, were not observed; it followed that the sphere of the fixed stars must lie far beyond the orbit of Saturn, instead of just beyond, as Aristotle had supposed. In the second place, when once the sphere of the fixed stars had been brought to rest, it was no longer necessary to suppose that the stars were all the same distance.

Though the Copernican theory was slow in gaining acceptance, because it was at that time such a revolution in thought, the fundamental ideas of the theory of Aristotle were gradually undermined. The careful observations by Tycho Brahe of the bright new star, which suddenly blazed forth in 1572, enabled him to prove that it was much more distant than the Moon, in the region, therefore, where according to Aristotle no change could take place. Tycho Brahe, from his observations of the great comet of 1577, concluded that it was moving round the Sun in a circular orbit outside that of Venus, and therefore also in the region where no change could occur. The Aristotelian notion that comets were exhalations from the Earth in its atmosphere was also disproved, and the idea of solid spheres was put an end to; as Kepler afterwards said, 'Tycho destroyed the reality of the orbs'. The next blow to the old ideas came when Kepler showed that the orbits of the planets were ellipses, having one focus at the Sun, so that the belief that the celestial bodies must have circular motions had to be abandoned.

The final death-blow to the system of Aristotle came early in the 17th century, when the recently invented telescope was turned on the heavens. The discoveries by Galileo of sunspots and the Sun's rotation, of the

SPHERES, ORBS, WHEELS and epicyclic movements played a large part in mediaeval cosmology. This woodcut, which shows the intrepid explorer inspecting the celestial clockwork from the edge of the Earth, is often reproduced in serious vein. In fact, it parodies the uncomfortable conclusions that followed from too naïve a view and indicates that more sophisticated ideas were required.

four major satellites of Jupiter which revolved around it as a parent body, and of the phases of Venus and Mars, firmly persuaded him of the correctness of the Copernican theory, which he vigorously championed in his writings. Galileo, by his refusal to accept statements merely on the authority of others and by his insistence on the necessity of appealing directly to observation, laid the foundation of rational scientific method.

In 1718 Halley found that some of the brighter stars were in motion, so that the stars could no longer be regarded as fixed. In 1783, William Herschel, from a study of the motion of a few of the brighter stars, proved that the Sun itself is in motion. The view that the stars extend outwards to great distances and that differences in brightness were due primarily to differences in distance had gradually come to be accepted. The Sun could therefore no longer be regarded as the centre of the Universe. So first of all the Earth and then the Sun had been displaced from occupying the proud position of the centre of the Universe.

The formulation by Newton of the hypothesis of universal gravitation; the first attempt to describe the structure of the Universe; and the results of modern observations, which have shown that space is occupied by a vast number of galaxies extending beyond the present limits of observation, are described in the article **Astronomy.** These results are of fundamental importance for modern cosmological theories.

Over 130 years ago, Olbers called attention to a paradox, which followed in a logical manner from certain initial assumptions which appeared to be plausible. The assumptions were:

1. That the stars are distributed on the average uniformly throughout space and that the average intrinsic luminosity of the stars is the same in all regions of space.

2. That the average spatial density and the average luminosity of the stars do not change with time.

3. That there are no large systematic movements of the stars but that their motions are in random directions.

4. That space is Euclidean or, in other words, that it has everywhere the properties with which we are familiar in everyday life.

Suppose a series of large spheres to be drawn around any arbitrary point as centre, so that the difference between the radii of consecutive spheres is constant. The volumes of the successive spherical shells are proportional to the squares of the radii of the shells. Because of the uniformity in the spatial density and average luminosity of the stars and because these quantities do not change with time (so that we are not concerned with the fact that in looking outwards into space we are looking backwards in time), each shell must contribute an equal amount to the intensity at the centre. As we can add shells without limit, the total intensity of light at the centre must be infinitely great, which experience shows not to be correct. Therein lies the paradox. The paradox still persists if, in place of stars, we submit galaxies, in the light of our current knowledge of the Universe. It can be shown that the paradox persists even if space is not assumed to be Euclidean, provided only that it is homogeneous. It is not even essential for space to be infinite; the paradox will still hold for a finite but un-bounded non-Euclidean space.

The assumptions on which the paradox is based require to be closely examined. All observational evidence is in support of the assumptions that the galaxies are distributed on the average uniformly through space and that the average intrinsic luminosity is the same in all regions of space. But we are on much less sure ground when we consider the further assumption that there has been no change with time in either the spatial density or the average luminosity.

If the Universe has a finite age, light can reach us at the present time only from galaxies that are within the distance that light can travel in the time that has elapsed since the Universe was born. The number of spherical shells in Olbers' argument would in that case be finite, instead of infinite, and the paradox would then no longer exist. If, however, we retain the assumption that the average spatial density and luminosity of the galaxies do not change with time, we must drop the assumption that there are no large-scale motions, for the paradox can be avoided if the contribution to the total intensity provided by the distant galaxies is much below the estimate of Olbers. This is possible, when the other assumptions are retained, only by one cause known to physics: the **Doppler** displacement of light. If the distant galaxies are receding rapidly, the radiation received from them is much reddened and is thereby reduced in intensity. If the velocity of recession is large enough, the reduction may be sufficient to reduce the total intensity of the radiation to a finite amount.

Thus there appear to be only two possible ways by which the paradox can be avoided. It is necessary to assume that the Universe has a finite age or that it is expanding; one or both of these is a feature of all modern cosmological theories.

MODERN COSMOLOGY started with the publication by Einstein in 1915 of his theory of general **relativity** and with the observations by Hubble in the 1920s of the **red shifts** and distances of galaxies. All major cosmological theories have used general relativity (somewhat modified in the steady state theory) to describe the Universe, and they have interpreted the red shifts of galaxies as due to expansion of the Universe. They also assume the Universe to be *isotropic* (broadly the same in every direction when viewed from one point) and *homogeneous* (broadly the same when viewed from different points at the same time). The recession of galaxies from our own galaxy is not incompatible with homogeneity as it might appear. A simple analogy is a flat, infinitely large sheet of rubber with a number of points on it. If it is stretched uniformly an observer at one of these points will see the other points recede from him. An observer at another point would similarly see all the points receding from him and neither observer could claim to be in a privileged position.

EVOLUTIONARY COSMOLOGIES. Using the theory of general relativity, a range of expanding isotropic homogeneous models of the Universe can be constructed. A particular model can be specified by values of the present-day mean density of the Universe and of the *cosmological constant*, Λ. The effect of the cosmological constant on a small scale is that of a force between any two bodies proportional to their separation. Observations of the solar system do not show any effects on planetary orbits and thereby set limits on the value of Λ. No other measurement of Λ has been made or attempted. A large quantity of matter could be in intergalactic space without having been detected, so the density of galactic material, if it were spread out over the whole Universe, is a lower limit on the total mean density. Since we do not know Λ or the mean density the choice of the model which best describes the Universe must be made after comparing predictions with observations.

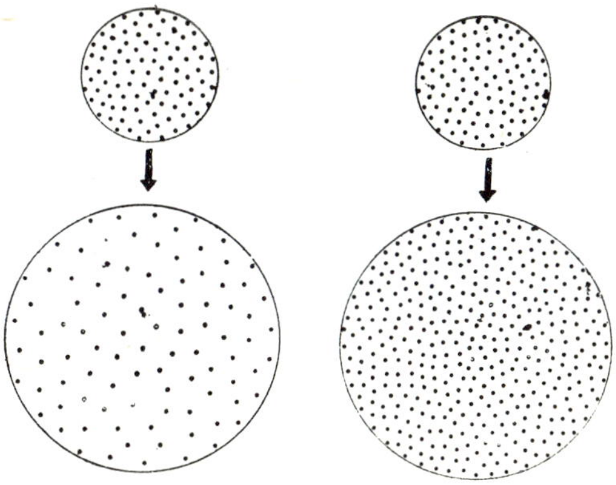

EVOLUTIONARY AND STEADY-STATE THEORIES of the Universe represented diagrammatically. The two top circles are portions of the Universe at the same time. with galaxies (*black dots*) scattered uniformly through them. Recession of the galaxies spreads them over a larger volume (*left*). The steady-state theory supposes that meanwhile new galaxies come into being so that, in spite of the expansion, the mean density of the galaxies remains unchanged. One of the chief tasks of modern astronomy is to find ways of testing these views with observational evidence.

GENERAL PROPERTIES. Most of the evolutionary models start some time in the past, typically ten thousand million years ago, with all the matter at a point, from which they expand; these are the *big bang* models. In some of these models the expansion is halted and reversed by gravitation, and contraction back to a point follows. The other models expand for ever, so that as time goes by the mean density becomes ever smaller. Nothing is known of what the Universe was like before the big bang or, if contraction to a point occurs, what it will be like afterwards.

The space of many of these models is curved in a way analogous to the surface of a sphere (see **Space, Geometry of**) and the relationship between the red shifts and distance moduli (see **Magnitudes**) of galaxies varies from one model to another.

The light from galaxies with large red shifts has taken a long time to reach us, so we see them at an earlier stage in the development of the Universe than we see nearby objects of small red shift. It is possible that galaxies evolve and change their absolute magnitudes in this time, therefore even if we know the absolute magnitudes of nearby galaxies we cannot assume that distant galaxies are the same. Nearby galaxies have a range of absolute magnitudes. The relationship between the apparent magnitudes and red shift of galaxies is thus affected by the possible evolution and intrinsic differences of the properties of galaxies as well as by the geometry of the cosmological model. Cosmologists have attempted to separate these effects when analysing the observations but so far without success.

RADIO SOURCE COUNTS. The vast majority of radio sources that have been optically identified are galaxies (see **Radio Galaxies**). If we assume that all sources are galaxies we can count how many, N, there are with *flux density* (i.e. the brightness at some frequency as we see it) greater than a value S. A graph of the logarithm of N against the logarithm of S can be drawn for the observable range of S and compared with theory. If evolution of sources is neglected all the models predict the same slope for large S with a flattening off as S becomes smaller. Observations by several groups of astronomers show a greater slope than predicted for large S although the graph does flatten off for small S as expected. One explanation is that in the past the sources had on average a sufficiently higher intrinsic intensity or a sufficiently higher concentration (over and above the effects of expansion). Since we have very little knowledge of the evolution of radio sources we are free to predict the evolution required to explain the counts.

BLACK BODY RADIATION. It is normally assumed in constructing cosmological models that the **background radiation** is **black body** at about 3° K. As the Universe expands this temperature falls, so at early times it was very high. The radiation at present has little interaction with the matter but earlier this was not true: the matter must have been at the same temperature as the radiation. One second after the big bang the temperature was 10,000,000,000° C. and the Universe consisted of neutrinos, photons, electrons, positrons, neutrons and protons. As the Universe expanded and cooled the neutrons and some of the protons combined to form helium. For a wide range of densities, within which we think the density of the Universe to lie, the amount of helium formed is almost the same, about 10 % by number of atoms. Measurements of the fraction of helium in a range of astronomical objects have been made and all give about 10 % by number, apart from a few exceptions which can probably be explained away.

Other elements are formed but only in negligible quantities. After all the reactions have occurred there remain hydrogen (whose nuclei are protons) and helium, which form into galaxies and stars, photons, which we see as the background radiation, and neutrinos.

STEADY STATE THEORY. This theory states that as well as being isotropic and homogeneous the Universe looks broadly the same at all times; it thus has neither beginning nor end. Since the galaxies are receding, and thus tending to reduce the mean density of the Universe, matter must be created to counteract this and keep the mean density constant. This matter is normally assumed to be hydrogen. The rate of creation depends on the mean density of the Universe but is a few atoms per cubic mile per year. This is so small that it cannot be directly measured. General relativity can be modified to include constant creation and a detailed steady state theory formulated. Unlike the range of big bang theories, there is only one steady state theory which makes unambiguous predictions to be compared with observation.

For the radio source counts the steady state theory predicts the same slope for large S as the evolutionary theories. Since a steady state Universe is broadly the same at all times we cannot attribute the observed slope to evolution or changes in the concentration of sources. It has been suggested that the difference is due to the **quasars** and that if these are not as distant as their red shifts suggest their contributions to the counts must be eliminated; theory and observation might then be brought into agreement.

Black body background radiation and helium are not produced cosmologically in a steady state Universe. An alternative explanation of the observed background radiation is given in the article on it. Predictions of the amount of helium that can be produced in stars (see **Nucleogenesis**) give about one-tenth of that observed so some other process must be invoked to produce the helium. Objects with the mass of thousands of stars have been suggested to be capable of producing helium in the observed quantities; these would have to exist shortly after the formation of a galaxy so that the helium can subsequently find its way into stars.

The steady state theory has thus explained many observations less well than big bang theories have, and few astronomers now consider it to be a possible model of the Universe. (H.S.J./P.R.O.)

COUNT-DOWN. The procedure by which a missile or space vehicle and the associated ground installations are prepared and checked before a firing. The count, first in hours and finally in minutes and seconds, is synchronized to the range timing system and proceeds backwards towards zero at the planned moment of lift-off. When there are a great many systems to be checked and prepared, both in the vehicle and on the ground, the count-down may last more than a week. At the other extreme, as in the case of an operational ICBM, it is very desirable to be able to fire the weapon with the shortest possible count-down. In practice, for a solid-fuelled ballistic missile whose target is pre-planned, the preparation time might be expected to be well under an hour.

CRAB NEBULA

In the last stages of the count when a large number of parameters have to be monitored and controlled in an exact sequence, control of the count is usually taken over by a computer-controlled automatic sequencer. If the count-down is a long one and the launch must occur within a short time-interval around the nominal time (see **Launch Window**), there may be planned intervals or *holds* in the count to permit some scope for rectifying unforeseen defects; usually there will be sufficient flexibility to correct minor defects without enforcing an unplanned hold.

CRAB NEBULA. A nebulosity in our galaxy of most unusual structure, with a central star formerly thought to be a **white dwarf,** but now considered by some astronomers to be a **neutron star.** The spectrum of the surrounding gases shows a **Doppler** shift which indicates that these gases are streaming outwards at a speed of about 1,000 kilometres per second, presumably as a result of a violent explosion in the past. By working backwards on the basis of the present rate of expansion the date of the outburst can be calculated, and the Crab Nebula has been identified as the débris of a **supernova** whose appearance in a position corresponding to that of the Crab Nebula was recorded in detail by Chinese astronomers in A.D. 1054.

The Crab Nebula is one of the most powerful radio transmitters in the sky. It has also been found to emit X-rays (see **X-ray Astronomy**). Most recently a **pulsar** has been found in the direction of the Crab and coinciding with the position of this pulsar there appears to be an optically varying object.

To the naked eye the nebula is a faint star in the constellation Taurus, on the edge of the Milky Way.

CRATERS, LUNAR and **MARTIAN.** See **Moon** and **Mars.**

CRITICAL ANGLE. The greatest angle of incidence for which refraction of a ray travelling from one medium into an optically less dense medium is possible. For greater angles of incidence the ray will be totally *reflected.* (See **Refraction.**)

CRITICAL FREQUENCY. The frequency at which radio waves transmitted vertically change from penetrating an ionized layer to being reflected back to the Earth's surface. Communication with bodies moving above the ionized layers must be made at frequencies above the critical frequency.

CRITICAL TEMPERATURE. Pressure raises the boiling point of liquids. Many gases can therefore be liquefied by pressure alone at temperatures at which they are still gaseous under normal pressure. If the gas is, however, above its *critical temperature,* no amount of pressure can liquefy it.

CUSPS are the pointed ends of a crescent. The cusps of the crescent Moon, Mercury or Venus always point away from the Sun.

D

DATE LINE. At any one moment, two calendar dates are effective simultaneously over different parts of the globe, separated by the *International Date Line.* The need for this will be seen from the following argument:

Let it be midnight exactly at Greenwich between, say, a Wednesday and a Thursday. Southend lies a few miles to the East, and is therefore a little nearer the rising Sun; the astronomical time at Southend is accordingly 0.03 a.m. on Thursday morning. At Whitehall, a shorter distance to the West, the time is 11.59 p.m. on Wednesday night. The confusion which a strict adherence to astronomical time would create is avoided locally by the use of **Zone Time,** but over greater distances it remains true that East of Greenwich it is Thursday and West of Greenwich it is Wednesday. On the opposite side of the globe it is noon. Noon on Wednesday or Thursday?

The Date Line resolves this difficulty by setting a definite though arbitrary limit to the areas over which the two calendar dates apply. The line follows the 180th meridian through the middle of the Pacific Ocean, diverging occasionally so as to include the whole of an island group on one side or the other. When crossing this line on a westerly course a day is 'lost', and the date must be advanced one day. The reverse is true for an easterly crossing. For one instant only each day, when it is midnight on the Date Line, does the whole world share the same calendar date.

DAWN ROCKET. An Earth satellite vehicle is often launched in an easterly direction so that the speed of *rotation* of the Earth's surface is added to the rocket's orbital velocity round the centre of the Earth, with a consequent economy in the weight of propellents required for a specific mission. In the same way, it is advantageous to launch an artificial *planet* at dawn in the direction of the Earth's *orbital motion*, so that the Earth's orbital velocity is added to the launching vehicle's orbital velocity round the Sun. This velocity of 18·5 miles per second is lost rather than gained in the case of a *dusk rocket*.

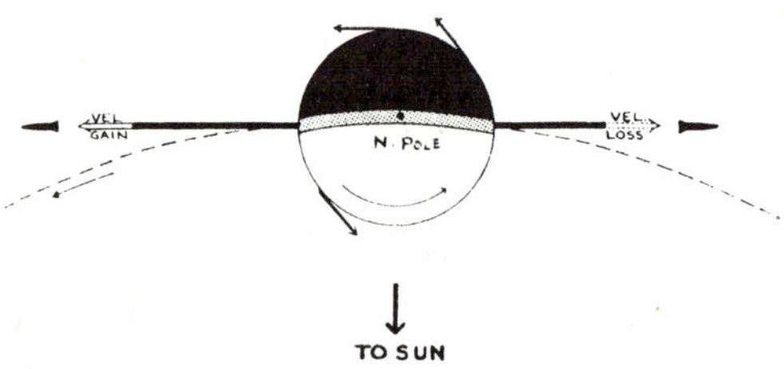

DAWN AND DUSK ROCKETS. Their velocity gain or loss is independent of the latitude of the firing point. The smaller arrows indicate that an Earth satellite vehicle can benefit from the Earth's rotation irrespective of the time of day; but the effect decreases away from the equator and is zero at the poles.

DAY. One *sidereal* day is the interval of time between two successive meridian passages of a star. One *mean solar day* is the interval between two successive meridian passages of the **Mean Sun.** A sidereal day is three minutes and fifty-six seconds of mean solar time shorter than a mean solar day. (See **Time Measurement.**) The *Civil Day* is based on the mean solar day, and is used for all ordinary purposes.

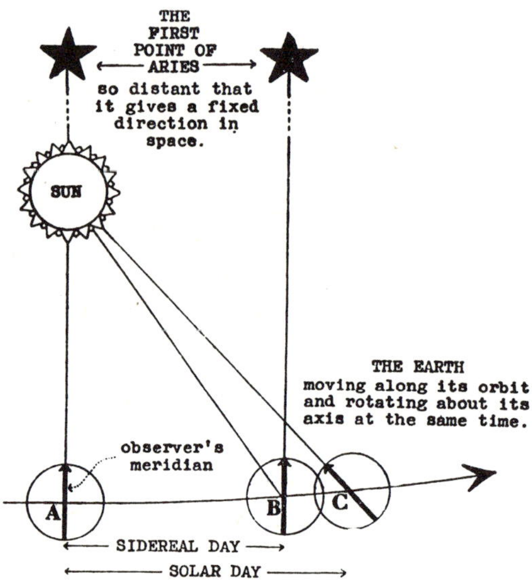

Tidal friction is very gradually slowing down the rotation of the Earth. This is causing the day to lengthen by approximately one thousandth of a second per century.

DAYTIME SHOWER. A shower of meteors coming from the direction of the Sun, and therefore not visible at night. (See **Meteor.**) One of the first results of the continuous watch by radio-echo methods was the discovery of the extraordinary activity of the daytime radiants during the summer months. From May to September there is a continuous succession of active streams, with hourly rates of the order 20 to 80. Accurate velocities have now been measured, and the streams are known to have small orbits of high eccentricity, only slightly inclined to the ecliptic. The *o-Cetids* have an inclination of 34°, a period of 1·5 years and an eccentricity of 0·91; the other three showers have much smaller inclinations. This is perhaps the most remarkable feature of these showers, because the meteors can intersect the Earth's orbit at both nodes. Thus the *Arietids*, travelling in a small orbit with period 1·6 years and eccentricity 0·49, are seen again at the end of July as the night-time *δ-Aquarids*; the *ζ-Perseids*, which occur at the same time, have an orbit of similar size but smaller eccentricity of 0·78, and are seen again in November as the Southern Arietid stream. Whipple's study of the November *Taurids* led him to the expectation of the return of this shower in daylight hours in July, and the discovery of the daytime Taurids showed the correctness of this view. The Taurids have a period of 3·2 years, and, like the night-time Taurids, are related to Encke's Comet. (J.G.P.)

DECLINATION. The declination of any point on the Celestial Sphere is its angular distance from the celestial equator, expressed in degrees and positive if the point is North of the celestial equator, negative if it is South of it. Declination corresponds to Latitude on the Earth (but see **Celestial Sphere**).

DEGENERATE MATTER. A tremendously dense form of matter that occurs in **white dwarf** and **neutron stars.**

In normal **atoms,** relatively large distances separate the nucleus from the electrons that encircle it. The atoms of degenerate matter have been stripped of their orbital electrons, and the nuclei are packed close together. This results in a form of matter in which a 'gas' of free electrons moves about between heavy, more sluggish, positively charged nuclei. The physical properties of degenerate matter are more like those of a *metal*, than a gas in normal conditions. Thus energy is transported to the surface from the interior of a degenerate star by conduction, rather than by radiation or convection (see **Stellar Energy and Interiors**).

The volume of a nucleus is an insignificant fraction of the volume of the whole normal atom, but virtually all the atom's mass resides in the nucleus. The density of degenerate matter, in which these nuclei are closely packed together, is therefore very high. In white dwarfs it can be about 36 million times that of water, so that 1 cubic centimetre would weigh over thirty tons at the surface of the Earth.

Although the nuclei have been stripped of their surrounding orbits of electrons, the number of negatively charged electrons now freely moving between the nuclei is exactly equal to the number of positive charges of the nuclei themselves. Thus degenerate matter is still electrically neutral.

Matter in the degenerate state can exist only under extremely high pressure, which in white dwarf and neutron stars is maintained by vast gravitational forces. The behaviour of a degenerate gas differs from that of an ordinary gas in one very important way. Whereas in a normal gas, composed of atoms and molecules, the temperature of the gas increases as the pressure increases, in the case of a degenerate electron gas the temperature remains constant, or may even fall, as the pressure increases. This strange behaviour gives rise to many remarkable properties in the structure of white dwarf stars. Neutron stars are composed of an even more extreme and dense form of degenerate matter, which is composed solely of the neutral elementary particles called neutrons.

DEGREE OF ARC. A unit for measuring angles. A rotating line sweeps through 360 degrees of arc in one complete revolution. Each degree is subdivided into 60 *minutes* of arc (60′), and each minute into 60 *seconds* of arc (60″). A right angle contains 90°.

DEIMOS. The outer satellite of **Mars.** It is even smaller than Phobos, with an estimated diameter of only 5 miles, and is thus a difficult object to observe even when Mars is near opposition.

From any point on the surface of Mars, Deimos remains visible for two and a half days at a time, and during that period goes through all its phases twice.

DENSITY. The density of a given substance is its mass per unit volume. Its numerical value depends on the units of mass and length that are employed. In the centimetre-gram-second system the density of water is unity, i.e. 1 cubic centimetre of water weighs 1 gram. The density of the Earth is about 5½, while stars range from less than a millionth to many millions.

DIP OF HORIZON. The angle by which the horizon appears below the true horizontal direction as viewed by an observer above ground level. (See diagram under **Altitude.**)

DISTANCE. Astronomical distances are expressed in the **astronomical unit** (A.U.), the **light year,** the **parsec** and in terms of **parallax.**

	miles (millions)	*Astron. Units*	*Light Years*
1 ASTRON. UNIT . . . . =	92·9	1	8·3 *light minutes*
1 LIGHT YEAR . . . . =	5,880,000	63,300	1
1 PARSEC . . . . =	19,150,000	206,000	3·26

DIAMETER. This is a measure of the size of a circular or spherical body. It is the length of the longest straight line that can be drawn within the body, and passes through its centre. The *apparent diameter* of a body is the angle subtended by the body's diameter at the eye of the observer. Thus, the Moon is rather over 2,000 miles in diameter, but its *apparent* diameter as seen from the Earth is about half a degree.

The apparent diameters of virtually all stars are too small for the disc to be seen or photographed with even the largest telescope. All stars should therefore appear as points in photographs, and variation in the sizes of their images is a purely photographic effect owing entirely to differences in brightness.

DIFFRACTION GRATING. An alternative to the **prism** for forming a **spectrum** from light containing different wavelengths. (See under **Spectroscopy.**)

DIONE. One of the satellites of **Saturn.** It is the fifth satellite out from the planet, and with a diameter of about 700 miles is the fourth largest. It completes one revolution round Saturn in rather less than 3 days.

DISTANCE MODULUS. A convenient method of expressing large distances. (See **Magnitude.**)

DOPPLER EFFECT. The apparent change of wavelength of light (or any other form of wave motion) when the source and the observer are in motion relative to one another.

The following analogy will illustrate the principle. Suppose that a ship is at anchor in a sea-way. If the waves are regular, a certain number of them will pass under the ship during a given interval of time. If the ship were moving directly into the waves, it would encounter more than that number during the same interval, and if it were moving with the waves it would encounter fewer. The frequency of the waves as observed from the ship would vary according to the ship's motion, and so would their wavelength (see **Wave Motion**).

Similarly in the case of light from a star, if the relative motion between the star and the Earth is such that the star may be said to be approaching the Earth (or the Earth approaching the star) a terrestrial observer encounters more light waves per second, i.e. the frequency of the light waves will seem to

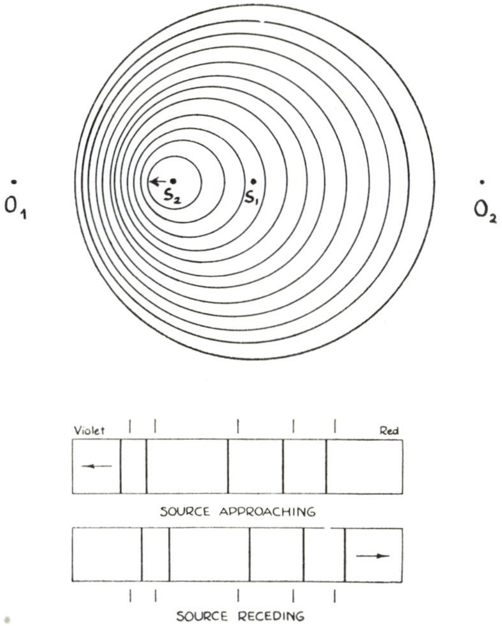

THE DOPPLER EFFECT. The upper diagram shows a source of waves moving from S_1 to S_2. The outermost circle represents the wave that was emitted when the source was at S_1; the innermost circle stands for the wave emitted from S_2. The waves encountered by an observer at O_1 will be much more closely spaced (i.e. their frequency will be higher and their wavelength shorter) than those encountered by an observer at O_2. For light waves, this shift of frequency is revealed by the spectrum. Five spectral lines are chosen in the lower diagram to illustrate this point; the marks above and below the spectra indicate the normal position of the lines.

have increased and the lines of the spectrum will have shifted towards the violet end. For a receding star there will be a similar shift towards the red side of the spectrum (see **Red Shift**). This displacement of the spectral lines can be accurately measured, and the relative velocity of approach or recession calculated from it.

In certain cases the Doppler effect may also be used to detect *rotation* of a star. If the rotation is sufficiently rapid, the relative velocity of the edge or *limb* of the star which is turning *towards* the Earth will differ from that of the limb which is turning *away* from the Earth. The spectral lines in the light from these two limbs will therefore be displaced in opposite directions along the spectrum, while the light from the central part of the star's disc will remain unaffected. This will bring about a broadening of the spectral lines (see **Spectroscopy**). This principle is also applied to some eclipsing binaries (see **Binary Star**).

Exactly the same method is used in **Radio Astronomy** to deduce relative velocities from changes in the frequency of radio emissions whose true frequency is already known from other considerations.

DOPPLER RADIO. A device for tracking the flight of a missile, based on the **Doppler effect.** A transmission is made from a ground station at a constant frequency. This reaches two or three other ground stations (*a*) directly, and (*b*) after having been picked up and re-transmitted by a repeater unit in the missile. At each of the receiving stations a Doppler shift will be found to have altered the frequency of the transmission received via the missile, and from this the velocity of the missile relative to the station can be calculated. Three such relative velocities suffice to determine the path of the vehicle, and the necessary computation by triangulation methods is done by an electronic computer at such speed that the results can be used for guidance of the missile.

DOUBLE STAR. See **Binary Star.**

DOUBLET. Originally used to describe a close pair of lines in a spectrum. In spectra of many atoms (e.g. sodium) the strongest series is a series of doublets. The term is also used for close pairs of energy levels in atoms that are related in a particular manner. (See **Spectroscopy.**)

DOVAP stands for a missile tracking method based on the **Doppler effect,** velocity and position. (See **Doppler Radio.**)

DRACONIDS. See **June Draconids** and **October Draconids.**

E

E LAYER. An ionized region of the Earth's atmosphere at a height of about 70 miles. It can reflect radio waves. (See **Ionosphere.**)

EARTH. The Earth, our home in space, is one of the group of inner planets. Apart from its exceptionally high mean density, 5·52 times that of water, it is quite unremarkable from an astronomical point of view.

ORBIT. The Earth revolves round the Sun at a mean distance of 93,000,000 miles, the perihelion and aphelion distances being respectively 91·4 and 94·6 million miles. The orbital eccentricity is 0·017, less than that of any other major planet except Venus and Neptune. The mean velocity is 18·47 miles per second.

ROTATION. The axial rotation period is 23 hours 56 minutes, and the inclination of the axis 23° 27′.

The rotation period is known with high accuracy, and in fact modern quartz clocks are better timekeepers than is the Earth. Small fluctuations in period occur. There is also the almost inappreciable lengthening of the day due to the tidal pull of the Moon. (See **Tidal Friction** and **Tides.**)

MASS, DIMENSIONS AND SHAPE. The rotation of the Earth sets up a centrifugal force which is greatest at the equator and zero at the poles, leading to a reduction in surface gravity at the equator of about ½ %, and to an equatorial bulge and corresponding polar flattening of the globe. The equatorial diameter is 7,926 miles; the polar 7,900, giving a polar compression of 1 part in 298. Other irregularities of shape are illustrated in the diagram.

The escape velocity is 7 miles per second. The Earth is the largest of the inner group of planets, though only very slightly superior to Venus. Its mass is 6,000,000,000,000,000,000,000 tons.

CONSTITUTION. Although we live on the Earth's surface, our knowledge of the inner composition of the globe is still scanty. Studies of earthquake waves indicate that there is a core of liquid iron or nickel-iron perhaps 4,000 miles in diameter, overlaid by a layer of stony material which is in turn overlaid by a layer of peridotite. The actual crust is less than fifty miles thick, and is made up largely of granite and volcanic rocks.

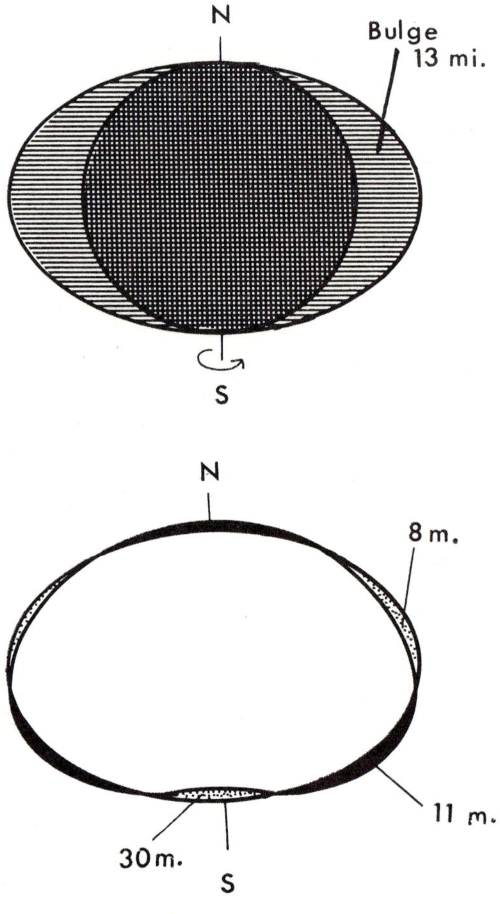

THE SHAPE OF THE EARTH. As a first approximation, this may be taken as a spheroid, and the upper diagram shows the equatorial bulge, greatly exaggerated. The lower diagram shows the 'pear-shaped' departures from spheroid form which have been detected by close study of satellite orbits. Solid black areas represent elevations of mean sea-level above the spheroid surface, and stippled areas indicate depressions, with heights given in metres.

(*After D. King-Hele.*)

Pressures inside the Earth are tremendous (10,000 tons to the square foot at only 25 miles). The central temperature is uncertain, but is probably only a few thousands of degrees. The Earth is unique in the solar system in that it possesses vast oceans.

ATMOSPHERE. Since the Earth's escape velocity is 7 miles per second, the planet has retained a considerable atmosphere, though our present mantle is probably not the original one. The bulk of the air is made up of nitrogen (78 %); oxygen accounts for 21 %, and other gases are present in smaller quantities.

The abundance of oxygen in the Earth's atmosphere has probably been built up gradually by green plants which absorb carbon dioxide and release oxygen. The composition of the atmosphere has thus conditioned the forms of life found on Earth, and has itself been conditioned by them. Burning of fossil fuels by man has increased the carbon dioxide content by 15 % in the last 70 years. (See **Atmosphere.**)

ORIGIN OF THE EARTH. Though many hypotheses have been advanced, we are still uncertain as to the manner in which the Earth came into being. The theory now favoured is that of von Weizsäcker, who supposes that it and the other planets were formed from material collected by the Sun during its passage through an interstellar cloud, but the whole question is still open. (See **Solar System, Origin of.**)

AGE OF THE EARTH. From measurements of radioactivity in accessible rocks the age of the Earth's crust has been estimated at over 3,000 million years. The modern estimate for the age of the Earth is 4,700 million years. (P.M.) (*Pictures on following page.*)

EARTHSHINE. The faint illumination often visible on the 'dark' side of the crescent Moon. Objects on the part of the Moon not illuminated by the Sun are distinguishable by the light which is reflected from the sunlit side of the Earth. When the Moon is new, the Earth as seen from the Moon is in full phase and appears thirteen times as large as the Moon appears to us; it also has a considerably higher **albedo,** and therefore earthshine on the Moon is far more intense than moonlight on the Earth. When the earthshine is especially prominent we can see 'the Old Moon in the New Moon's arms'.

ECCENTRICITY is a quantity which defines the general shape of a curve (or **orbit**) of the kind known as **Conic Sections.** The eccentricity of an ellipse lies between 0 and 1, and the closer it is to 1, the more elongated the ellipse.

ECLIPSE. An eclipse occurs when one celestial body obscures another by passing in front of it. During an eclipse of the Sun, the Moon passes between the Sun and the Earth. During an eclipse of the Moon, the Earth is placed between it and the Sun and casts its shadow on the Moon. Planets occasionally eclipse stars, and the Moon does so constantly, but this phenomenon is always called an *occultation.* In certain **binary stars** the components eclipse each other as one moves across the line of sight from the Earth to the other. On rare occasions one of the inferior planets (Venus and Mercury) travels across the Sun's disc. This also is technically an eclipse, but as only a very small part of the Sun's surface is obscured the term *transit* is used for this.

An eclipse is said to be *partial* if part of the eclipsed body remains visible throughout; it is *total* if during one stage none of the eclipsed body can be seen; and it is *annular* (i.e. 'ring-shaped') if the eclipsing body is not large enough to block the light from the whole of the eclipsed body, so that during the middle of the eclipse only the central portion is obscured, leaving the entire rim to be seen.

(Strictly speaking, of course, the Moon stays visible even when it is eclipsed, if only in dark outline. Moreover there can never be an annular eclipse of the Moon, because the diameter of the Earth's shadow at the distance of the Moon is still considerably greater than the diameter of the Moon itself.)

When a source of light that is not a point-source casts a shadow of an object, this shadow consists of two parts: the *umbra* or total shadow is the region completely cut off from the light, while the *penumbra* remains partly illuminated. This can be verified by holding a pencil above a sheet of paper under an ordinary light bulb. As the pencil is raised the *penumbra* broadens and the darker *umbra* in its middle shrinks and finally disappears altogether. When the Moon's

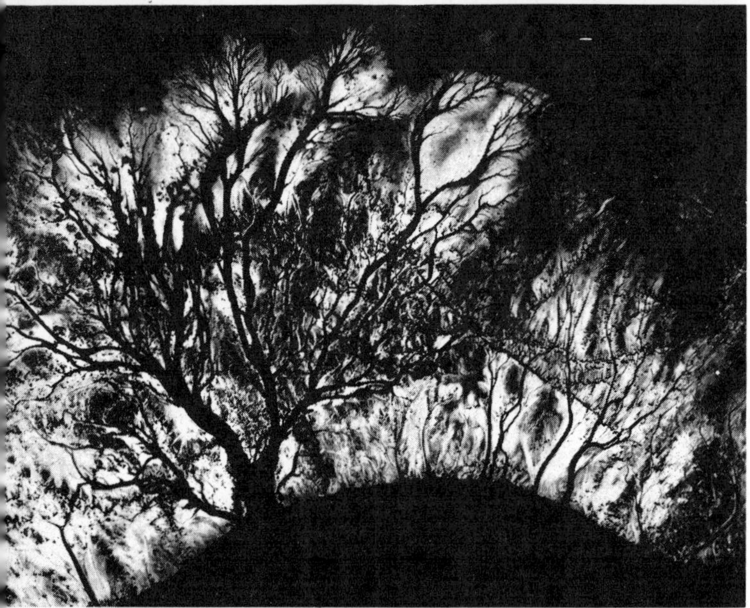

THE SURFACE OF THE EARTH presents some unfamiliar aspects when seen from even a moderate height. Studies of pictures such as those reproduced here are helpful in the interpretation of satellite and rocket camera photographs taken from considerable heights.

Top: an Alaskan volcano, seen from an aeroplane flying above cloud level.

Middle: the *Burning Tree* – a famous shot of the Colorado River delta in the Gulf of California by Fairchild Aero Survey. The branches of the 'tree' were formed as the river successively choked itself off by forming deposits of brilliant sand and light-coloured shoals.

Bottom: Tanzania coastline from 7,500 feet, near Kilwa. The open sea is at top left, the shore runs diagonally across the picture from bottom left, sometimes fringed with a narrow white line of beach, and between them lies a belt of reefs, sandy shoals and weed-covered rocks.

Top: DIZFUL EMBANKMENT, IRAN. Bare rocks, dried river beds, and evidence of stratification, erosion, folding and faulting.

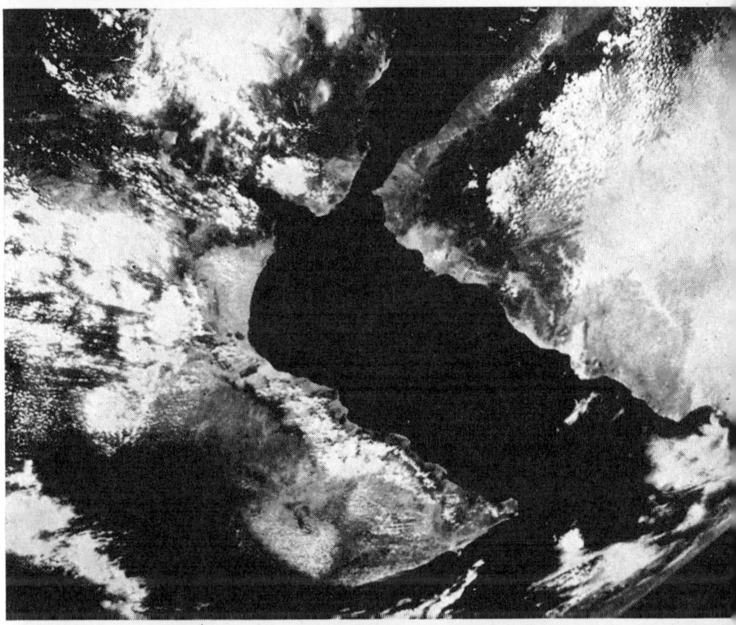

Middle: THE RED SEA from *Gemini II.* The narrow part is at Bab el Mandeb, between Aden and Yemen on the right, and Ethiopia, Somalia and Djibouti on the left. The peninsulas at Aden harbour are clearly visible.

Bottom: THE EARTH from *Apollo 8* on its way home from the Moon. The North Pole is at 11 o'clock, South America lies at the centre, the U.S.A. at upper left. A small part of the bulge of West Africa appears along the sunset terminator at upper right.

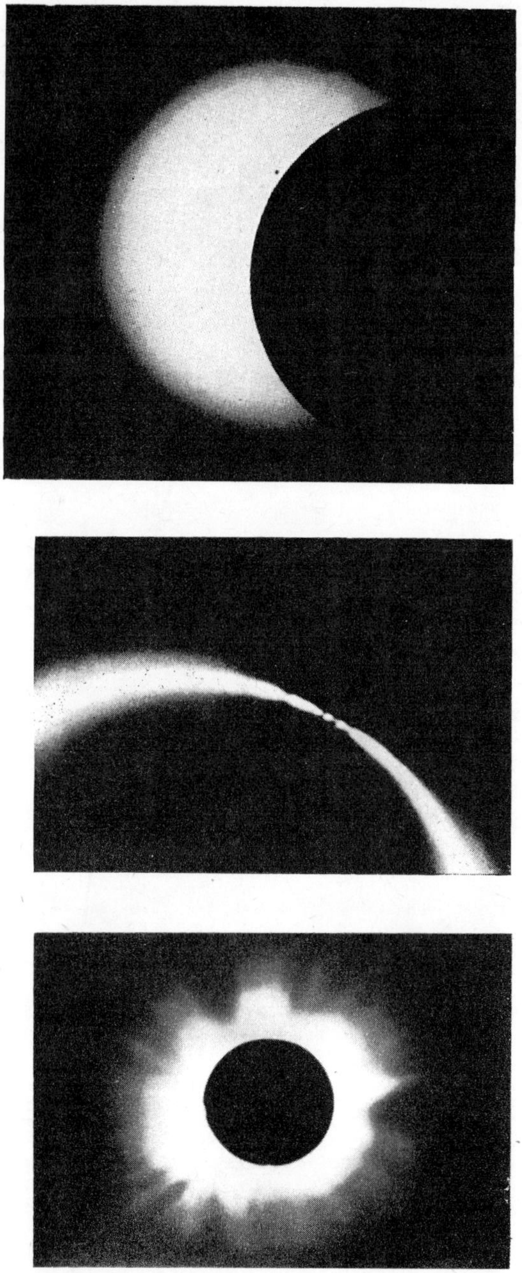

SOLAR ECLIPSES. Top to bottom: (1) A partial eclipse, with a sunspot close to the Moon's shadow. (2) Baily's Beads appear briefly just before an annular eclipse. (3) Photograph of a total eclipse of 1937 taken at 25,000 feet in Peru, showing the corona, coronal streamers and irregularities in the Moon's outline.

shadow travels across the Earth, an observer in its penumbra will see a partial or annular eclipse of the Sun and one in its umbra will see a total eclipse.

Eclipses of the Moon can be seen simultaneously from all points on the Earth where the Moon is above the horizon, but they are of no great interest to astronomers. Eclipses of the Sun, on the contrary, have been of the highest value, chiefly because they allowed an examination of the Sun's corona and other phenomena which until the invention of the **coronagraph** was not possible at other times. For this reason, many expeditions have been sent to the narrow and often remote strips over which the tip of the Moon's umbra would pass during a solar eclipse, and all too frequently months of preparation have been foiled by a few clouds during the crucial moments of totality.

The nearer the Moon is to the Sun, the shorter the cone of its full shadow. When the Earth–Moon system is near perihelion this cone is generally too short to reach the Earth at all even if the Sun, the Moon and the Earth are exactly in line. A total eclipse is then impossible, and an annular eclipse will be observed.

A total eclipse begins at *first contact*, when the limb of the (inevitably *new*) Moon appears to touch the edge of the Sun's disc. There follows the partial stage during which the day grows dim as more and more of the Sun is obscured. Just before *second contact*, the beginning of totality, the contours of the leading edge of the Moon break up the vanishing crescent of the Sun into patches of light called **Baily's Beads.** They disappear again almost at once, and the *corona* flashes into view. A bright halo surrounds the hidden Sun, traversed by the coronal streamers and perhaps the vivid red arch of a *solar prominence*. The brighter stars shine in a dusky sky, and possibly also a comet near its perihelion. (See also **Sun.**)

After seven and a half minutes at the most the sequence of events is reversed. At *third contact* the Sun begins to re-emerge, at *fourth contact* the last of the Moon draws clear of its edge.

Eclipses permit very accurate determinations to be made of the Moon's motion and the **perturbations** that affect it. They have also made it possible to confirm a prediction contained in the Theory of **Relativity** that

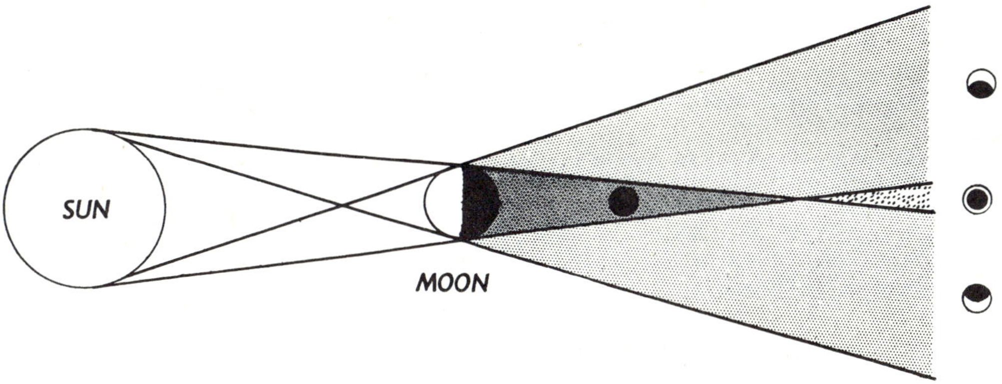

Above: THE MOON'S SHADOW. No part of the Sun's disc is visible to an observer within the dark cone of the Moon's total shadow or *umbra*, as indicated by the black circle. The other circles show how the Sun would appear from various positions within the half-shadow or *penumbra*.

Below: THE MOON IN ECLIPSE photographed over Tokyo at ten-minute intervals, beginning on the left. A telephoto lens was used, and this makes the Moon appear larger and higher in the sky than would be the case with an ordinary lens.

light is bent towards a gravitating body. To test this point, it was necessary to photograph stars whose light passes close to the Sun before reaching the Earth, and then check for any apparent change in the star's position. This was done successfully during an eclipse in 1919, when the displacement was found to agree almost exactly with the predicted amount.

It is not difficult to calculate the dates and paths of eclipses forwards or backwards over many centuries. Many historical events whose records mention an eclipse have thereby been dated.

Every 18 years and $11\frac{1}{3}$ days the sequence of eclipses repeats itself almost exactly. This interval is called a **Saros** cycle, and has been known since the days of the Chaldeans. (M.T.B.)

ECLIPTIC. The intersection of the plane containing the Earth's orbit with the **celestial sphere.** It can also be defined as the apparent path of the Sun among the stars.

As the Earth moves in its orbit the position of the Sun relative to the star background changes, making one complete circuit of the sky in a year. It is of course immaterial that these background stars cannot be seen at the same time as the Sun, since their positions on the celestial sphere are known. The

Sun lies in the plane of the Earth's orbit, and so its apparent path marks the ecliptic in the sky.

As the plane of the Earth's equator is inclined to that of its orbit at an angle of about $23\frac{1}{2}°$, the ecliptic is inclined to the **celestial equator** by the same amount. The points of intersection of the ecliptic and the celestial equator are occupied by the Sun at the *equinoxes*; the Sun reaches its farthest North of the equator at the *summer solstice*, and its farthest South at the *winter solstice*.

The other major planets of the solar system, and most of the asteroids also, have orbits lying in nearly the same plane as that of the Earth. They can therefore never be seen far from the ecliptic.

The constellations through which the ecliptic passes are called the Signs of the **Zodiac.** (R.G.)

ELECTRIC PROPULSION systems offer the best hope of achieving really high rocket velocities, and will therefore play an important part in interplanetary journeys. They can provide low thrust for long periods and can therefore accelerate a vehicle, once it is in orbit, to speeds which theoretically could approach that of light, but they would do this gradually and cannot provide the rapid acceleration (and deceleration) needed for lift-off or landing. In practice they are likely to be used in combination with chemical and **nuclear propulsion.**

These systems can be of the arc-jet type in which a propellent is heated in an electric arc and expanded through a nozzle as in the case of a nuclear thermal rocket, or can rely on the acceleration of a beam of charged particles or plasma to achieve thrust. Very high **exhaust velocities** are possible for **plasma** or **ion rockets,** and though thrust/weight ratios are likely to remain poor because of the weight of the electric power source, the resultant very large **specific impulse** means that a much higher proportion of the total space craft weight can be payload than is the case with chemical or nuclear thermal rockets. Arc-jet or electrothermal rockets have been operated at specific impulses of around 1,600 sec., or about double present levels for nuclear thermal rockets, and there seems some prospect that a system combining certain of the characteristics of an arc-jet and a plasma rocket, in a way which is not well understood theoretically, could achieve an exhaust velocity of about 100 km./sec. or a specific impulse of about 10,000 sec. The simplest kind of electrothermal rocket is a 'resisto jet' in which a flow of hydrogen is heated as it passes along an electrically heated duct.

Electrostatic or ion rockets use an electric field to accelerate a beam of charged particles. The particles need not be ions, and should not have too small or too large a mass: light ions require currents of thousands of amperes, heavy ions require very high voltages. If ions of only one sign are being expelled from a vehicle, the latter will rapidly acquire a very large charge of opposite sign and will soon attract the beam so powerfully that no further ions can be expelled. It is necessary therefore to neutralize the beam just downstream of the motor by injecting sufficient electrons. In the early days of work on ion engines it was not known whether sufficiently exact neutralization was possible, but doubts have now been dispelled by successful Russian and American tests of ion engines in space. The thrust of an ion rocket depends only on the propellent flow rate and the power contained in the beam, and is independent of the molecular weight or particle size of the propellent. The particle charge and mass do, however, determine the relationship between beam current and motor voltage, and also the exhaust velocity and hence specific impulse. Typical ion rockets might cover the specific impulse range from about 3,000 to something under 20,000 sec.

Plasma propulsion systems differ from ion rockets in that at least part of the beam kinetic energy is derived from acceleration in a magnetic field. If crossed magnetic and electric fields exist in the motor, charged particles of opposite sign will move in spiral paths of opposite sense but in the same direction. Space-charge effects prevent the separation of positive and negative ions even if these are of such different mass that separation might be expected, and as a result of the electrical neutrality of the beam there is no tendency for the beam to blow apart, an effect which limits the beam intensity in an ion rocket. This means that a plasma rocket could be made with smaller cross-section than an ion rocket of the same thrust. Although efficiencies are still too low for worth-while applications to be in sight, the long-term prospect for plasma systems is very promising, leading eventually to a controlled thermonuclear plasma rocket. (C.W.M.)

ELECTROMAGNETIC RADIATION. Electromagnetic waves are caused by periodic fluctuations in space of electric and magnetic fields. These fields act in directions perpendicular to each other and to the line of propagation of the electromagnetic waves. Energy can be radiated through a *vacuum* in the form of electromagnetic waves; this form of energy transport is called a wave motion purely because it obeys the basic *mathematical laws* governing a wave motion propagated by particles in a *material* medium. Radiated heat, light, radio waves, X-rays, and gamma-rays are all forms of *electromagnetic radiation*, or electromagnetic waves of different **wavelengths;** a detailed list is given under **Electromagnetic Spectrum.**

In the absence of matter, all electromagnetic waves travel with the speed of light, which is almost exactly 300,000 km. per second, a distance equal to seven and a half times the circumference of the Earth. They can be *refracted*, as light is refracted or 'bent' by a lens, and *reflected*, as in a mirror, and their direction of travel may be curved by a gravitational field.

Electromagnetic waves are created when an atom gives up energy by relaxing the orbits of one or more of its electrons, or by changes involving its nucleus (see **Atom**). They can also arise through the motion of electrically and magnetically charged matter.

There are occasions when they do *not* obey the mathematical rules of wave motion and behave in every way as if they were fast-travelling *particles*. The point is that, whatever happens, electromagnetic energy does follow certain rules and laws in a consistent way, and it is as unimportant to be able to *visualize* its nature as the precise shape of chessmen is unimportant to an understanding of the game of chess.

ELECTROMAGNETIC SPECTRUM. The full range of wavelengths of **electromagnetic radiation.**

The character of this radiation changes markedly as the wavelength decreases. The longest waves are known as radio waves, with wavelengths of several kilometres, while the shorter ones include gamma-rays and X-rays. The whole spectrum is shown in the diagram.

The use of the word *spectrum* here extends the idea of the spectrum of light which a

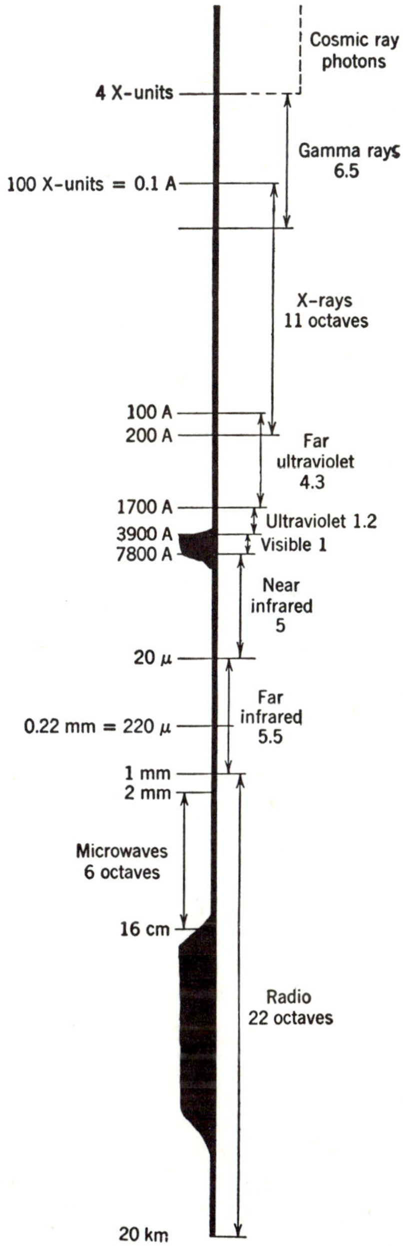

THE ELECTROMAGNETIC SPECTRUM plotted on a logarithmic scale. The thickness of the solid black areas represents the extent to which the Earth's atmosphere is transparent to the corresponding wavelengths. (*After Menzel*)

spectroscope can produce, but there is no single apparatus that can deal similarly with *all* forms of electromagnetic radiation.

ELECTRON. A fundamental particle that exists in every **atom.** The mass of an electron is about 1/1840 of that of a hydrogen atom, and is therefore much less than the masses of the other common particles, the *proton* and the *neutron*, which are almost equal to the hydrogen atom. The electron, however, possesses an electrical charge as great as the charge of a proton, but negative whereas the proton's charge is positive.

The electron is the only particle outside the nucleus, or heavy central body, of an atom. It can be separated from the atom to which it belongs by friction, heat, radioactivity or any other form of energy which promotes ionization. It is then called *free*.

Although the number and distribution of electrons in an atom largely determine the latter's properties, the electrons themselves are not to be regarded as material particles like hard little spheres: they are not *made* of anything. Their effects can nevertheless be observed when they are in the free state. The familiar electric current is simply the flow of electrons along a wire or other conducting body. Electrical conductors are materials whose atoms offer very little resistance to an electron flow. Although an electrical current travels with the speed of light, the electrons constituting it move relatively slowly, at a speed of the order of millimetres per second. In the same way, if a valve is turned on at one end of a long pipe containing water, the water begins to flow out of the other end practically at once even though it moves quite slowly in the pipe.

Electrons are responsible for the opacity of stars, which might be expected to be transparent as they are made of gas. In stars at temperatures similar to that of the Sun, light of any wavelength may be absorbed by bound electrons in atoms. Light with energy greater (i.e. wavelength less) than a certain critical amount is capable, not of causing an electron to change its orbit, but of removing it from the atom altogether. In doing this, the light is absorbed and may be re-emitted with a different wavelength. Light coming from deep down in a star is thus absorbed and re-emitted at a higher level, so that we see no light from the interior. In stars at very high temperatures, light is scattered by electrons which are already free, and this is sometimes the most important source of opacity. (R.G.)

ELEMENT, CHEMICAL. One of over 100 substances which cannot be split up into simpler substances by chemical means. Elements combine chemically to form *compounds*, and the properties of a compound usually differ entirely from those of its constituent elements.

Of the known elements, about eighteen do not occur naturally and have been produced artificially in the laboratory. Many others are comparatively rare. The most common elements of the Earth include hydrogen, oxygen, nitrogen, carbon, silicon, sulphur, chlorine, iron, nickel, magnesium and aluminium. In addition to most of these, helium and to a lesser extent titanium are common in stars.

All neutral atoms of an element are alike. The radiation emitted by atoms gives rise to spectra (see **Spectroscopy**) from which they can be identified when chemical tests cannot be applied. Most elements have stable atoms, but some, like radium or uranium, undergo radioactive *decay* and change into other elements, the change being accompanied by a release of energy and particles that are fragments of an atom. (See **Atom** and **Nucleogenesis.**)

ELEMENT, ORBITAL. One of six quantities necessary to define an **orbit.**

ELLIPSE. See **Conic Sections.**

ELLIPSE, TRANSFER. The path by which an interplanetary vehicle can transfer from an **orbit** about one body into an orbit about another. It is part of an ellipse; the point on this ellipse farthest from the first body should lie in the orbit about the Sun of the second body for the sake of fuel economy. An acceleration is needed to embark on the transfer ellipse. By combining careful timing with a suitable acceleration or retardation upon reaching the second body's orbit, the vehicle could then either land on it, follow it, or circumnavigate it and return to the first body. (See **Planetary Probe.**)

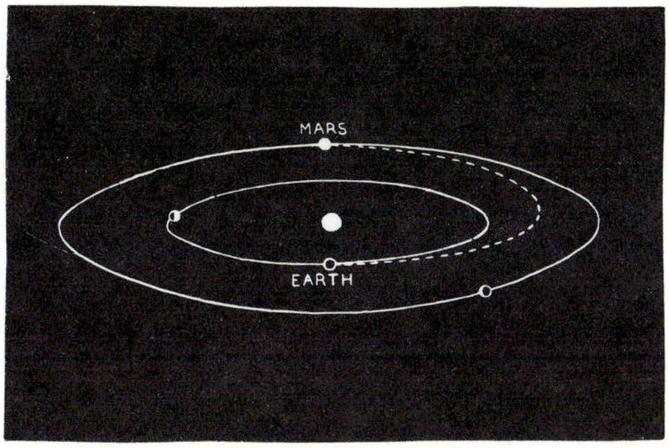

TRANSFER ELLIPSE from Earth to Mars. During the transfer both planets move (counter-clockwise in the diagram) along their orbits.

ELONGATION. (See diagram under **Conjunction.**) The difference in **Right Ascension** between an inferior planet (Mercury or Venus) and the Sun. It is usually measured in degrees of arc. The possibility of observing a planet depends on the darkness of the sky when the planet is at a reasonable altitude, i.e. high enough not to be excessively affected by atmospheric vagaries. The sky darkness depends on the distance of the Sun below the horizon, and the elongation is thus an indication of the visibility of the planet. Superior planets are rarely observed when they appear close to the Sun, but the elongation of the inferior planets is of importance because it can never be great – not more than 47° for Venus, or 28° for Mercury. When these planets have elongations East of the Sun they can be seen in the evening sky; western elongations correspond to morning appearances.

EMISSION SPECTRUM. A spectrum consisting of bright lines, bands or a continuum emitted by a hot source. (See **Spectroscopy.**)

ENCELADUS. One of the smaller satellites of **Saturn.** It is the third nearest to the planet; its diameter is about 300 miles and it takes 33 hours to describe one revolution in its orbit.

ENCKE'S COMET has the shortest revolution period of all the periodic comets – 3·3 years. Its perihelion lies within the orbit of Mercury, while its aphelion lies well out in the asteroid zone. The comet was discovered in 1786, and since 1819 it has been observed at each return to perihelion except that of 1944. Searches with the large telescopes on Mount Wilson before the returns of 1951 and 1954 resulted in its recovery seven months and ten months before perihelion respectively, while the comet was still in the outer half of its orbit. On the latter occasion the comet was barely of the twentieth magnitude at the time of its recovery. Nearly a year later, at perihelion, it exceeded sixth magnitude.

ENCKE'S DIVISION. A minor gap in **Saturn's** outer ring.

ENERGY. The capacity for doing work. There are several kinds of energy to be considered.

KINETIC ENERGY or energy of motion. Any moving body possesses kinetic energy in proportion to its mass and to the square of its velocity:

$$\textit{kinetic energy} = \tfrac{1}{2}\,\textit{mass} \times \textit{velocity}^2.$$

Examples: a cannon ball does work when it strikes a wall by breaking and moving the stones; a flywheel can continue to turn over a motor after the power supply has ceased; the kinetic energy of a meteor is dissipated in the form of heat, light, sound and radio waves as it is slowed down on entering the atmosphere.

POTENTIAL ENERGY. A body has potential energy by virtue of its position relative to a force that is acting on it. Examples: an object raised above the ground – the force acting on it is that of gravity, and if the object is allowed to fall it acquires kinetic energy and can therefore do work; a loaded spring – if it is released, it can do work by moving a body or driving a mechanism such as a clockwork; a piece of iron near a magnet – if it is permitted to move, it too will acquire kinetic energy, and it may pull another body with it.

RADIANT ENERGY. This is the energy inherent in electromagnetic waves. An outstanding example is the radiated energy of the Sun, which when it reaches the Earth does work by warming and ionizing the atmosphere, by heating the ground and evaporating water, by effecting the build-up of starch and sugars in green plants, and in countless other ways.

CHEMICAL ENERGY. More often than not this form of energy requires a small initial investment of work before it becomes apparent. Examples: the heat of a match applied to a sheet of paper will release chemical energy as heat of combustion; a small spark can initiate the explosion of a mixture of hydrogen and chlorine leading to the evolution of heat of combination.

NUCLEAR ENERGY. Of all the above forms of energy it is true to say that, *in a closed system, the total amount of energy remains constant*, no matter what changes may take place in the system. Nuclear energy, however, stems in part from the transformation of matter into energy. The quantity of energy that can be obtained from the destruction of a given amount of matter is given by the equation:

$$Energy = mass \times (velocity\ of\ light)^2.$$

ENERGY LEVEL. The distribution of the electrons in an atom among the various orbits they can occupy.

If an electron jumps from one orbit into another, smaller one, it loses energy which is emitted as radiation of a particular wavelength. An electron can also absorb radiated energy by jumping into a greater orbit. The first event will lower, and the second raise the energy level of the atom in which it occurs. The lowest energy level of an atom is called its *ground state*. (See **Atom** and **Spectroscopy.**)

ENERGY, ORBITAL. The sum of the potential and kinetic energies of a body revolving in an **orbit** about another body.

EPHEMERIS. A table giving the predicted positions of some body in the solar system for a succession of future dates.

EPOCH. In astronomy, an arbitrarily chosen moment in time to which measurements of position are referred. Owing to the **precession** of the Earth's axis, the equator shifts slowly, and with it the celestial equator. This causes the **Right Ascension** and **Declination** of a star to vary continuously, as these quantities depend upon the position of the celestial equator (see **Celestial Sphere**). The lines of Right Ascension and Declination drawn across star maps are therefore strictly correct only for the date or *epoch* for which they were drawn. The epochs commonly used are 1855, 1900, 1920 and 1950; and a position measured from an atlas is not completely specified unless the epoch is also quoted, so that suitable corrections may be made if necessary.

EQUATION OF TIME. See **Mean Sun.**

EQUATOR. The equator of a rotating sphere is the line joining the points on the surface which are equidistant from the two poles of the axis of rotation. The plane in which the equator lies cuts the axis at right angles in the centre of the sphere. At the equator, centrifugal force and rotational velocity are greater than anywhere else on the surface of the sphere. (See also **Celestial Equator.**)

EQUINOX. A moment at which the Sun's centre crosses the **celestial equator** as seen from the Earth's centre. During its annual journey around the **ecliptic,** the Sun crosses the celestial equator twice. There are therefore two equinoxes each year: the *vernal equinox* occurs about March 21 and marks the official beginning of spring, while the *autumnal equinox* near September 22 marks the beginning of autumn. At those dates, day and night are of roughly equal length all over the Earth.

EROS. The first discovered of the minor planets which come within the orbit of Mars; as it is of some size, with a longest diameter of 15 miles, it has proved very useful for determining the value of the astronomical unit. (See **Asteroids.**)

ESCAPE VELOCITY. The minimum velocity which will enable an object to escape from the surface of a planet or other body without further propulsion.

The escape velocity of the Earth is just over seven miles per second, or 25,000 m.p.h. Once a body moving upwards has exceeded this speed, its own momentum will carry it away from the Earth in a hyperbolic **orbit,** which may presently turn into a heliocentric one, and it will never return under the influence of gravity. This is equally true of a heavy missile and of a light particle, such as an atmospheric molecule. In the upper reaches of an atmosphere, where the density is low and the temperature may be high, the gaseous molecules are likely to be in considerable agitation, and their 'mean free path' (i.e. the average distance they can travel without being deflected by collision with another particle) is likely to be long. A molecule will therefore sooner or later acquire an outward velocity exceeding the escape velocity and will then be lost to the atmosphere. It follows that many bodies with low escape velocities must more or less quickly be denuded of any atmosphere they may at some stage possess.

Escape velocity does not allow for air resistance, but in all practical cases the velocity will not be attained until heights are reached at which air resistance is so low that it can, for this purpose, be disregarded.

A few calculated escape velocities are given below, in miles per second:

Earth	7·0
Moon	1·5
Mercury	2·6
Venus	6·3
Mars	3·1
Jupiter	37·0
Callisto	0·9

EUROPA is the smallest of the four major satellites of **Jupiter,** having a diameter of 1,950 miles. It revolves in its orbit in about 3½ days.

It is a good reflector of light, and appears almost as bright as the larger **Io.** The escape velocity of only 1·3 miles per second indicates that it is incapable of retaining an atmosphere.

EXHAUST VELOCITY. The velocity with which the propellent gas leaves a rocket motor. In the case of all rockets this is a function of the amount of energy transferred to the gas stream and of the molecular weight of the gas; the efficiency of the nozzle and the pressure upstream of it also affect the value achieved.

For chemical rockets the highest specific impulse is obtained from propellents which react to produce a large *heat of reaction* and give combustion products with a low mean molecular weight. Liquid oxygen and kerosene give a molecular weight of products (water and carbon dioxide) of somewhat under 40, but hydrogen and oxygen produce only water with a molecular weight of 18, and combinations even more difficult to handle, such as hydrogen and fluorine, will yield still lower molecular weight products and will give twice the specific impulse available with oxygen and hydrocarbon fuels.

Exhaust velocity is related to specific impulse by the equation:

$$\textit{exhaust velocity} = \textit{specific impulse} \times g,$$

where g is the acceleration due to gravity.

Theoretical values in the region of 12,000 ft./sec., representing a theoretical specific impulse of around 400 seconds, are now possible for chemical rockets, though the actual value achieved in practice is somewhat less for a variety of reasons, including pressure losses in the nozzle and combustion problems caused by the need to ensure adequate cooling by having a fuel-rich mixture in the flame zone adjacent to the combustion chamber walls and so a cooler flame than that theoretically possible.

For nuclear and electrical rockets no such theoretical limits exist. The exhaust velocity achievable with a plasma or ion rocket is a function only of the electric field strength which can be produced and controlled, the mass of the ions used and the space available for accelerating them. Of these only the mass of the ions is a fixed quantity. Similarly for a nuclear rocket in which the heat from the nuclear reactor is transferred to a *working fluid* of minimum molecular weight (say, hydrogen), the limiting parameter is the rate at which heat can be transferred, or indirectly the temperature which the reactor core can sustain. This may have a practical limit at the present time, but higher working temperatures should be possible eventually. Specific impulses at least an order of magnitude better than chemical rockets should be achieved with nuclear propulsion systems in the foreseeable future. Values of the order of 10,000 are already attainable with electrical thrusters, but the actual value of thrust for a given weight is very low indeed. Specific impulse of 10,000 sec. corresponds to an exhaust velocity of about 320,000 ft./sec.

EXOSPHERE. The outermost region of the **Atmosphere of the Earth.**

EXPANSION OF THE UNIVERSE. A term which has been used to describe the apparent spreading apart of the **galaxies,** which constitute the largest units in the Universe. The rate of recession of each galaxy from our own is consistent with an expansion in all directions of the Universe as a whole and not merely of the matter in it. (See **Red Shift** and **Cosmology.**)

EXTRAGALACTIC NEBULAE. An obsolete name for **galaxies.**

EYEPIECE. The **lens** system at the viewing end of a **telescope.** The eyepiece is in effect a microscope through which the image formed by the telescope objective is examined. Eyepieces, like microscopes, can have a variety of magnifying powers, and generally several are provided with one telescope to give a number of different magnifications.

The commonest form of eyepiece is named the Huyghenian, after the astronomer Huyghens; it consists of two convex lenses spaced a short distance apart. Although it is satisfactory for use with most refractors it is generally unsuitable for reflecting telescopes on account of their relatively short focal lengths.

F

F LAYER. An ionized layer at an altitude of about 150 miles in the Earth's atmosphere. (See **Ionosphere.**)

FACULAE. Unusually bright patches in the solar photosphere. (See **Sun.**)

FIELD. The region over which the effects of a force such as gravity or magnetism are appreciable. At any point within it the field has a certain strength and acts in a certain direction. For instance, the field of gravitational attraction of the Earth at a point at ground level acts vertically downwards with a force corresponding to an acceleration of 32 feet per second per second.

FIELD OF VIEW. The area of the sky which is visible at any one time in a **telescope.** The term is often abbreviated to 'field'. The diameter of the field depends on the telescope and on the eyepiece employed, and is usually quoted in minutes or degrees of arc. The diameter of the photographic field of view of the 200-inch Hale telescope is about 10′ of arc; this means that one photograph covers one-millionth of the visible hemisphere of the sky.

FILAMENTS, SOLAR. Dark linear markings shown on the Sun's disc by a spectroheliograph. (See **Sun.**)

FINDER. A small telescope enabling a larger one to be quickly trained on to specific objects.

A large telescope generally has a **field of view** of less than one degree of arc, and it is frequently difficult to locate an astronomical object in such an instrument. On the other hand it can be readily found in the large field of view of the finder. The latter is mounted on the large telescope and adjusted so that it points in the same direction. If a star is brought to the centre of the finder field, it will be in view in the main telescope.

FIREBALL. A very bright **meteor.**

FIRST POINT OF ARIES and **FIRST POINT OF LIBRA.** See **Celestial Sphere.**

FLARES, SOLAR. Brilliant eruptions of light from glowing hydrogen in the solar chromosphere in the neighbourhood of sunspots. (See **Sun.**)

FLASH SPECTRUM. The spectrum of the solar chromosphere observed at the beginning and end of a total solar eclipse, when the layers of the Sun below the chromosphere are still hidden from view by the Moon. Its name derives from the suddenness of its appearance and its brief duration. (See **Sun.**)

FLYING SAUCER. Journalistic phrase that has been applied to various objects reported to have been seen in the sky by casual observers. Meteorological balloons, refraction phenomena, searchlight beams playing on clouds and Venus seen in daylight are among the commoner causes of such reports. Photographs of 'flying saucers' supposed to have landed are cool impostures.

The term has also been used as a nickname for some experimental aircraft designs. (M.T.B.)

FOCUS. See **Conic Sections.** For optical focus, see **Lens** and **Telescope.**

FORBIDDEN LINES. These lines sometimes appear in the spectra of very tenuous gases and arise from normally rare transitions between energy levels in **atoms.** They have been observed in the spectra of the uppermost parts of the Earth's atmosphere, the solar corona, and galactic nebulae. (See **Spectroscopy.**)

FRAUNHOFER LINES. The absorption lines in the spectrum of the Sun. The term is also used rather loosely for the absorption lines in the spectra of other stars. Fraunhofer himself catalogued over 750 lines, and over 26,000 have subsequently been recorded in the parts of the solar spectrum which have so far been accessible to observation. (See **Spectroscopy.**)

FREE FALL. The condition of unrestricted motion in a gravitational field.

On the Earth's surface the ground restricts the downward motion that would otherwise take place. In a rocket or satellite coasting without power, the rocket and all its contents are equally under the influence of **gravity,** irrespective of the distance from the Earth, and the entire rocket is in a state of free fall. It may still be moving upwards under its own momentum, but from the moment it begins to coast it is technically falling.

In these circumstances, the organs of balance in the human ears do not function, and controlled muscular movements are difficult until some practice has been acquired (see **Space Medicine**). Many of the most ordinary actions become complicated or impossible: everything that is not fastened tends to float about, and liquids behave rather like balloons filled with air and drift out of open vessels at the slightest local disturbance.

Conditions inside the rocket are then *as if* there were no gravitational field acting, but this is only because there is *no difference* between the effects of gravity on the rocket, on objects inside it and on their 'supports'. (See **Weightlessness.**)

In other words, free fall does not depend at all on the absence of gravity, but only on the absence of all accelerating or retarding forces other than gravity. Except for the retarding force of air resistance, a cricket ball would be in free fall from the moment it leaves the thrower's hand, even while it is still rising. An object resting on a table is not in free fall because the support of the table opposes the action of gravity. The Earth is in constant free fall about the Sun.

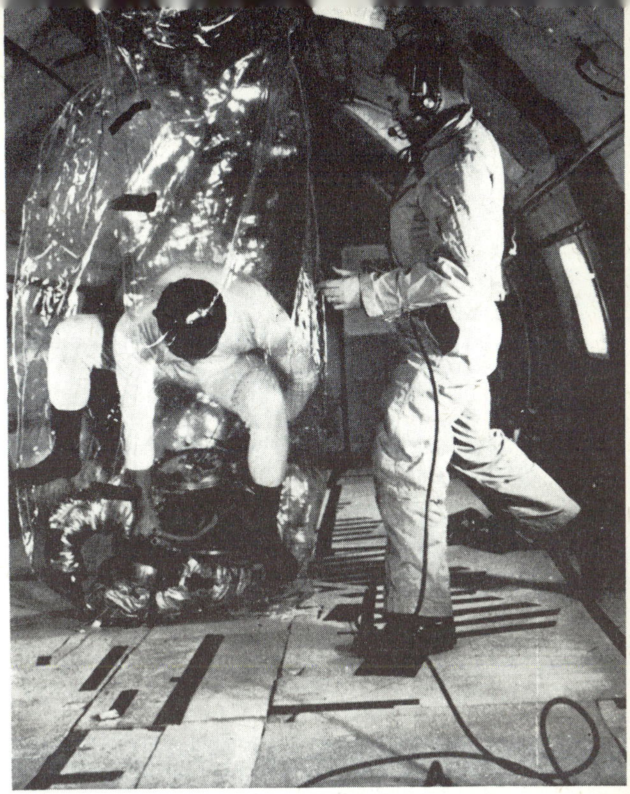

SUITING-UP IN FREE FALL. An Apollo astronaut is floating inside a 'don-and-doff' bag which is designed to provide protection in a depressurization emergency until he can put on his pressure suit. The equipment, and the difficult task of manipulating it under conditions of weightlessness, are being tested aboard a coasting aeroplane.

FREQUENCY. The frequency ν of a wave motion is the number of cycles it completes in one second. The frequency ν of **electromagnetic radiation** in space is related to its **wavelength** λ by the formula $\lambda\nu = c$, where $c = 300{,}000$ km. per second is the velocity of light.

G

g. A symbol for the force of **gravity** at the Earth's surface. This force will accelerate any body in free fall by 32 feet per second per second, and this **acceleration** is equal to 1 g. This unit can be used indiscriminately to express any acceleration or gravitational force. For instance, a rocket whose speed increases by 96 feet per second every second is said to be accelerating at 3 g. For the effects of acceleration on human beings, see **Space Medicine.**

G LAYER. An ionized layer at a height of 300–400 miles in the Earth's atmosphere.

GALACTIC ABSORPTION. The absorption of light by interstellar matter in the **Galaxy.** Large particles block light of all wavelengths equally and cause an overall lowering of the intensity of the light. Small particles whose dimensions are comparable with the wavelength of light cause a reddening, as they scatter the blue end of the spectrum more than the red. This is quite different from the reddening due to the **Doppler effect,** which *displaces* the entire spectrum of a receding light source to the red and alters the positions of the spectral lines without changing their relative strengths. Absorption reduces the background brightness of the continuous spectrum more in the blue than in the red, but does not displace the lines.

Owing to the concentration of dust in the galactic plane, the reddening and absorption are most noticeable when we look along this plane from the Earth at distant objects: no extragalactic nebulae are visible near this plane, the whole of their light having been absorbed before it reaches us.

GALACTIC CLUSTER. A loose aggregation of stars occurring in a spiral galaxy. (See **Galaxies.**)

The number of stars in a galactic cluster is no more than a few hundred, and their arrangement is usually haphazard, with perhaps a slight concentration towards a centre; the contrast with **globular clusters** is very striking. Over seven hundred galactic (or *open*) clusters have been found in our Galaxy, and many more in other spirals.

NGC 2682 (Messier 67), an open cluster in our galaxy, in the constellation of Cancer.
(*Mount Wilson – Palomar*)

The galactic clusters of the **Pleiades** and the Hyades in the constellation Taurus are obvious to the unaided eye; the Double Cluster in Perseus is a splendid object in a small telescope.

Galactic clusters lie in the spiral arms of galaxies and participate in galactic rotation, and belong to Population I. (See **Galaxy.**)

GALACTIC NEBULAE. Clouds of interstellar matter whose presence is revealed either because they are illuminated by a bright star or because they noticeably weaken the light

NGC 3587, a planetary nebula in the Great Bear constellation. The hot central star is well shown. (*Mount Wilson – Palomar*)

from stars in a particular region of the sky. The former are *bright* and the latter *dark* nebulae. Both types are found in our own and in other spiral galaxies.

The bright nebulae in the Milky Way are usually closely associated with stars. The light from the stars is reflected by the nebula. If the stars are very hot their ultraviolet radiation can be absorbed by the nebula, causing it to heat up and emit light of its own. An excellent example of an irregular nebula of this type is the **Orion Nebula** illustrated overleaf.

The *planetary nebulae*, which are a type of bright nebula, are usually symmetrical shells of gas round extremely hot stars. These shells are thought to have been thrown off the star, late in its life.

Some planetaries appear ring-shaped. This is an effect of perspective, the denseness of the shell appearing greatest when we look through the edge of the apparent disc and least when we look through its centre. The shell often appears of different sizes when viewed in light of different colours: the same emission lines are not all excited in the same layers.

In many parts of the sky, particularly in the Milky Way, there are dark patches where few stars are visible, e.g. the **Coal Sack.** The *dark nebulae*, relatively dense clouds of interstellar matter, obscure the light from stars in the direction of these patches. They are not sufficiently near any stars to be illuminated, so they appear dark.

Some of the dark material appears in the form of dark globules which many astrono-

THE COAL SACK and, to the right and above it, the Southern Cross.

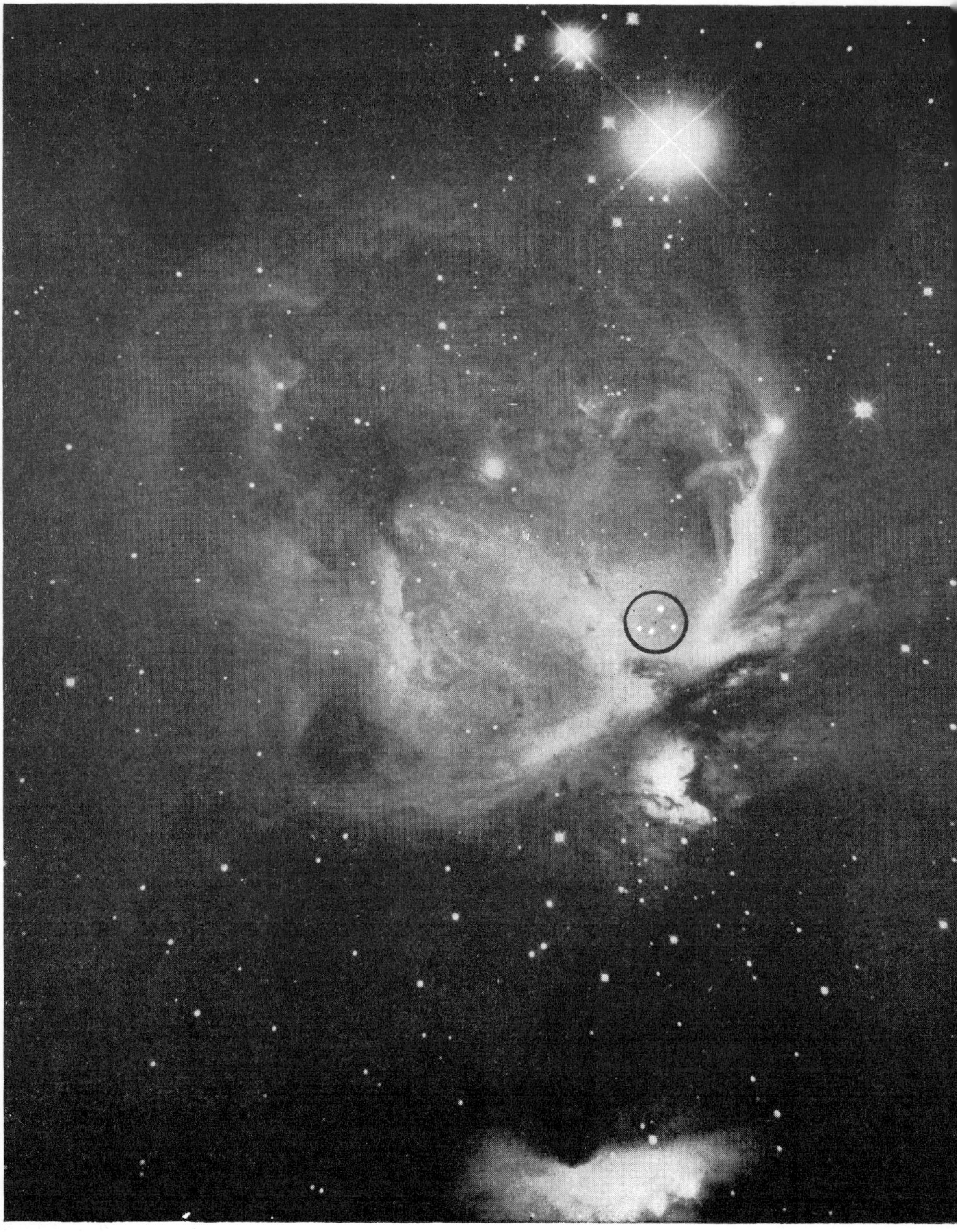

THE HORSEHEAD NEBULA, a region of dark absorption and glowing emission nebulosities near the star Zeta Orionis. A number of foreground stars are seen which lie between the Earth and the almost opaque dark areas. The picture is shown here with South towards the right.

Right: A view of a wider field surrounding the Horsehead Nebula.

Left: THE GREAT NEBULA IN ORION, a vast mass of gas illuminated chiefly by high-temperature stars within the black circle. The naked eye sees it as the middle star in the Sword of Orion, but a long exposure reveals that the nebula covers an area of the sky larger than the Moon's disc. (*Yerkes*)

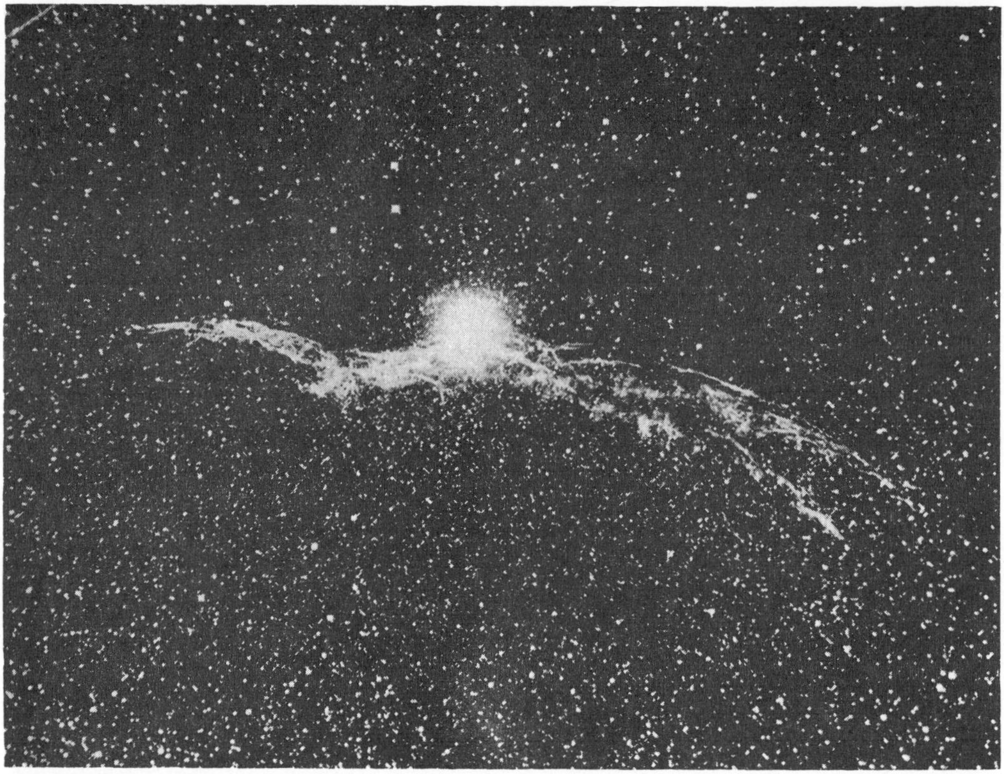

GALACTIC NEBULAE. *Above:* Filamentary nebula in Cygnus. *Below:* Nebulosities in Monoceros, part of the Milky Way. On the left, a small section of a 48-inch Schmidt plate, with the curious Cone Nebula in the centre; on the right, a 200-inch photograph of the Cone, a wedge of dark obscuring matter seen with a bright star at its apex.

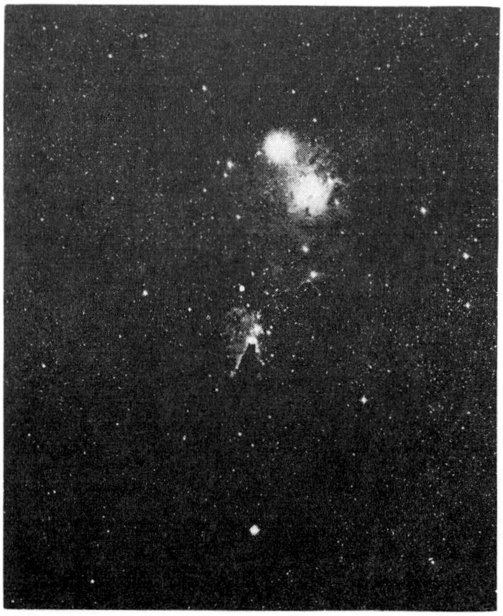

NGC 4594, the 'Sombrero Hat' galaxy – a vast spiral system of some 100,000,000,000 stars, seen edge on. (*Mount Wilson – Palomar*)

mers believe to be the initial stages of star formation. We can account for the formation of these dark globules by the **radiation pressure** of starlight (which acts only *towards* the nebula owing to the latter's darkness) which drives the dust particles in the nebula together. In this way the globule can double its mass in something like 100 million years. This continues until the gravitational attraction within the globule is large enough to bring about a very rapid evolution to the initial stages of a star.

GALAXIES. These are vast systems of stars and other celestial objects which exist outside our own Galaxy. They were formerly called extragalactic nebulae.

Three extragalactic nebulae are visible to the naked eye: the two **Magellanic Clouds,** which appear as luminous patches in the southern sky, and the **Andromeda Galaxy,** visible as a hazy, faint object. A telescope reveals many more; their distribution seems non-uniform, none being visible in the vicinity of the Milky Way, but this is probably a result of **galactic absorption.** In directions pointing out of the galactic plane vision is relatively unobscured by interstellar dust, and photographs taken with large telescopes show as many galaxies as stars.

The brighter galaxies have been assigned numbers in various catalogues. Thus the Andromeda Galaxy is M 31, being number 31 in Messier's catalogue; there are also the New General Catalogue (NGC) and the Index Catalogue (IC). As well as galaxies these catalogues include star clusters and galactic nebulae, all of which appear nebulous in small telescopes.

TYPICAL SPIRAL GALAXIES. Those on the left have roughly spherical nuclei, and are denoted by an S. The three on the right are all *barred spirals* ('SB') in which the arms emerge from the ends of a straight bar through the centre. The small letters a, b and c indicate progressively more open spirals. (*Mount Wilson – Palomar*)

TYPES OF GALAXIES. Our own Galaxy is thought to be a typical *spiral*. These consist of a bright ellipsoidal central nucleus from which spiral arms emerge tangentially at two diametrically opposite points to form a *disc*, and die out after about one complete turn. The relative importance of the nucleus varies and the spirals are classified as *Sa*, *Sb*, or *Sc*, where an *Sa* galaxy has an important nucleus, an *Sc* has a small nucleus and prominent arms, and an *Sb* is intermediate between these two. The nucleus shows clearly as a bulge when the galaxy is seen edge-on, as is NGC 4565 in the photograph. The spiral arms consist of relatively young Population I stars (see **Hertzsprung–Russell Diagram**) and the nucleus of old Population II stars; both rotate. Not sharing in the rotation, and distributed around the nucleus, is a halo of Population II stars and a hundred or so **globular clusters.**

Rather different from the spirals are the *barred spirals*. These are characterized by a bright nucleus with a bar running through it, from the ends of which the two spiral arms extend part of the way around the nucleus. They are classified as *SBa*, *SBb*, or *SBc*, where the *a*, *b*, or *c* indicates the relative importance of the nucleus as for the normal spirals.

Elliptical galaxies are structureless agglomerations of stars. They look like and are like the nuclei of spirals, consisting entirely of stars belonging to Population II, and being free of dust. They are classified by *E* followed by an integer ranging from 0 for those that appear circular to 7 for the most spindle-shaped. Intermediate between the elliptical and spiral (or barred spiral) galaxies are the *SO* (or *SBO*) galaxies which have a large nucleus and a small disc but show no visible spiral arms.

Irregular galaxies have no obvious symmetry. They probably contain objects of

NGC 4565, a spiral galaxy seen edge on. Interstellar matter in the spiral arms obscures part of the nucleus.

FULL EARTH, photographed by a weather satellite in synchronous orbit, stationary 22,300 miles above the mouth of the Amazon. South America is near the centre; the southern U.S.A. at top left; the western bulge of Africa at upper right; and Great Britain on the edge of the cyclonic depression at about 1 o'clock.

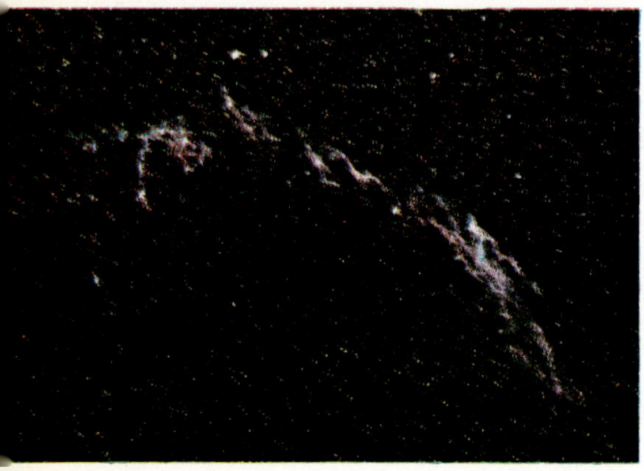

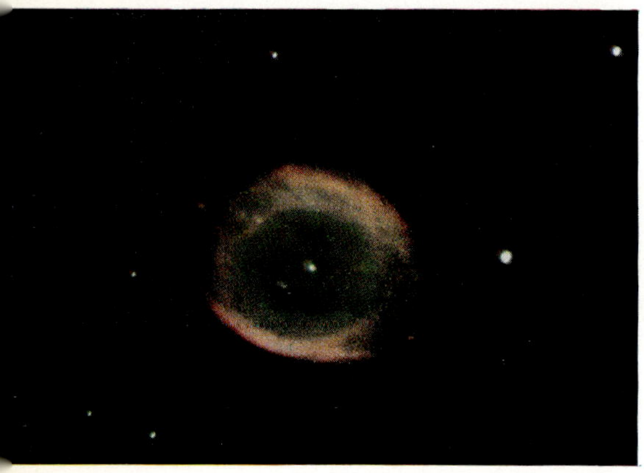

NGC 6781

Planetary Nebula in Aquila

This nebula is similar in principle to the Ring Nebula below and to the Dumb-bell (facing p.241). Our solar system would fit into this hollow sphere of gas some 20 million times.

NGC 1432

The Pleiades

The six bright stars of the Seven Sisters, a galactic cluster. Even without binoculars one can often see eight more. See p.214.

NGC 6992

Veil Nebula in Cygnus

Gaseous remnants of a supernova explosion about 50,000 years ago, made to glow by irradiation from nearby hot stars in the Milky Way. Another nebulosity in *Cygnus* is shown on p.118.

NGC 1976

The Great Nebula in Orion

A black-and-white picture appears on p.116. The colour photograph shows how we should see it if the pupils of our eyes were the size of cartwheels.

NGC 6720

Ring Nebula in Lyra

A shell of gas fluorescing under ultraviolet radiation from the hot, blue star at the centre. Temperature drops and colour reddens from the centre outwards.

NGC 2024

The Horsehead Nebula

A region of dark absorption and glowing emission nebulosities, also illustrated on p.117.

SIX GALACTIC NEBULAE

photographed at Mount Palomar Observatory. The human eye cannot perceive colour in very faint light. The camera can collect light over minutes or hours and reveal the true colours.

INFRARED VIEW OF MEXICO near the mouth of the Colorado River, taken from *Apollo 9* at 140 miles. Healthy green foliage shows up in reddish hues in the irrigated farmlands along the river and on the slopes of the San Pedro Martia and Juarez Mountains. The white ribbon-like shapes are salt lakes. Reflections from silt indicate currents and shoals in the estuary. The area covered includes that of the *Burning Tree* picture on p.102.

both populations. The Small Magellanic Cloud is a good example.

A large galaxy may contain over a hundred thousand million individual stars.

CLUSTERS OF GALAXIES. Many galaxies belong to clusters which have up to a thousand members. Our own Galaxy is one of the *local group* of about twenty galaxies, including the Andromeda Spiral and its two elliptical companions as well as the Magellanic Clouds.

A rich cluster in the constellation *Coma* is notable for the absence of any galaxies showing spiral arms or interstellar dust. The density of galaxies in this cluster is so great that nearly all must have suffered many collisions with other galaxies. In such a collision the stars, which are tiny compared with the spaces between them, would pass through unharmed; but the interstellar dust and gas would undergo real collisions and be swept out of both galaxies, which could not then continue to form Population I stars from it.

DISTANCES AND MOTIONS OF THE GALAXIES. The spiral systems in the local group contain, among other types of star, Cepheid **variables,** supergiants, and quite frequent **novae.** The absolute magnitudes of

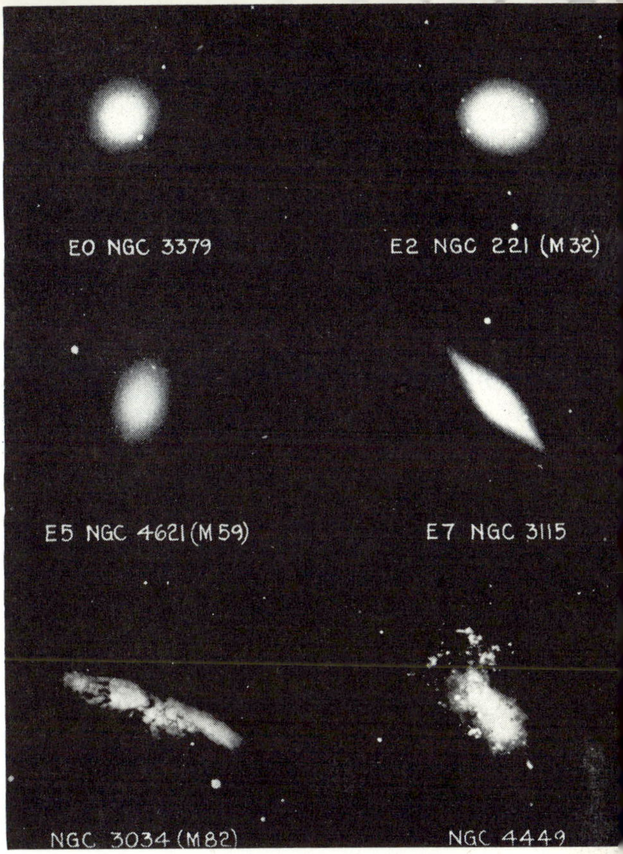

ELLIPTICAL AND IRREGULAR GALAXIES. Some of the foreground stars from our own Milky Way can be seen in front of the elliptical galaxies. (*Mount Wilson – Palomar*)

Below: A possible course of the evolution of galaxies.

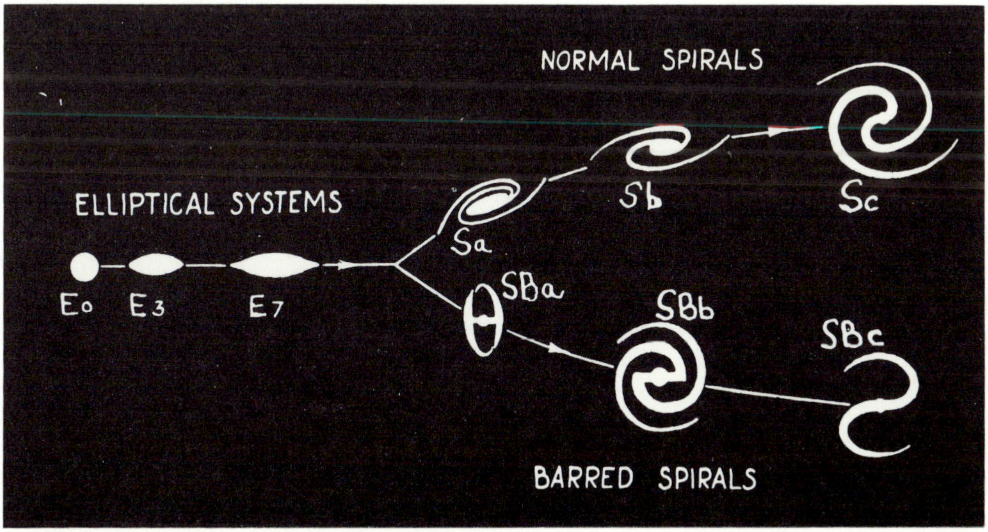

NGC 5128 galaxy in Centaurus. A source of radio noise and one of the strangest objects ever discovered. This was first thought to be two galaxies in collision but many astronomers now think it to be a single exploding galaxy.

GALAXIES IN SPACE. A group of four stellar systems in the Great Bear constellation. The beautifully formed spiral of Messier 81 is near the centre of the picture; the others are elliptical galaxies, with the top one seen more or less edge on. This group is about 7,000,000 light years distant.
(*National Geographic – Palomar Sky Survey*)

NGC 1300, a typical barred spiral galaxy in *Eridanus*.
(*Mount Wilson – Palomar*)

MESSIER 64, a spiral galaxy in *Coma Berenices*, photographed with an eight-hour exposure. (*Ritchey*)

THE WHIRLPOOL GALAXY, Messier 51, in *Canes Venatici*. One of its two spiral arms reaches out towards its unusual companion.
(*Mount Wilson – Palomar*)

A CLUSTER OF GALAXIES in Coma Berenices. Each of the objects with fuzzy outlines in this picture is a galaxy containing many millions of stars; the local foreground stars of the Milky Way can be distinguished by their firm, circular outlines. Our picture is an enlargement of part of a photograph taken with the 200-inch Hale telescope; in the minute portion of the sky covered by it, over a hundred galaxies can be traced, but at a distance of about 80 million light years many of them are too faint for reproduction in print. Two galaxies in collision are at lower left.

these are known, and by comparing them with the apparent magnitudes we can estimate the distances. The Andromeda Nebula turns out to be 575,000 parsecs away.

The nearest cluster of galaxies apart from the local group is in the constellation *Virgo*. The very brightest stars in it, of absolute magnitude −7 and −8, can just be photographed as individuals, and again comparison with the measured apparent magnitude yields the distance, in this case eleven million parsecs. The distances of more remote clusters of galaxies are estimated by assuming that their brightest member *galaxies* (and not *stars* in a galaxy) are similar to the brightest members of the Virgo cluster.

Even the nearest galaxies are too far away to show any **proper motion,** but they show **red shifts.** This is normally interpreted as being due to expansion of the Universe and is one of the bases of modern cosmology.

ROTATION OF GALAXIES. In every spiral galaxy for which measurements have been made the rotation is such that the spiral arms trail behind the nucleus. Their spectra show the relationship between rotational velocity and distance from the centre. They demonstrate that the middle parts of the spirals rotate as rigid bodies, but the outer regions lag behind more and more. The nucleus of M 31 revolves once in about 25 million years.

A GROUP OF GALAXIES IN LEO. Four more dissimilar individuals would be hard to find: one is elliptical, one a normal spiral, and two are very different examples of barred spirals. The arms of one of the latter are drawn out, probably by the gravitational attraction of its neighbour.
(*Mount Wilson – Palomar*)

For spirals the periods range from 10 to 80 million years. Ellipticals have shorter periods.

The Sun shares the rotation of our own Galaxy. It is about 8,000 parsecs from the galactic centre, a little beyond the radius of maximum rotational velocity. Moving at about 225 km. per second (and carrying the entire solar system with it) it will complete one revolution about the galactic centre in about 200 million years. Comparisons with M 31 show the Milky Way galaxy to be very similar in character, but slightly smaller and also rather less massive.

EVOLUTION OF GALAXIES. The first suggestions were that galaxies evolve from ellipticals to spirals. More recently evolution from spirals to ellipticals has been suggested. Some astronomers think that, depending on how fast it is rotating, either a flattened spiral galaxy or a more globular elliptical galaxy is formed initially and that evolution from spirals to ellipticals or vice versa does not occur. A possible course of evolution is shown in the diagram on page 121.

We form our theories concerning the evolution of galaxies on the assumptions (*a*) that the same physical laws apply to them all, regardless of their location in space; (*b*) that we are observing a representative sample. These assumptions are not necessarily correct. If, for instance, there were 'dead' galaxies emitting virtually no radiation, we could *ipso facto* not observe them directly.

GALAXY, THE. The spiral system of stars of which the Sun is a member.

Observation of the night sky in the absence of moonlight, even with the naked eye, leads to the conclusion that space does not seem to be uniformly populated with stars. An irregular streak of luminosity, known as the *Milky Way*, runs right across the sky. A telescope reveals that this luminous band contains myriads of stars. We live in a disc-shaped aggregation of stars, and when we look at the Milky Way we are looking along the plane of the disc and thus seeing a great thickness of star-populated space; in other directions we are looking out of the disc into the relatively empty space beyond.

In our attempts to discover the form of our own Galaxy we are severely handicapped by being inside it: we are rather in the position of an ant trying to map a large lawn from a fixed position amongst the blades of grass composing it. An additional difficulty is the **galactic absorption** which prevents us from seeing far into the Milky Way. The dark matter which causes the absorption can be seen in many spiral **galaxies.**

Observation of **globular clusters** shows that they occupy a roughly spherical volume with a centre 8,000 parsecs away in the direction of the constellation *Sagittarius*; this is also the *galactic centre.* Photographs taken in infra-red light, which penetrates the obscuring matter in the galactic plane better than visible light, reveal very dense star clouds probably on the edge of the nucleus.

The very brightest stars, those of **spectral classification** *O* and *B*, and clouds of ionized hydrogen (H II regions, see **Interstellar Matter**) can be seen at great distances. Counts of them reveal three spiral arms in the vicinity of the Sun, where the concentration of stars is greater than average. The velocities of these *O* and *B* stars can be measured by observing the **Doppler effect** in their spectra and they show that stars in the solar neighbourhood are rotating about the galactic centre in circles, like planets around the Sun, with velocities about 200 kilometres per second. The stars nearer the galactic centre than the Sun move faster and get ahead of the Sun while those farther out fall behind. Knowing the distance of the galactic centre we may calculate that one revolution is completed in about 200 million years and that the total mass of the galaxy is about a hundred thousand million times that of the Sun. This makes our Galaxy rather less massive and smaller than the **Andromeda Galaxy.**

Neutral hydrogen emits radio waves with a wavelength of 21 cm. (see **Radio Astronomy**). The clouds of hydrogen can be observed in all parts of the Galaxy, since radio waves are

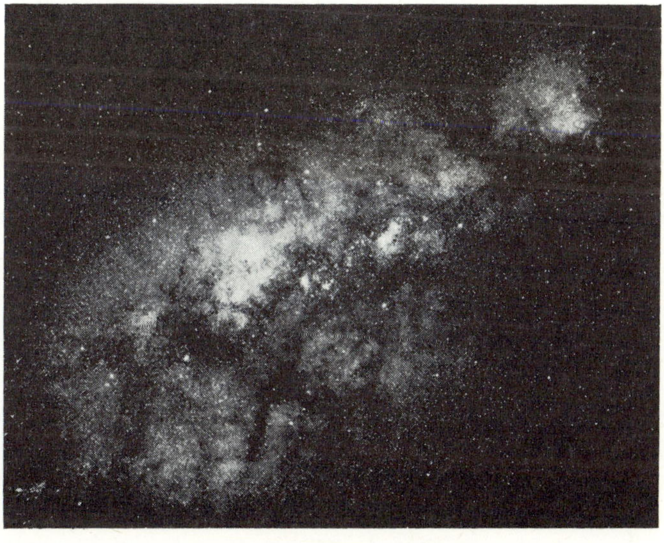

Part of the MILKY WAY in Sagittarius. These dense star clouds lie in the direction of the galactic centre.
(*Mount Wilson*)

not affected by galactic absorption. In most directions it is possible to calculate the distance of the hydrogen, and astronomers find that the hydrogen is concentrated into arms. Near the Sun these are in the same places as those discovered optically.

Thermal radio waves come from a source, *Sagittarius A*, in the direction of the galactic centre. This source is believed to be actually at the centre.

The halo is also a source of radio waves; similar emission has been detected coming from the Andromeda Galaxy. The stars in the halo move in elongated orbits often highly inclined to the galactic plane which take them both far from and near to the galactic centre. When they are in the vicinity of the Sun their velocities around the galactic centre are much less than the Sun's and so they have high velocities relative to the Sun; they are thus known as *high-velocity stars*. Among the most studied groups of these stars are the RR Lyrae **variable stars;** their absolute magnitudes are well known and so their distances can be found once their apparent magnitudes have been measured.

THE MILKY WAY is what we see when we look from the Earth in any direction that lies within the galactic plane. It forms a band that circles the sky, and represents an edge-on view of those parts of our spiral galaxy that are sufficiently distant for the dense concentration of stars to become apparent. This picture shows the entire band flattened out.

Both old Population II stars and younger Population I stars are found in the Galaxy (see **Hertzsprung–Russell Diagram**). The Population I stars and the dark absorbing matter (*dust*) are confined to the plane of the Galaxy and the stars rotate about the centre in circular orbits like the Sun. The Population II stars form a much more nearly spherical system, being concentrated in the nucleus, but the globular clusters and the galactic halo are also composed of these stars. The *halo* is a vast ellipsoidal, almost spherical, distribution of stars whose radius is considerably larger than the distance of the Sun from the galactic centre.

GALVANOMETER. An instrument for measuring weak electric currents.

GAMMA-RAY. A very high-energy **photon.** Gamma-rays are electromagnetic waves of the shortest known wavelength, from 10^{-11} to below 10^{-14} cm. They are produced by the disintegration of the nuclei of atoms such as radium.

GAMMA-RAY ASTRONOMY. The detection of **electromagnetic waves** of extremely short wavelength, beyond the **X-ray** region of the spectrum, emitted by celestial sources. Cosmic gamma-rays cannot be detected at the surface of the Earth since gamma radiation is strongly absorbed in the atmosphere. Experiments can be carried out with instrumentation only in high-altitude balloons or in **artificial satellites.**

A diffuse background radiation of gamma-rays has been detected, with a uniform distribution of intensity over the whole of the celestial sky. This observation was made by instruments carried on the lunar probes Ranger 3 and Ranger 5 while on their journey to the Moon. At present only one positive detection of a discrete gamma-ray source has been made, and this has not been confirmed by further observations. However, the experimental techniques of this, the most recent branch of observational astronomy, are still relatively crude and exceedingly difficult to perform.

It is expected that astronomers will learn much about the state and structure of the Universe as this branch of astronomy develops, for processes which are predicted to give rise to the emission of gamma-rays play a part in several of the present theories of the structure of the Galaxy and of **cosmology.** Thus the steady-state theory of continuous creation of matter and theories requiring the presence of **anti-matter** in the Universe both demand the production and emission of gamma-rays, with distinctive, and possibly observable properties.

GANTRY. A metal framework housing a vertical missile prior to launching. It provides shelter for the missile, and contains platforms at different levels for testing and servicing. When the missile is about to be launched the gantry is often rolled aside.

GANYMEDE. The brightest and most massive of the satellites of Jupiter, though in size it seems to be slightly inferior to Callisto. The diameter is 3,200 miles, and the mass over twice that of the Moon. As it has an escape velocity of 1·8 miles per second, Ganymede might be expected to retain a tenuous atmosphere; but so far none has been detected. Surface details can be observed with large instruments. (See **Jupiter.**)

GEGENSCHEIN (*lit.* 'counter-glow'). See **Zodiacal Light.**

GEIGER-MUELLER COUNTER. An instrument for detecting charged particles and high-energy radiation, e.g. γ-rays.

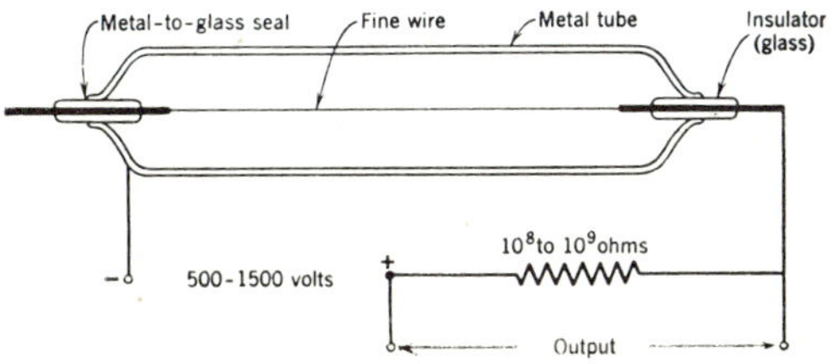

The basic parts are a metal tube, filled with gas at low pressure, and a metal wire which runs through the centre of the tube. A fairly high voltage is applied to the wire and the tube, which are insulated from each other. When a fast particle or a γ-ray enters the tube, **ions** are formed by its violent collisions with atoms of the gas. By virtue of their charge, the ions are attracted to the wire or tube, and their collisions with other atoms *en route* furnish more ions. The resulting stream of charged ions is equivalent to the flow of a small current whose existence betrays the entry into the tube of the original particle or ray. The current is sufficient to give a loud click in a telephone receiver connected in the circuit, or it may be amplified to operate a recorder or counting mechanism. Many modifications of the tube counter have been evolved for specialized purposes. In particular, it can be screened so that only particles from a certain direction are registered, and two or more counters can be placed one behind the other to time the speed of a particle which traverses them all in turn.

GEMINI PROJECT. The U.S. two-man manned space flight programme. This was the follow-on to the initial manned flights in the *Mercury* programme, and its most significant achievement was to demonstrate the feasibility of the rendezvous techniques which were crucial to the success of the **Apollo** lunar landings and **Skylab** missions.

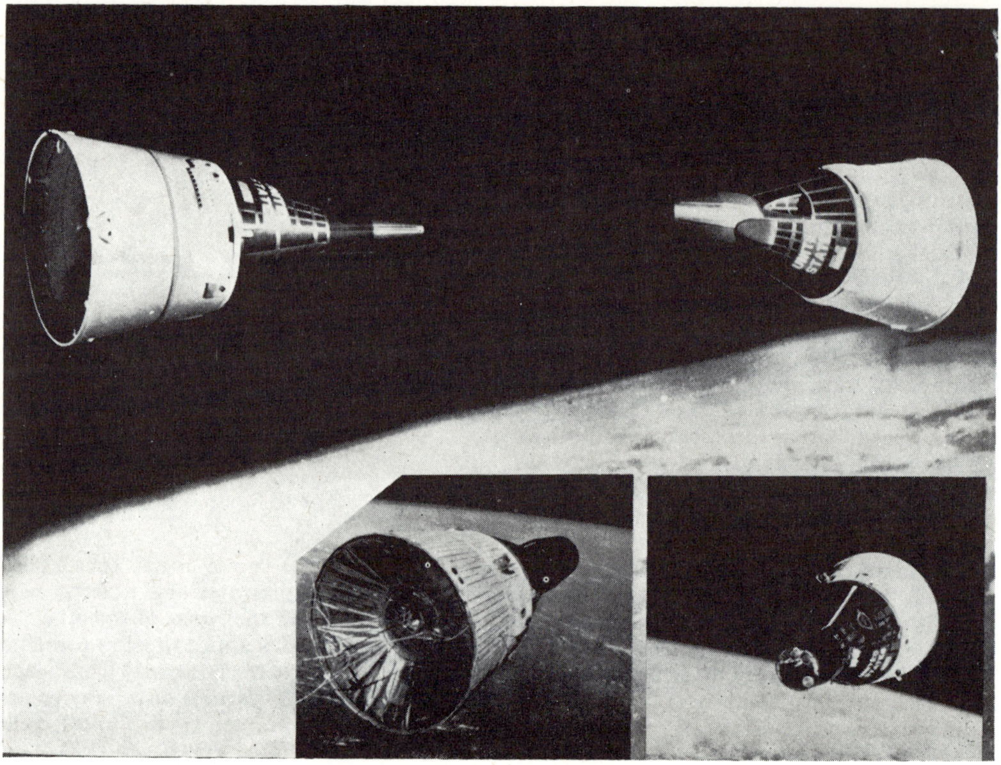

GEMINI 6 AND 7 RENDEZVOUS in orbit. An artist's impression of the meeting, and (*insets*) photographs taken from the two vehicles of each other. Gemini 7 was launched eleven days before Gemini 6 and remained in orbit for a fortnight. Each carried two astronauts.

Below:
A comparison of the *Mercury*, *Gemini* and *Apollo* capsules for one, two and three astronauts respectively.

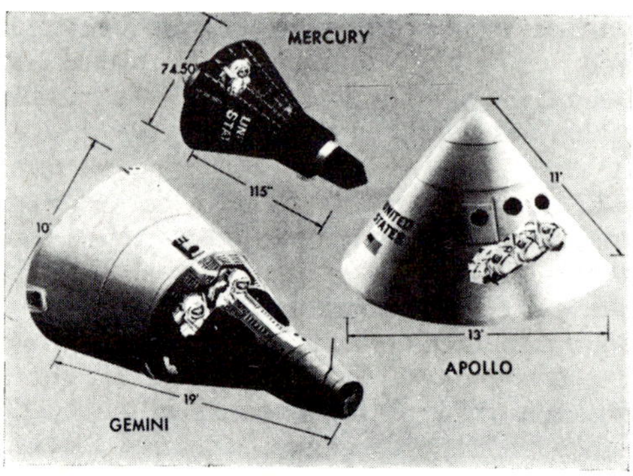

Nine two-man missions were flown, including several rendezvous and docking manoeuvres. In addition, although the Soviets were first with a spaceman outside the capsule protected only by a space suit, the Gemini team built up a total of almost 12½ hours of extra-vehicular activity. Experience of this kind has been very useful in the development of hardware for the Apollo flights.

There were some technical problems, but the programme was generally very successful. The Titan 2 launch vehicle, a man-rated version of an ICBM, performed faultlessly on every occasion, and enabled the launch to take place at the right time for the rendezvous missions to achieve orbits co-planar with the previously launched **Agena** target vehicles.

GEMINIDS. One of the richest of **meteor** showers, with maximum activity about December 12.

GEODESY concerns itself with the study of the **Earth's** dimensions, elasticity, mass and local variations of gravity.

GEOGRAPHIC POLE. One of the two points where the Earth's axis of rotation cuts its surface. All geographic **meridians** meet at the poles. Since the **magnetic poles** are at a considerable distance from the geographic poles, compass needles do not as a rule point exactly towards the geographic poles.

The Earth's axis of rotation differs slightly from its axis of symmetry. The difference causes a wandering of the precise location of the geographic poles in cycles of 432 days, over a range of about 60 feet, and hence corresponding small changes in latitudes all over the Earth.

A far more gradual and sustained change arises from the movement of the Earth's crust relative to the axis of rotation. (During the last 150 million years both England and North America have moved from near the equator to their present positions, but are almost stationary now.) The motion of land masses in relation to the poles has been called *polar wandering*, although that expression is putting the cart before the horse.

GIACOBINIDS. An alternative name for the **October Draconid** meteor shower.

GIANT STARS. These are stars whose diameters are considerably greater than those of most stars of similar surface temperature in the neighbourhood of the Sun. (See **Hertzsprung–Russell Diagram.**)

GIBBOUS (of Moon or planet). The phase when more than half the disc appears illuminated, i.e. between 'half Moon' and 'full Moon'.

GLOBULAR CLUSTER. A dense ellipsoidal aggregation of stars. Globular clusters occur in our own Galaxy and have been discovered in nearby spiral **galaxies.** The brightest globular cluster of our own Galaxy is ω Centauri, visible to the naked eye as a fourth magnitude object in the southern hemisphere.

SPACE WALK taken by Edward White during the third orbit of *Gemini IV*, June 1965. He is tethered to the 25-foot umbilical line and is holding the jet gun unit which assists his movement. A camera is strapped to his arm.

There are four other such naked-eye globular clusters including M 13 in the constellation Hercules, 47 Tucanae, M 22 in Sagittarius and M 5 in Serpens. Many others can be seen with small telescopes, but an instrument with an aperture of at least ten inches is required to show a cluster well resolved into separate stars, and the very faintest of these objects are detectable only with the most powerful telescopes.

The Andromeda Galaxy, which is very similar to our own Galaxy, has about 200 known globular clusters distributed roughly spherically around the nucleus. Over a hundred clusters are known in our Galaxy and it is estimated that another hundred are hidden from us by **galactic absorption.** Those visible appear mainly in the direction of the constellations *Scorpius* and *Sagittarius*, and, assuming that these too have a spherical distribution, they indicate that the centre of

A GLOBULAR CLUSTER, Messier 3 in the constellation Canes Venatici. (*Ritchey*)

the galaxy lies in that vicinity. The globular clusters do not take part in the rotation of the Galaxy as a whole, but move in eccentric orbits often at high inclinations to the galactic plane.

The number of stars in a globular cluster is difficult to estimate, as only the brighter ones can be seen; even the most laborious work with the world's largest telescope, the 200-inch on Mount Palomar, has not made it possible to discern in a cluster individual stars less bright than absolute magnitude 6. A conservative estimate places the total number at a hundred thousand and some clusters may have more than a million. The distances of many clusters have been found by examining the RR Lyrae **variable stars** in them. These measurements show that the centre of the distribution of globular clusters, which we identify with the galactic centre, is about 8,000 parsecs away and show that the absolute magnitudes of whole clusters range from −5 to −10.

The full diameters of globular clusters range from 7 to 120 parsecs. Most of the stars are concentrated towards the centre in a core which in most clusters does not exceed 5 parsecs across. In this core there may be as many as 30,000 to 40,000 stars with a density thousands of times greater than in the solar neighbourhood. The ellipsoidal shapes of the clusters indicate that they are rotating.

The brightest stars in these clusters are red, instead of blue as is the case for stars near the Sun. Studies of their **Hertzsprung–Russell diagrams** and of **stellar evolution** show these clusters to be among the oldest objects in the Galaxy.

G.M.T. Abbreviation for **Greenwich Mean Time.**

GRAVITATION. The force which exists between all particles of matter everywhere in the Universe. The force is one of attraction, and between any pair of bodies it is proportional to the mass of each and inversely proportional to the square of the distance between them.

(Thus by doubling the distance separating two bodies, their mutual attraction is reduced to one quarter its original value; by doubling the mass of one body and trebling that of the other, the gravitational force between them is made $2 \times 3 = 6$ times as great as before.)

The gravitational field of a body extends throughout the Universe. It becomes rapidly weaker and finally negligible with increasing distance, but never quite ceases. It is therefore wrong to suggest that any place, however far away, is beyond the Earth's gravity, although outside a certain range the gravitation due to some other body may greatly exceed that due to the Earth.

There is a fundamental difference between *mass* and *weight*. The mass of a body is the amount of matter in it, regardless of where the body is placed. Weight is the force of gravity exerted on a body, and depends not only on mass but also on the strength of the gravitational field at the place where we measure the weight. Confusion sometimes arises because the same units are employed to express both weight and mass: the mass of a body is equal to its weight *at the surface of the Earth*. Thus, on the Earth, a cubic centimetre of water has a mass of 1 gram and weighs 1 gram; on the Moon, the water still has a mass of 1 gram, but its weight is only 1/6th of a gram, because the Moon's gravitational attraction is 1/6th of that of the Earth.

If a body in a gravitational field is free to move it will fall, i.e. it will be accelerated. We can state the strength of a gravitational attraction in terms of the acceleration it imparts to a freely falling body (see **Acceleration**).

The nature of gravitation is still a mystery. Certainly there can be no question of 'shielding' against it by means of screens as we can in the case of radiation.

GRAVITATIONAL COLLAPSE. Any collection of matter will collapse under self-gravitation unless the collapse is halted by non-gravitational forces; these include *magnetic forces*, *rotational forces* and *pressure forces*. Since magnetic and gravitational forces keep step with one another during the collapse, it seems unlikely that magnetic forces *alone* can halt a gravitational collapse; if gravitational forces are initially larger, they will always be larger and collapse will proceed unchecked. Similarly pressure forces, which will develop through compression, appear to be inadequate if the mass of the object is large enough. Rotation, however, is much more complicated: it is certainly true that, if **angular momentum** is conserved, rotational forces could halt the collapse, but it appears likely that even if a small magnetic field is present, angular momentum will be transferred away from the body via this magnetic field and collapse could proceed. Although this problem has not been resolved, it is certain that *some* angular momentum transfer *must* take place as, without it, stars would not be formed!

Classically, if one is able to accumulate matter indefinitely, the **escape velocity** at the surface of the object will eventually exceed the speed of light. This conclusion is still valid in General **Relativity.** Light emitted from this collapsing object will show a **red shift** as viewed by a distant observer, and as the radius decreases the light is shifted more and more. This continues until the body reaches the critical *Schwarzschild radius*, when the red shift is infinite. A strange consequence of relativity is that, to an outside observer, a body collapsing towards its Schwarzschild radius *never* reaches it, since the visible motion slows down and then stops: however, if one were sitting on the surface of the object, the time of collapse would be finite.

Gravitational collapse has been proposed as an energy source for **quasars** and **radio galaxies.** This is essentially due to the fact that gravitational energy, generally unimportant in ordinary stars, depends on the square of the mass of the object and, hence, becomes important for objects having masses greater than that of one hundred thousand suns. Further, whereas nuclear conversion of hydrogen into heavier elements (see **Stellar Energy**) yields an energy of about 1 % of the rest energy of matter, one can, theoretically at least, obtain up to eight-ninths of the rest energy of matter by gravitational collapse. It remains an open question, however, whether this energy can in fact be released from the body. (T.H.M.P.)

GRAVITATIONAL LOSS. The loss of efficiency of a rocket escaping from the Earth occasioned by the time required for **escape velocity** to be reached. If the rocket could attain escape velocity instantly, it would require not more than the theoretical minimum of energy to escape. But an actual rocket cannot do this; it is limited by the power of its motors. For each second that the rocket spends in accelerating, the gravitational attraction of the Earth retards it by about 32 feet per second (roughly 20 m.p.h.). Even if the period of acceleration were as short as 100 seconds (corresponding to an improbably great average acceleration of over 10 g.) it would involve a loss of $20 \times 100 =$ 2,000 m.p.h., which is not negligible even in comparison with the Earth's escape velocity.

A simple analogy may be drawn with a man running *up* an escalator which is moving *down* with constant speed. If he ran so fast that he got to the top instantaneously, he would have climbed no more steps than if he had gone up a flight of stairs of similar height; but the longer he spends on the escalator the more steps he will have had to climb before reaching the top.

GRAVITY, ARTIFICIAL. The centrifugal force which may be used to simulate gravity in manned rockets and space stations.

Long periods of **free-fall** conditions may have harmful effects on the human system and will certainly be a considerable inconvenience. If the inhabited spaces are rotated about a suitable axis, the resulting centrifugal force will be in all respects equivalent to a gravitational field directed outwards from the axis of rotation. The strength of this field at any point increases in proportion to that point's distance from the axis of rotation, and to the rate of spin. Some rather ambitious designs for space stations envisage the living accommodation disposed in a ring round the rotating axis of the station. One revolution in three minutes would furnish gravitation equal to that on the Earth's surface at 128 feet from the axis. However, rotation can set up physiological disturbances of its own, and for technical reasons it may prove simpler to rotate crew resting compartments within a station rather than the whole station.

GREAT BEAR (*Ursa Major*), also known as The Plough or Big Dipper, is one of the most prominent constellations in the northern sky. From the left, the stars in the diagram are Benetnasch, Mizar with the faint Alcor ('The Test' for normal vision) close by, Alioz, Megrez, Phecda and the two Pointers, Merak and Dubhe.

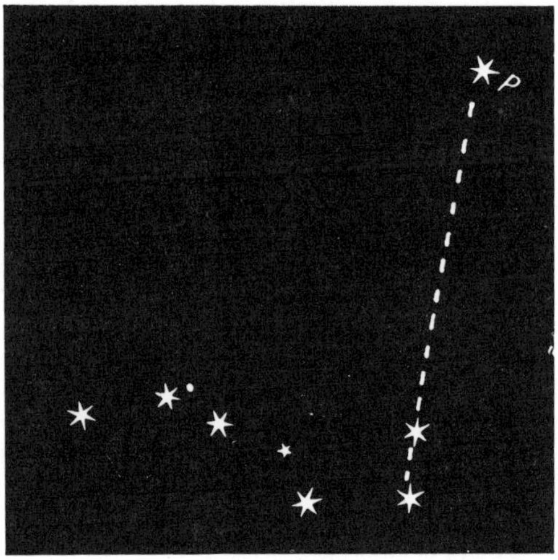

About five times the distance between the Pointers along the line joining them lies the Pole Star, which always appears within 1° of true North.

GREAT CIRCLE. A circle on the surface of a sphere which has the same centre as the sphere. The shortest route across the surface of a sphere between two points on it lies along the Great Circle through those points.

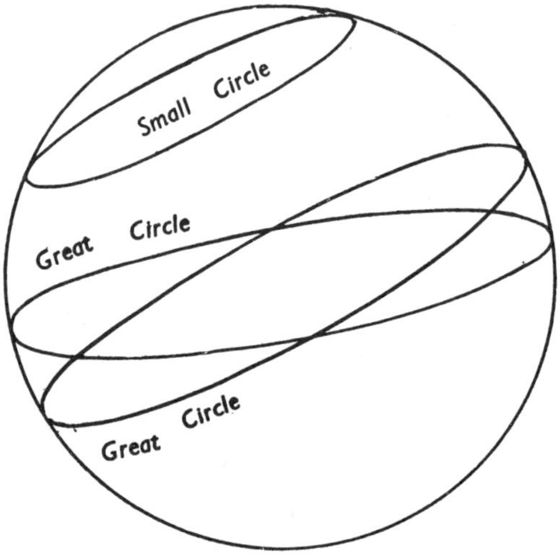

The term is applied to the Earth as though it were a perfect sphere. The equator is a Great Circle, and each **meridian** of longitude is half a Great Circle. By contrast, parallels of latitude other than the equator are *Small Circles.*

GREENWICH MEAN TIME is the time at Greenwich according to the **Mean Sun,** and is a standard to which all observations can be referred.

GREENWICH MERIDIAN. The line of longitude which bisects the now obsolete Airy **transit circle** of the **Royal Greenwich Observatory.** This line is used as the arbitrary zero of longitude measurements on the Earth's surface, and all other longitudes are reckoned in degrees East or West from it.

GREGORIAN CALENDAR. The calendar now in use. (See **Julian Calendar.**)

GUIDED MISSILE. A projectile whose trajectory can be modified after launch to compensate for errors which would otherwise result in failure to achieve the required result.

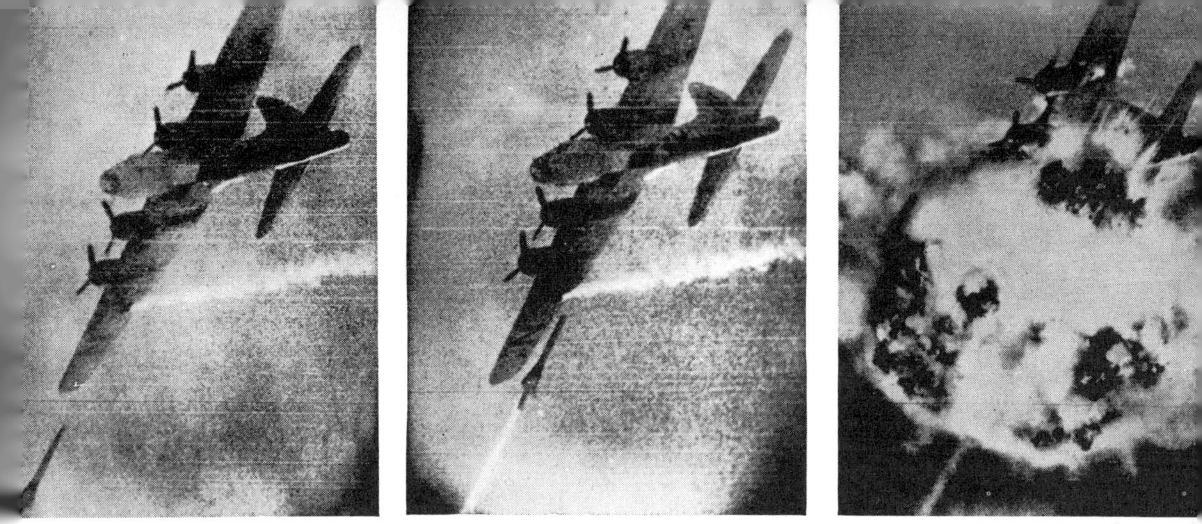

Above:
A guided missile destroying a four-engined bomber during trials.

Guided missiles are classified according to launch environment and intended target, and within certain categories also by range. For example, an *Air to Surface* missile is launched from an aircraft against a ground (or possibly seaborne) target. Such a missile might also be described as a stand-off bomb, and is used either when the target is too heavily defended to risk exposure of the aircraft, or when the required accuracy in weapon delivery cannot be achieved with conventional bombs. The roles of *Surface to Air*, *Air to Air* and *Surface to Surface* weapon systems will be readily deduced.

Only surface to surface (SSM) missile systems normally include hardware which has a potential space application, and though there are some SSMs which operate wholly in the atmosphere and have aerodynamic controls and air-breathing propulsion, most are ballistic systems employing rocket vehicles. A ballistic missile is one which is guided only during powered flight, and afterwards coasts to the target on a ballistic trajectory. There can be no re-use of such a vehicle, so the cost per pound of payload delivered is high. So high, in fact, that except at the shortest ranges, only nuclear warheads make it worth while. Ballistic missile systems are normally classified by range (which in essence means the velocity attainable with the required payload) in ascending order of range as Short Range (SRBM) up to a few hundred miles range, Intermediate Range (IRBM) up to about 3,000 miles and Intercontinental above this (ICBM). A sub-group of the ICBM class includes weapons placed in orbit and capable of being called down to strike any target within a belt of latitude which is governed by the inclination of the orbit to the equatorial plane. Such weapons

Below:
A Poseidon missile carrying a dummy multiple independently-targetable re-entry vehicle (MIRV) warhead being launched from a submerged submarine. A very tall antenna from the launching vessel sticks out of the water on the right. The Poseidon has a greater range and payload than the Polaris missile, and is now operated in over 30 U.S. nuclear-powered submarines.

(*U.S. Navy photograph*)

are prohibited by international Space Treaties – a small sacrifice, since they are in any case highly vulnerable during their waiting phase.

A ballistic missile is a guided missile only during its short powered flight phase. The guidance system is therefore concerned only with achieving the correct velocity at cut-off. In principle this can be done using range and rate measurements and commands from suitably placed ground stations, but the ground tracking system might be interfered with in some way and the number of missiles which can be fired in a given period is limited by the available tracking facilities. A better approach, from the military point of view, is to make the missile measure its own performance and compute the trajectory required from the pre-set target coordinates and its present position and velocity – if possible without using any external references. This is feasible using **Inertial Guidance.**

Two philosophies can be effectively applied. We may select a trajectory before lift-off and constrain the missile to fly along it with a pre-planned velocity programme which will in general require a degree of thrust control (*tracer guidance*), or we can start off with a planned trajectory and allow the missile to modify it, taking account of its actual flight performance, and computing its own cut-off point (*computed cut-off guidance*).

A variety of refinements and modifications are possible to improve the final part of the trajectory, particularly for shorter-range missiles. Infra-red and magnetic sensors, devices for homing along enemy radar beams, etc., can set off proximity fuses or activate controls which, with a short final burst from a rocket motor or by aerodynamic means, refine the approach path. One of the greatest problems is that of recognizing dummy targets.

Although one may deplore the existence and the need for such weapon systems, there is a consoling thought: it is not the military who have exploited advances in space techniques, but rather the scientific and exploratory space programmes which have benefited from military requirements. Both aspects contribute to latent dangers (see for instance **Space Pollution**), but the first artificial satellites used rockets which had been originally developed for warlike purposes. We owe carpets to the Crusades, and the conquest of

Following page:

THE HERTZSPRUNG–RUSSELL DIAGRAM – the key to the evolution of stars.

The diagram proper is the graph occupying the central portion of the colour plate. A star finds its position on the diagram according to its **absolute magnitude** (*vertical scale on the left reading from* -5 *to* $+10$) and its **spectral class** (*horizontal scale at the top reading from O to M*). The magnitude scale refers to visual and not bolometric magnitude.

Once the position of a star has been plotted, its approximate *temperature* may be read off from the second horizontal scale at the top in which the figures indicate thousands of degrees Centigrade; and its *emission of visible light* compared with that of the Sun, according to the vertical scale on the right. Approximate size may be judged from the radius of curvature, which for giants and supergiants is drawn to scale on the right and left respectively. White dwarfs are too small for their diameters to be visible on this scale. The proper place in the diagram for the supergiants and giants is shown by the asterisks of matching colour in the appropriate region of the diagram. The position of the Sun lies at the point in which the horizontal white line through the centre crosses the Main Sequence.

W = Wolf-Rayet stars

N = old novae and nuclei of planetary nebulae

C = Cepheid variable stars

T = T Tauri stars

The Hertzsprung Gap is the region between the Giants and the Main Sequence.

For a full explanation please refer to the article on the **Hertzsprung–Russell Diagram.**

JUPITER FROM CALLISTO, the largest of its twelve moons, with Ganymede and two other moons. It has been suggested that Callisto is a sort of 'celestial snowball', with ice and frozen gases overlaid on a rocky core with occasional outcrops. The scene depicted illustrates this theory. (*Painting by David Hardy, F.R.A.S.*)

Supergiants to scale
O
B
A
35
25
11
SUPE
-5
W
MAIN SEQUENCE
0
N
5
10

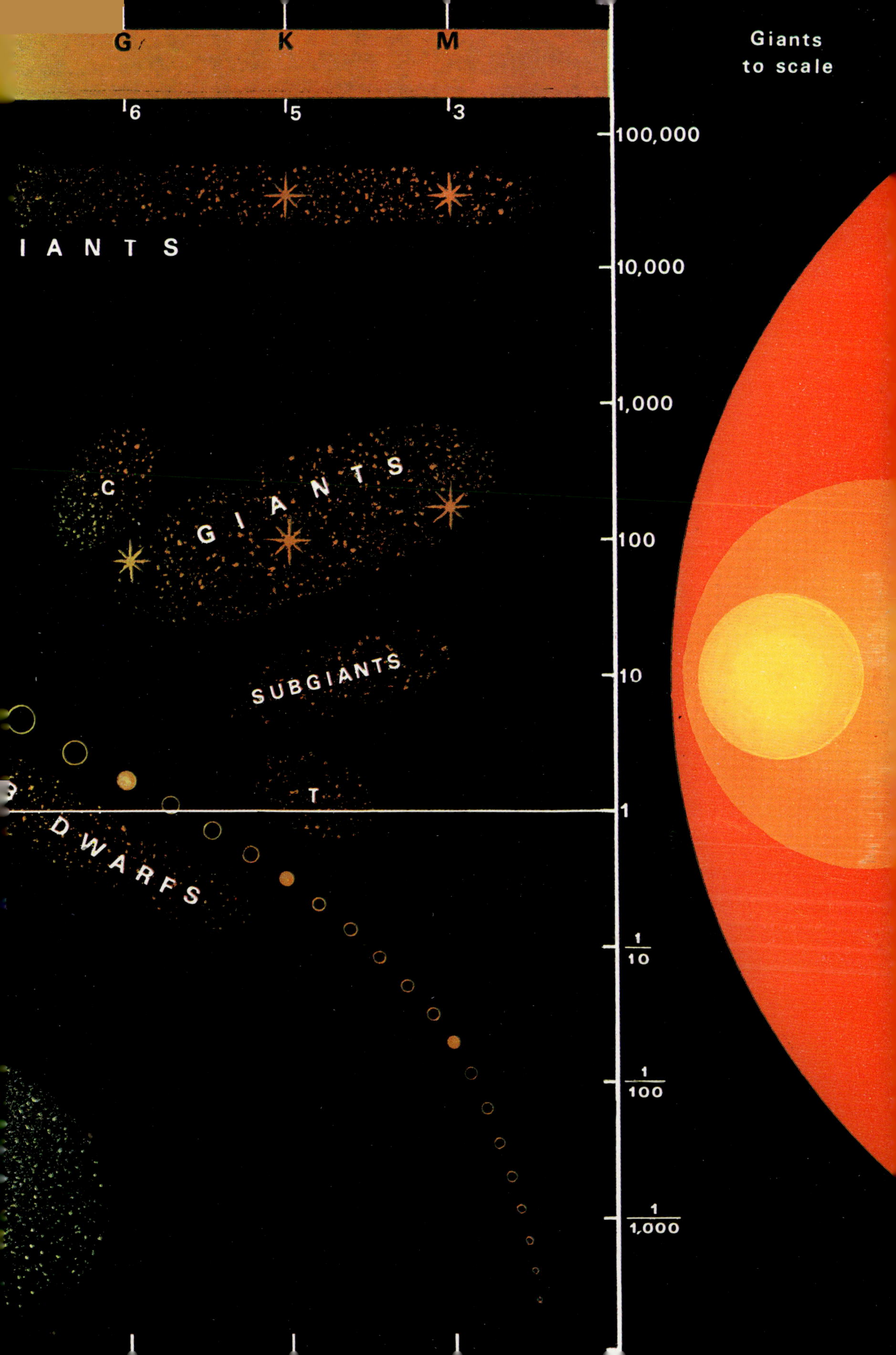

G
K
M
Giants
to scale
6
5
3
100,000
IANTS
10,000
1,000
C
GIANTS
100
SUBGIANTS
10
T
1
DWARFS
1/10
1/100
1/1,000

THE TRIFID NEBULA (NGC 6514) in SAGITTARIUS. It contains barely detectable dark spots in the pink portions, the Globules of Bok, which are believed to be stars in the process of condensing out of the surrounding dust and glowing gases. (Mount Wilson-Palomar.)

the Moon less to Apollo than to Mars. (C.W.M.)

GYROSCOPE. In its simplest form a gyroscope consists of a heavy wheel or disc which can be made to spin. Its axle is supported by a system of concentric rings or *gimbals* in such a way that the outer ring may be tilted or turned without exerting any significant forces on the wheel. Such an arrangement, known as a *free gyroscope*, will *tend to keep its axle pointing in the same direction* in space regardless of the movement of the outer gimbal. If the axle is aligned along the direction parallel to the Earth's axis the free gyro will behave as a compass. A free gyro suitably connected to a control system can be used for a variety of purposes, such as stabilizing ships in rough weather, but if the gimbals are constrained other valuable properties of the gyroscope can be exploited. If the axle is tilted at a constant *rate* about an axis in the plane of the wheel, the axle will exert a force in a direction at right angles to the direction of tilt which is exactly proportional to the rate of tilt. This type of constraint is employed in the instrument known as a *rate gyro* which is a vital component of missile and aircraft *autopilots* and *autostabilizers* and is widely used in all control systems requiring accurate *rate* feedback where this type of instrument is appropriate. If a *force* is applied to one end of the axle in a plane parallel to the wheel, and the other end is held fixed, the axle will *precess* at a rate proportional to the force, so once a mechanism is constructed to exploit this phenomenon we have a device which will produce a *rate* output for a *force* input. A little thought will show that here is a device with just the properties needed to integrate acceleration. If the force is an inertial one due to the acceleration of the instrument, the output will be proportional to the integral of acceleration or, in other words, to the velocity. This principle is used in the integrating gyro-accelerometers commonly used in **Inertial Guidance** systems.

H

HALLEY'S COMET. This, perhaps the most famous of all the **comets,** was also the first whose return was predicted. Edmund Halley, observing it in 1680 and applying Newton's new laws of gravitation, computed its orbit and identified it with the great comets of 1607 and 1531. He did not live to see its return in 1758, within a few weeks of the time he had predicted.

Appearances of Halley's comet have been traced back in historical records to 240 B.C. The average period is 77 years, but **perturbations** by the outer planets may cause this to vary by 20 years. It passed through **aphelion** in 1948 and is now moving towards the Sun again. Its next return will be in 1986.

In 1910 its 37-million-mile tail extended more than half-way across the sky as a faintly luminous band not unlike the Milky Way.

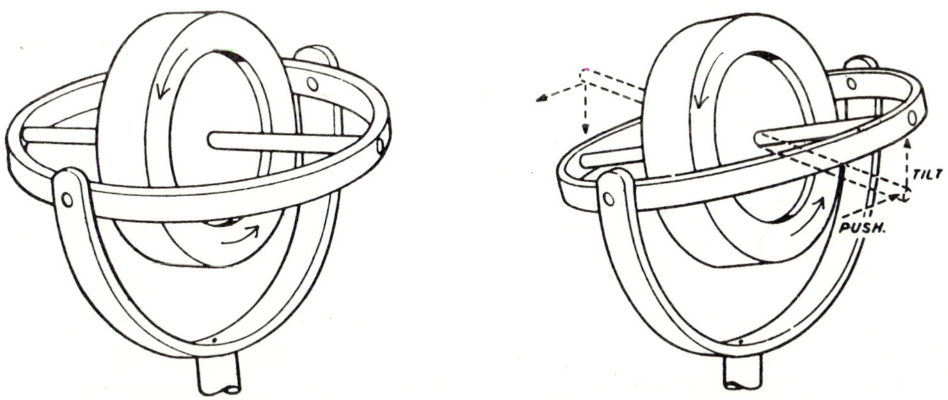

THE GYROSCOPE. The axle of the rapidly rotating wheel is mounted in gimbals which themselves can rotate around the vertical shaft. A push against the axle tilts it at right angles to the direction of the push.

HALLEY'S COMET AND VENUS during the comet's last appearance in 1910. The star images are elongated because the telescope followed the comet during exposure. A series of further photographs of Halley's Comet will be found in the article on comets. (*Slipher*)

On May 21 the Earth actually passed through part of this tail, but no effects whatsoever could be observed, and even large telescopes were unable to detect distinct bodies in either the tail or the head of the comet, nor was there a measurable diminution in the brightness of the Sun or any chemical change in the atmosphere.

For details of the orbit, see under **Comet.**

HALO. Name given to optical effects sometimes seen near the Sun or the Moon, due to the refraction and reflection of their light by minute particles of ice suspended in the air. These particles tend to assume the same attitude while falling, so that the planes of the ice crystals are parallel to each other and affect the light that strikes them in a systematic manner. The result may take several forms: coloured discs of light called *parhelia* or 'sun dogs' may appear on either side of the Sun and 22° away from it, joined by a ring of light, the halo; there may be a fainter, outer ring, and the parhelia may be absent. Haloes occur mostly in high altitudes, but lunar haloes are not uncommon generally on clear, frosty nights. Parhelia have started more than one 'flying saucer' report.

HARVEST MOON. The rising of the Moon is delayed from day to day by an average of 52 minutes, but this delay varies somewhat throughout the year. It is as little as 13 minutes near the time of the autumnal equinox (in northern latitudes). For several nights the full or nearly full Moon then rises at approximately the same time early in the evening, and is called the Harvest Moon.

HAYFORD SPHEROID. The solid figure whose proportions are the same as those of the Earth. It is a slightly flattened sphere, but the flattening amounts to less than 0·5 % of the diameter, and would not be detected by the naked eye in a scale model. The Earth's diameter is 7,927 miles at the equator and 7,900 miles from pole to pole. The equatorial bulge is due to centrifugal force, and is far more noticeable in the case of Saturn and Jupiter. It fluctuates in size to a minute extent under the influence of tidal forces set up in the Earth's interior by the Sun and Moon. Recent modifications are described under **Earth.**

HEAT is the energy of the random motions of molecules in a substance. The amount of heat contained in a body depends upon its temperature, and upon its specific heat. *Radiant* heat is **electromagnetic radiation** at wavelengths between those of infra-red and the shortest radio waves.

Heat can be transferred from one body to another in one or all of three ways:

1. By *conduction* across the area of contact between the two bodies;
2. By *convection*, in which heat is first given up to a fluid which then moves and conveys the heat to another place – the cooling of a car engine by the circulation of water is an example of convection;
3. By *radiation*, which can take place through a vacuum. The Sun heats the Earth's surface by radiation.

When two bodies are in contact and neither adds heat to the other, then both are at the same temperature. (See also **Specific Heat, Temperature,** and **Absolute Zero.**)

HEAVISIDE LAYER. A region of the **ionosphere.** Heaviside inferred its existence from the propagation of radio waves round the Earth.

HECTOR. An **asteroid** of the **Trojan** group.

HELIUM. An inert chemical element. The usual form of the helium **atom** consists of a nucleus of two protons and two neutrons, with two electrons outside it. The electrons are very tightly bound to the nucleus, and the atom is correspondingly difficult to ionize: it therefore forms no chemical compounds, and the spectral lines of ionized helium are seen only in the spectra of the very hottest stars, at temperatures far above those required to ionize most other elements. The abundance of helium in the Universe is second only to that of hydrogen, and the processes of energy generation in stars involve formation of helium from hydrogen. (See **Stellar Energy.**)

Helium was discovered as a constituent of the Sun before it was found on Earth. A bright yellow line was observed in the Sun's **flash spectrum** in 1868 by the astronomers Lockyer and Janssen; the line was not given by any substance known at that time. It was named helium from the Greek *helios*, meaning Sun. In 1895 a substance which gave the helium spectrum was found to be evolved on heating the rare mineral cleveite; subsequently it was detected in many mineral springs, and in natural gas.

Helium is a gas with about four times the density of hydrogen but a quarter that of air; it remains gaseous at atmospheric pressure down to a temperature hardly five degrees above absolute zero. It is produced by the radioactive decay of atoms such as radium. Owing to its low density the random motion of its atoms is rapid; helium can slowly escape from the Earth's atmosphere, as a small proportion of the helium atoms in the upper atmosphere have velocities in excess of the **escape velocity** of the Earth.

HERMES. An **asteroid,** only about a mile across, which passed within half a million miles of the Earth in 1937 and has not been seen since. No other asteroid has ever been so close, but an even nearer approach by Hermes is possible.

HERTZSPRUNG-RUSSELL DIAGRAM. Named after its inventors, this diagram is a graph on which the absolute **magnitudes** of stars are plotted against their spectral types. It is of fundamental importance in modern astronomy.

In order to know the absolute magnitude it is usually necessary to find the distance. (This is not true of all stars, for instance Cepheid variable stars have known magnitudes; so do a few other particular types of stars, but the statement remains true for the vast majority.) As the only measured distances, in the days of Hertzsprung and Russell, were those which were susceptible of measurement by **parallax,** and therefore small as astronomical distances go, it was natural that the stars represented on the diagram should be those relatively close to the Sun. Also, attempts to measure parallax had been largely confined to those stars which were likely to be near, that is, the bright stars. Relatively faint stars close at hand were not measured. Thus the effects of observational selection were enormous – all the stars on the diagram were near the Sun, and bright stars were grotesquely over-represented.

The modern diagram is shown in the colour plate between pages 136–137 (*caption on page* 136). (Fig. 1.) It reveals striking regularities in the properties of the stars. Most of these stars are represented by points which lie on a narrow band diagonally across the graph, indicated by the broken black line, while those which do not give points on this line are nevertheless confined, with very few exceptions, to quite distinct areas on the graph. The line representing the majority of stars is known as the *Main Sequence.* Those stars which lie above this line are brighter than main sequence stars of the same spectral class and hence of similar temperature. In order to give more light at the same temperature they must have considerably greater surface areas and are therefore called *giants.* Most of the giants belong to a small region of the diagram, called the *red giant region* because their *low* surface temperatures make them appear red. Stars lying very far above the main sequence are *supergiants,* while those only slightly above are called *subgiants.* Similarly those stars which are below the main sequence are fainter than main sequence stars of the same temperature, so they have much smaller surface areas and are called *dwarfs*. Most of the dwarfs belong to the **white dwarf** region because their *high* surface temperatures make them appear white. Stars only just below the main sequence are called *subdwarfs*. The typical positions of other groups of special types of star are also shown in the diagram: **Wolf-Rayet stars, Cepheid** and T Tauri **variable stars,** old **novae** and the nuclei of planetary nebulae (see **Galactic Nebulae**). The methods used to obtain the observational data for placing a star on the diagram are described under **Star.**

The *Hertzsprung Gap* is the region on the diagram between the main sequence and the red giants, where there are very few stars. The existence of this gap and the fact that very few stars lie outside the named regions are important in the theories of **stellar evolution.**

During the last two decades our knowledge has been greatly increased by the observation of stars in **globular clusters** and **galactic clusters.** Many of these are close enough for fairly faint stars in them to be examined, and there is the advantage that we know all the stars in a cluster to be at virtually the same distance from us. We have, therefore, only to measure the apparent magnitudes and spectral types of the stars in these clusters, draw the Hertzsprung–Russell diagram and correct all the magnitudes by the same amount, the **distance modulus.** This can be done by observing the Cepheid or RR Lyrae variable stars in the cluster, which have known *absolute* magnitude.

There is a tendency today to use **colour index** instead of spectral class; the two are closely related, and the colour index is far more readily determined. A diagram using colour indices is often called a colour-magnitude diagram, but it is fundamentally the same thing as the Hertzsprung–Russell diagram.

The Hertzsprung–Russell diagram for one of the globular clusters is shown in detail in Figure 2. It is at once evident that the star population of the globular cluster is completely different from that of the neighbourhood of the Sun. Figure 3 shows the colour-magnitude diagram of the same cluster, M3, drawn with lines representing the main groupings of stars shown in Figure 2, superimposed on the diagram for stars near the Sun. The only parts of the graphs showing any agreement are those below about the fourth absolute magnitude: here the M3 stars lie nearly on the Main Sequence. Stars below the sixth magnitude doubtless exist in M3, but were not bright enough to observe – Figure 2 shows that even these were fainter than the 21st apparent magnitude. Above absolute magnitude 4 there are two sudden corners in the diagram and the line rises and turns towards the side of increasing colour index, i.e. the brighter stars are redder. The brightest are red stars of absolute magnitude −3, in striking contrast to the brightest main sequence stars which are blue and two or three magnitudes brighter. The nearly *horizontal branch* of the colour-magnitude diagram shows that the blue stars in M3 are uncommon and relatively faint. A gap occurs in the horizontal branch at zero absolute magnitude: it is found that all the RR Lyrae variable stars (which are omitted in the diagram) fall exactly in this gap, and none outside, while no stars other than these variables occur in the gap. Evidently some cause of instability exists which affects all stars of the particular colour range and magnitude which correspond to the gap.

Figure 4 shows a composite Hertzsprung–Russell diagram for several galactic clusters with the globular cluster M3 for comparison. There is a point on each of the main sequences where they turn off to the right. This *turn-off point* is different for the different clusters. The brighter stars use up their energy faster (see **Stellar Energy**), so those stars in the cluster which were on the main sequence above the turn-off point have evolved off the main sequence in the lifetime of the cluster. Stellar evolution theory tells us how fast stars of a given brightness evolve, so the age of a star cluster can be found from its Hertzsprung–Russell diagram.

The Hertzsprung–Russell (or colour-magnitude) diagrams show that there are two fundamentally different types of stellar population: one, called *Population I*, occurs near the Sun and in galactic clusters, and another, *Population II*, in the globular clusters. Observations of our own and neighbouring galaxies confirm this sharp division (see **Galaxy, The**). Population I stars, which include Cepheid variables and supergiants, are found to be intimately associated with interstellar dust and occur in the spiral arms of **galaxies.**

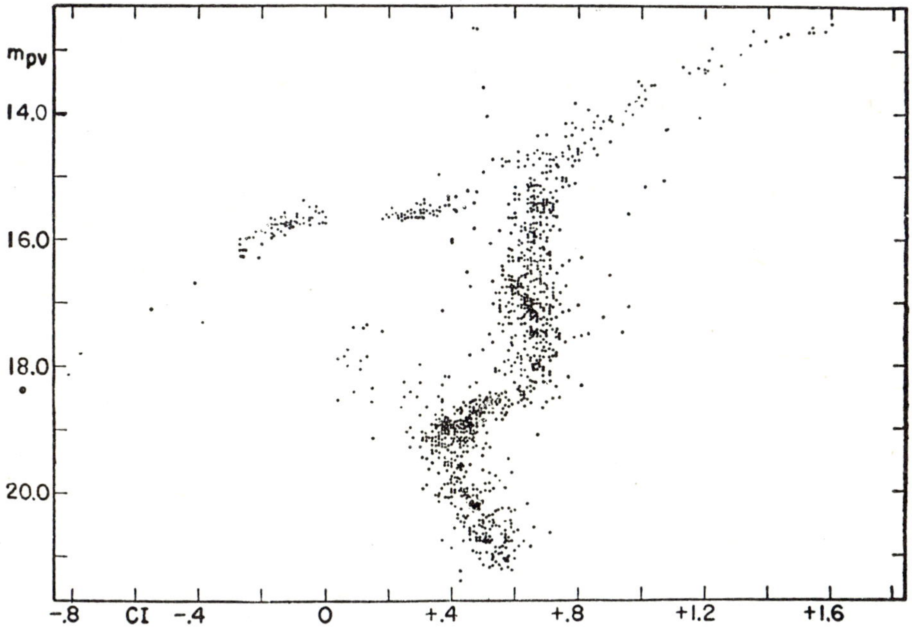

Fig. 2

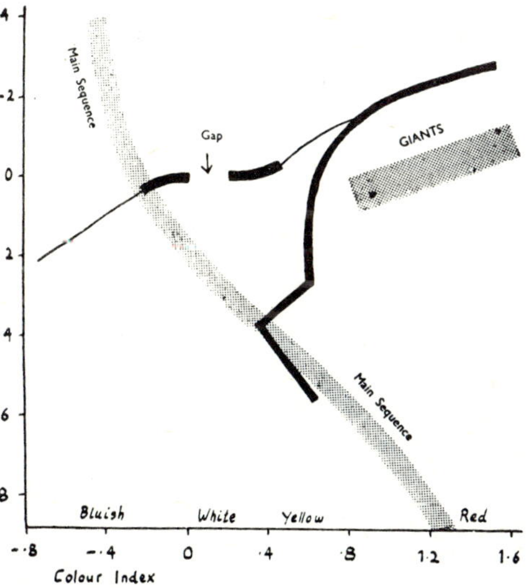

Fig. 3

For explanations please see text.

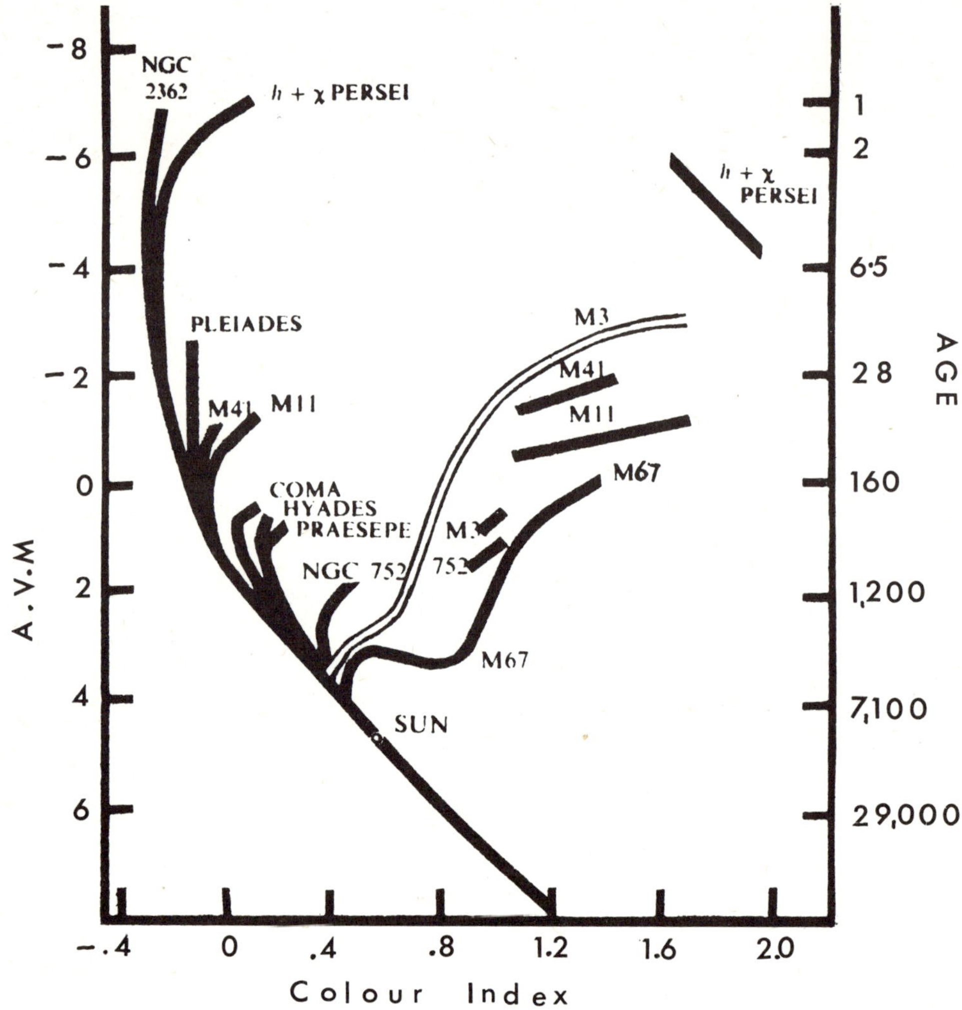

Fig. 4: a composite H–R diagram of various open clusters. The left-hand scale is that of Absolute Visual Magnitude; the scale on the right gives age in millions of years, but applies to the turn-off points only. (*After Sandage.*)

Population II stars form not only globular clusters but also the nuclei of spiral galaxies, and occur in regions free from interstellar dust. They constitute, in addition, the nearly spherical cloud of stars which surrounds the nucleus of such a galaxy and which, like the globular clusters, does not take part in the general galactic rotation. Most of the stars in a spiral galaxy are concentrated in the nucleus, so Population II stars are more common than Population I, in a ratio of the order of ten to one.

In the Hertzsprung–Russell diagrams for galactic clusters the main sequence includes early spectral type stars; whereas in globular clusters there are few such stars, and the main sequence turn-off point is at a fainter luminosity and a lower surface temperature. It is believed that the brighter main sequence stars in the globular clusters have already evolved, which suggests that the globular clusters are older. There are consistent differences in the abundances of the *heavy elements* (the elements other than hydrogen

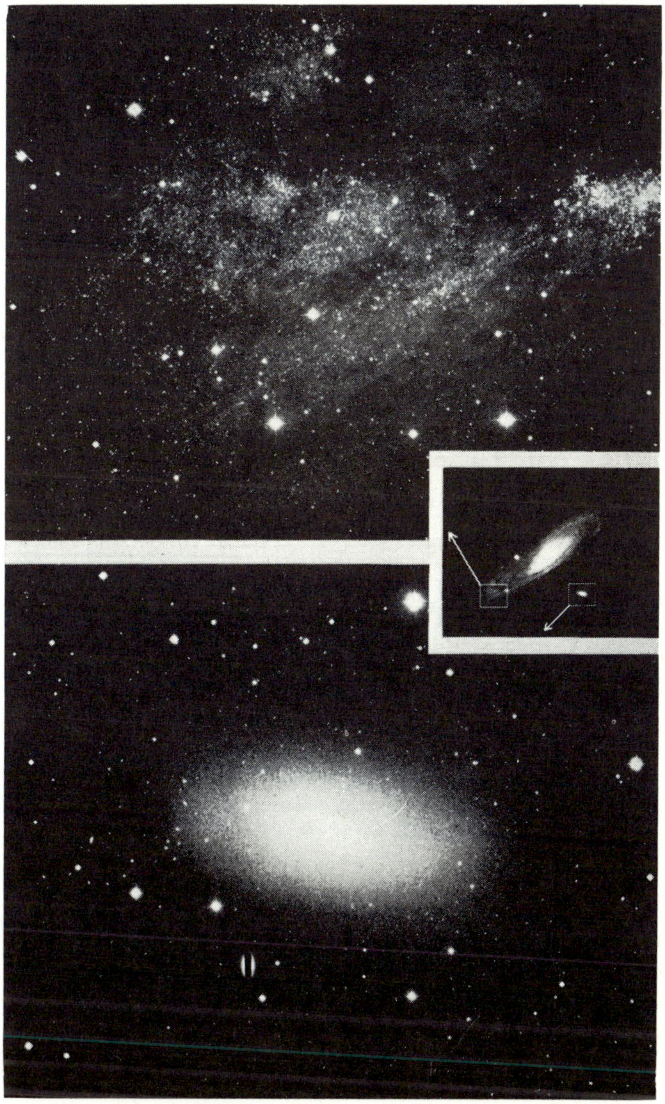

STELLAR POPULATIONS I and II. *Top:* the Andromeda galaxy, photographed in blue light, shows giant and supergiant stars of Population I in the spiral arms. The hazy patch on the right is composed of unresolved Population II stars. *Bottom:* NGC 205, Andromeda's companion galaxy, photographed in yellow light, shows stars of Population II. The brightest stars are red and 100 times fainter than the blue giants of Population I. – The very bright individual stars scattered over both pictures are foreground stars belonging in our own Milky Way system.
(*Mount Wilson – Palomar*)

and helium) in the stars of the two Populations. The abundance of heavy elements in Population I stars is about 2 % by mass, but it is only 0·02 % by mass in Population II stars. Stars are believed to condense from **interstellar matter** and to synthesize elements during their evolution which are later returned to the interstellar medium. (See **Nucleogenesis.**) Stars condensing later will condense from a medium richer in heavy elements; hence stars with a larger abundance of heavy elements are expected to be younger. This is a further indication that Population II stars are older than Population I stars. Population I stars are even now forming from the interstellar dust among them; Population II stars cannot now be forming as there is no such dust in their vicinity.

The Hertzsprung–Russell diagrams are of great importance in the study of stellar evolution. They provide a wealth of observational data for comparison with theory. The reason why only certain regions of the diagram are occupied by stars, in particular the main sequence and the red giant region, is that nearly all stars spend most of their lifetimes in these two regions. In fact most of a normal star's life is spent on the main sequence where the star converts hydrogen into helium in its central regions. When the central hydrogen is exhausted the star evolves quite quickly into a red giant. These topics are dealt with in greater detail under **Stellar Evolution** and **Stellar Energy.** (R.G., J.A.J.W.)

HIDALGO. An **asteroid** discovered in 1920. It has the largest known asteroid orbit, with aphelion as far from the Sun as Saturn. The orbit has exceptionally great inclination (42·6°) and eccentricity (0·66), and resembles that of a comet – it is quite closely similar to that of Tuttle's Comet. Long exposures with the 100-inch telescope on Mount Wilson have, however, failed to reveal any trace of coma.

HIGH-TEMPERATURE BELT. A region of the Earth's **atmosphere** at an altitude of about 35 miles. The temperature of the belt is about 80° C. compared with −60° C. at 15 miles and −30° C. at 55 miles. The high temperature of the layer is probably associated with the presence of ozone, a form of oxygen having three, instead of two, atoms in its molecule.

HOUR ANGLE. The hour angle of a celestial object is the time which has elapsed since its **meridian passage.** If the object has not yet crossed the meridian, i.e. is still in the eastern part of the sky, its hour angle is negative.

HUBBLE'S CONSTANT. See **Red Shift.**

HYDRAZINE. A powerful reducing agent ($H_2N\text{–}NH_2$) sometimes used as a propellent in rocket motors, as are some of its derivatives such as *hydrazone*.

HYDROGEN. The chemical element with the simplest possible **atom** in which one electron moves around one proton. Isotopes (*heavy hydrogen*) contain one or two neutrons in addition. Hydrogen is a gas at ordinary temperatures and is only 1/15 as dense as air.

The Universe consists mainly of hydrogen, which provides the vast majority of the atoms both in stars and in interstellar matter.

HYDROGEN PEROXIDE. A chemical sometimes used as an oxidant in rocket propulsion. Its molecule contains two atoms of hydrogen and two of oxygen. One oxygen atom can be split off and used to oxidize the fuel, leaving a water molecule behind. This reaction liberates a good deal of energy apart from the combustion of the fuel, and hydrogen peroxide has been used alone as a monopropellent, but it has gone out of favour because of its tendency to explode spontaneously.

HYDROPONICS (*nutriculture*) is the process by which plants are grown in a liquid medium without the use of soil or any other supporting material. All the essential elements for plant growth are contained in the nutrient solution in which the plants are grown. Control of the acidity, aeration and temperature of this solution promotes growth which is often more rapid than that of the same plants in soil.

The technique was first used as a research tool in the 17th century, but since that time has found a considerable commercial application. It is of interest in astronautics because use can be made of the photosynthetic reactions of the plants to absorb carbon dioxide from the cabin atmosphere and replenish the oxygen used. The plants may also be harvested to supply food, i.e. a waste product can be converted into a weight-saving asset, and an oxygen–carbon dioxide–starch–oxygen cycle can be set up. (See **Space Medicine.**)

Algae are the most efficient plants in these respects and research is being conducted to find the most suitable species. *Chlorella* has been the subject of intensive investigation and has been tested in a number of laboratory-scale closed ecological systems. It may prove necessary to include more than one species in a practicable scheme to combine the advantages of the gas exchange reactions with the nutrient value of the cultured cells. At present, the system is handicapped by the large weight of water required for an installation capable of supplying even in part the needs of a space vehicle's occupants.

HYPERBOLA. See **Conic Sections.**

HYPERBOLIC VELOCITY. The velocity of a body describing a hyperbolic **orbit** in the gravitational field of another body. Any velocity greater than the **parabolic velocity** will cause a body to move in such an orbit, and it will then never return to the same vicinity.

HYPERION. The eighth satellite of **Saturn.** It is one of the smaller satellites, having a diameter of perhaps 200 miles. It takes three weeks to complete one revolution in its orbit.

I

IAPETUS. The ninth satellite of **Saturn.** Its diameter is uncertain; some recent estimates put it at almost 2,000 miles, rather greater than was previously thought. Iapetus is interesting because it is much brighter near western elongation than near eastern, which indicates that all parts of its surface are not equally reflective.

ICARUS is perhaps the most interesting of the minor planets. It was discovered accidentally by Baade in 1948, and named after the Icarus of Greek mythology who flew with artificial wings made of feathers and wax; when he came too close to the Sun the wax melted and he fell to his death. No other asteroid known has a period as short as 409 days nor an orbital eccentricity as large as its 0·83. It comes 17,000,000 miles closer to the Sun than any other while receding to a distance of 183,000,000 miles at aphelion so that it is well beyond Mars. The motion of Icarus is of great interest because its orbital characteristics make it a favourable object for determining the relativistic shift of perihelion. During the close approach to Earth on June 14, 1968, Icarus passed within 4,258,000 miles, and there were many widespread rumours perpetuated even in serious-minded newspapers that a collision of Icarus and the Earth was possible which led to an official question being asked in the House of Commons as to whether this were true! Icarus was detected on June 13, 1968, using radar techniques for the first time. Estimates of its diameter vary between 600 metres if it were completely metallic or 1,200 metres if it were like a stony meteorite. The radar results indicated that in shape it is an elongated object with three unequal axes, somewhat reminiscent of a peach stone. If, as is likely, it spins with a period of rotation less than about two hours, then centrifugal force at the surface near the equator is greater than the almost negligible gravitational pull; any loose object placed on that part of the surface would therefore immediately drift off into space. (See **Asteroid.**)

INERTIAL GUIDANCE. An inertial guidance system as used in a missile is able to measure the accelerations of the vehicle in all directions by means of **accelerometers** and hence, by means of *integrators*, to determine the vehicle's velocity and direction, the distance travelled and its altitude. The target coordinates are fed into the system before firing and it is able to deduce the geographic position of the vehicle and the distance and direction to the target throughout its flight. A correction can then be automatically made to put the missile on to its correct course. Such a system is completely self-contained and is invulnerable to enemy jamming; no ground facilities are required and no radiation is emitted which would assist detection by the enemy.

A typical system comprises three accelerometers mounted to measure North–South, East–West and vertical accelerations with their associated integrators. These are all mounted on a gyro-stabilized platform to maintain them in a fixed position despite changes in vehicle attitude and position. The servo-systems required to maintain the platform in its position must be exceptionally accurate and sensitive as all errors are cumulative and quite a small error in acceleration measurement can produce a large error in distance covered. The data are fed into a computer which works out the necessary correction and passes a signal to the main propulsion motor, the subsidiary rocket motors for pitch, roll or yaw control, or the aerodynamic controls if the vehicle is still within the atmosphere.

For very long range missiles (intercontinental) various special corrections must be introduced. For example, the Earth is not a perfect sphere but is slightly flattened at the poles to the extent of 13 miles in the radius of 4,000 miles. A further correction must be made for the fact that the Earth is revolving.

INFRA-RED brightness temperatures of the surface of the Moon in the crater Copernicus, in absolute degrees, obtained from observations during an eclipse and superimposed on an oblique photograph taken by Lunar Orbiter II. (*Shorthill and Saari, Boeing Laboratories*)

This system of guidance is best suited to static targets whose positions are well known.

INFRA-RED. The region of the **electromagnetic spectrum** with wavelength between about 7,800 Å and 1 mm., including most of the wavelength range of radiant heat. Much work has been carried out on the determination of planetary temperatures from observations in the infra-red and, with the availability of more sensitive detectors and sophisticated computer techniques, it has been possible to explore the lunar surface in much finer detail.

Further information on absorption of radiation by the cool layers of gas surrounding some irregular **variable stars** is also available from infra-red analyses.

INTERFEROMETER, STELLAR. An instrument first used by the physicist Michelson in 1920 to measure the apparent diameters of stars as seen from the Earth. It relied upon the fact that light waves travelling along very nearly parallel paths can cancel each other. The diameters of seven stars were obtained, all smaller than 0·06 seconds of arc. Although the design was very ingenious the measurements were very tedious and difficult to carry out, and the instrument has not been used since 1920.

Another interferometer, working on similar principles, has been constructed at Narrabri, New South Wales, Australia, and started operations in 1963. This instrument is designed for permanent use, whereas Michelson simply added some equipment to the 100-inch telescope at Mount Wilson Observatory, and it is capable of measuring much smaller angles and fainter stars.

Interferometers are also used in **radio astronomy.**

INTERPLANETARY MATTER in its widest sense includes comets, meteors and small zodiacal dust particles, but the term refers chiefly to the very tenuous gas of neutral and ionized hydrogen which pervades the solar system (and also forms part of **interstellar matter**), clouds of solar protons and electrons, and streams of other particles ejected from the Sun which create their own complex magnetic fields and the 'solar wind'.

Dust particles in interplanetary space must be constantly undergoing evolutionary changes since they are affected by the solar gravitational field, by solar **radiation pressure,** by sputtering due to solar corpuscular radiation and by the interaction between electrically charged particles and the interplanetary gas plasma. The source of the interplanetary dust could well be material ejected from comets as they approach perihelion passage or are disturbed by passing close to a planet.

INTERSTELLAR MATTER. Material which exists between the stars in a galaxy, either in gaseous form or as minute solid particles. The total mass of interstellar matter in a spiral galaxy is believed to be about one-tenth of the total mass of the stars.

The article on **galactic nebulae** describes the appearance of interstellar clouds, both dark and bright. Interstellar matter exists everywhere in the spiral arms, not only in these denser clouds. Certain lines in the spectra of binary and other stars do not move with the rest of the spectrum as the stars revolve; these lines do not originate in the stars themselves but in the interstellar matter between us and the stars. Close examination often shows the lines to be multiple. The multiplicity is caused by **Doppler shifts** that arise from the streams of interstellar matter moving at speeds of a few astronomical units per year relative to each other. Most of the interstellar matter is hydrogen, at an average density of the order of one atom per cubic centimetre, which is deduced from spectroscopic observations made by radio telescopes at a wavelength of twenty-one centimetres. (See **Radio Astronomy** and **Spectroscopy.**)

The number of galaxies visible per square degree of sky falls off towards the galactic plane, and practically none can be seen within 5° of it because the light from them is obscured by interstellar matter in the galactic plane.

A considerable fraction of interstellar matter consists of hydrogen. Other neutral atoms, ions and complete molecules whose presence has been detected include those of calcium, sodium, potassium, titanium, cyanogen, carbon and sodium hydrides and, surprisingly, formaldehyde.

Radio astronomy measurements show also the existence of an interstellar magnetic field. It is very weak but, nevertheless, it has important effects on the motion of interstellar gas clouds, the formation of stars and the motion of **cosmic rays.**

Beside the single atoms and molecules in interstellar space there are also 'grains', apparently made of a graphite core coated with ice. The size of these grains is about one ten-thousandth of a millimetre; this is deduced from the fact that they scatter blue light more than red, causing an overall reddening of the transmitted light. The light from distant stars is polarized, which can be accounted for if the grains are aligned by the interstellar magnetic field.

The interstellar matter is not in general in thermal equilibrium, so it does not behave as a **black body**, hence **Stefan's Law** cannot be used to calculate its temperature. Near hot, early spectral type stars, the interstellar matter is ionized by **ultraviolet radiation** and has a *kinetic* temperature (a temperature associated with the energy of motion) of about 10,000° K.; these are called H II regions. Away from stars the matter is not ionized and the kinetic temperature is about 100° K.; these are H I regions. The temperature of the grains may be much lower than that of the gas.

Intergalactic matter has been detected between galaxies that are closely associated. Very slight amounts of dark obscuring matter may exist between unrelated galaxies.

INTRA-MERCURIAN PLANET. Though it is now known that Mercury is the closest to the Sun of the major planets, it was formerly believed that a still closer planet existed; it was even given a name (Vulcan), and was thought to have a diameter of 1,000 miles, with a mean distance of 13,000,000 miles and a sidereal period of 19¾ days.

Le Verrier, whose calculations led to the discovery of Neptune, computed an orbit for Vulcan on the basis of some irregularities in the motion of Mercury. Various observations of the elusive body were reported; a French amateur, Lescarbault, stated that he had seen Vulcan in transit, and in 1878 two Americans, Watson and Swift, believed that they had recorded several previously unknown bodies by observing the neighbourhood of the Sun during the total solar eclipse of that year. However, the theory of relativity has cleared up the irregularities in the movements of Mercury, and it is now certain that Vulcan does not exist. It is not impossible that some minute asteroids may revolve at a mean distance from the Sun which is less than that of Mercury, but such bodies would be very difficult to detect. (P.M.)

IO. The innermost of the four large satellites of **Jupiter.** It is slightly larger than the Moon, having a diameter of 2,310 miles, and revolves once in its orbit in 42½ hours.

ION. An **atom** which has lost or gained **electrons.** A positive ion is an atom which has lost one or more electrons and is therefore positively charged, and a negative ion is an atom which has gained one or more electrons and is negatively charged. For example, a proton is a positive hydrogen ion (a hydrogen atom which has lost its one electron). Ions differ from the corresponding neutral atoms in their general properties. They are very reactive chemically, and move in an electric field towards an electric charge of sign opposite to their own. A stream of ions whose charges are of the same sign is equivalent to the flow of an electric current.

A positive ion is formed by excess absorption of radiation by an atom, causing detachment of an electron. All atoms form positive ions, and many singly or multiply ionized atoms are present in stars as well as free electrons. The symbols Ca^{+}, Ca^{++} represent singly and doubly ionized calcium respectively; frequently these are referred to as Ca II, Ca III so that Ca I represents a neutral atom of calcium. The capture of a free electron by a positive ion is called *recombination* and is accompanied by emission of radiation.

A negative ion is formed by attachment of an electron to a neutral atom; few atoms in fact can form negative ions. The most important negative ion of astrophysical interest is that of hydrogen, represented by H^{-}, which is a dominant source of absorption of radiation in the solar atmosphere. (See **Spectroscopy.**)

ION ROCKET. A rocket propelled by the recoil from the ejection at high speed of electrically charged particles. It derives its energy from solar radiation, or from a nuclear power plant.

The maximum velocity which a rocket with conventional propellents can reach is limited by its **mass ratio** and **exhaust velocity.** The ion rocket design could provide exhaust velocities bordering on the speed of light, but such a rocket would accelerate very slowly and would spend weeks or months working up to high speeds. It must therefore be placed into an orbit about the Earth by ordinary step-rockets before its electrical propulsion can become effective.

The principle is very simple: either solar energy from large mirrors or solar batteries or a nuclear reactor is employed to drive a generator, whose output creates a strong electrical field. An easily ionizable substance (e.g. the metal caesium) supplies the **ions** which the electrical field repels so strongly that they are ejected with very great velocities.

In pushing the ions backwards, the rocket drives itself forwards. In one design which has been thoroughly worked out for a rather large payload, the rocket would accelerate in its orbit at 1 millimetre per second per second, so that after 1 hour its speed would have increased by 13 km. per hour; each day it would add 311 km. per hour, and would spiral away from the Earth; after 5,000 years it could theoretically reach half the speed of light. Needless to say, such a speed would inevitably carry it far beyond the solar system, and the Sun's radiation could not remain effective. On the other hand, a journey to Mars would take 260 days and could be made on solar energy alone.

The theoretical basis of the ion rocket is quite sound, but its weak thrust renders it impracticable until ways are found of creating extremely strong electric fields without incurring unacceptable weight penalties.

IONIZATION. The formation of **ions,** or the degree to which an **atom** is ionized. Ionization of gases can be promoted by high-energy radiation such as light or ultraviolet rays, or by collisions of particles in thermal agitation. (See **Spectroscopy.**)

IONIZATION GAUGE. An instrument for measuring low pressures in gases. The strength of the electric current that flows between two terminals in a gas-filled tube depends upon the number of **ions** which are formed, and this in turn depends upon the pressure. The gauge is calibrated so that readings of the current also indicate the corresponding pressures of the gas.

IONOSPHERE. A region of the Earth's atmosphere extending roughly from 40 to 500 miles above the ground, merging into the *exosphere* above it.

Radio waves are expected to travel in nearly straight lines, and early experimenters were surprised that they could be received not only beyond the horizon of their transmitters but across thousands of miles round the Earth. An ionized layer in the atmosphere was postulated to account for this: such a layer would have the power of reflecting radio waves.

The ionosphere is now known to consist of *several* layers of ionized gases, whose altitudes are shown on the comprehensive diagram under **Atmosphere of the Earth.** The D region is only weakly ionized, and the chief reflectors are the E and F layers above it. During the night the F layer is single, but at sunrise it splits into two layers which drift apart and reunite at dusk. The E and F layers are sometimes named the Heaviside–Kennelly and Appleton layers respectively, after early investigators. Between them they reflect all radio waves which have wavelengths greater than five to thirty metres, the exact minimum value being variable. The E layer probably contains ionized oxygen, and the F layer ionized nitrogen as well.

The Sun exerts a controlling influence on the ionosphere. Its ultraviolet radiation ionizes the rarefied gases of the upper atmosphere, and the electron density there varies in step with the 11-year sunspot cycle and the 27-day period of the Sun's rotation (see **Sun**). Solar flares have profound effects, and cause short-wave radio fade-outs over the sunlit side of the Earth: the intense emission of X-radiation from a flare penetrates the E layer and ionizes the lower strata where the air is denser. Radio waves are then absorbed instead of being reflected.

Flares are also responsible for magnetic disturbances. When a flare occurs there is a sudden magnetic disturbance called a *crochet*, followed about a day later by a magnetic storm as the charged particles emitted by the flare arrive. The time delay enables us to calculate the speed of these particles as about 1,000 miles per second. They also cause **aurorae.**

When meteors become incandescent in the F layer they produce large numbers of ions; this causes noticeable momentary increases in the intensity of radio reflections.

ISOTOPE. One of two or more forms of an element which differ in the number of neutrons contained in the nuclei of their **atoms,** but not in their chemical characteristics. For instance, **hydrogen** has three isotopic forms with 0, 1 and 2 neutrons respectively in their atomic nuclei. Some elements have as many as twelve isotopes, several of which may be radioactive.

J

JANUS. The tenth satellite of **Saturn,** discovered in 1966 by A. Dollfus. It is the closest of the satellites, and is very difficult to observe.

JULIAN CALENDAR. The calendar instituted by Julius Caesar in 45 B.C. It was similar to the present system except that *every* fourth year was a leap year. This gave an average length of 365¼ days to the civil **year,** about 11 minutes longer than the tropical year. If this discrepancy had been allowed to build up indefinitely the civil year would have fallen progressively farther behind the seasons, the spring in the Northern Hemisphere beginning in February, January, December. . . . When Pope Gregory XIII rectified the situation in 1582, ten days of accumulated error were dropped from that year; in the *Gregorian calendar* only such hundredth years as are divisible by 400 are leap years. 1900 was, therefore, not a leap year, but 2000 will be. The Gregorian calendar was not adopted in England until 1752, when the error was eleven days: the day after September 2 was called September 14. This occasioned some riots; people believed that their lives had been shortened by this Popish scheme of omitting certain dates, and crowds in Westminster chanted: 'Give us back our eleven days!'

The error in the Gregorian calendar is less than 1 day in 3,000 years.

JULIAN DATE. The system of time reckoning used in most astronomical calculations. Its only unit is the day: shorter intervals are expressed in decimals of a day. The starting point or **Epoch** of the Julian Period was January 1, 4713 B.C., and the date is measured in days from then. The Julian date corresponding to January 1, 1960, is 2,436,935. The Julian day starts at noon, twelve hours later than the civil day. To avoid confusion the civil date is often altered by half a day. Thus A.D. 2,436,934 commenced on 1960, January 0·5.

JUNE DRACONIDS. A **meteor** shower with maximum activity near June 28.

JUNO. One of the largest and brightest of the **asteroids,** having a diameter of 120 miles; it was the third minor planet to be discovered.

JUPITER.

Nomenclature of the zones and belts.

SEB = South Equatorial Belt.

STB = South Temperate Belt.

SSTB = South South Temperate Belt.

Similarly for the northern features.

JUPITER is by far the largest and most massive of the planets and may be considered the most important member of the Sun's system. Despite its great distance from the Earth, it appears as a brilliant object in our skies, outshining all other planets apart from Venus and – on rare occasions – Mars. Unlike Mars, it comes to opposition almost every year, having a mean synodic period of 399 days.

ORBIT. Jupiter's mean distance from the Sun is 483 million miles, the perihelion and aphelion distances being respectively 460 and 507 million miles. The orbit is inclined at 1°.3 to the ecliptic. The orbital velocity is about 8 miles per second. Owing to its great mass, Jupiter has a very marked effect upon the motions of other members of the solar system, including the asteroids and comets.

DIMENSIONS AND MASS. Jupiter is appreciably flattened at the poles, the equatorial and polar diameters being 88,700 and 82,800 miles respectively. This compression is obvious in any small telescope. Yet although Jupiter's vast globe could contain 1,312 Earths, its mass is only 318 times that of the Earth. The density is relatively low, only 1·34 times that of water, and at once indicates that Jupiter is a world totally unlike our own. The surface gravity is 2·64 times that of the Earth, and the escape velocity is 37 miles per second. This high value means that landing upon the surface would be impossibly dangerous, even in the absence of other hazards.

TELESCOPIC APPEARANCE. Jupiter is a fascinating object in even a small telescope. The general hue of the disc is yellow, and there is much detail to be seen. The straight belts parallel to the equator are most conspicuous, and there are also spots and other more complex features. Yet the surface details are not permanent. They alter constantly, and it is unusual for any particular feature to persist in recognizable form for more than a few days. Even the belts vary in prominence and structure. In 1968, for instance, the south equatorial belt was conspicuous, whereas by 1969 it had become very obscure.

One or two features have longer lifetimes. Most famous of these is the Great Red Spot, which first became prominent in 1878, although indicated on drawings made much earlier. Another long-lived feature was the South Tropical Disturbance, first seen in 1901 and last recorded with certainty in 1940. In late 1955 a fresh disturbance was reported in this zone, and yet another in 1966, but neither feature persisted.

ROTATION. The inclination of the equator to the orbit is only 3°.1. The rotation is extremely rapid; Jupiter has in fact a shorter 'day' than that of any other planet, 9 hours 50 minutes 30 seconds in the case of the equatorial zone. However, this period is not constant for all parts of the planet, which is proof – if proof were needed – that the surface is not hard and rocky. The quick rotation means that the spots and other features can be seen to drift almost perceptibly across the disc.

It has been found convenient to adopt a system of nomenclature for Jupiter based upon the differing rotation periods. System I, bounded by the north border of the south equatorial belt and the south border of the north equatorial belt, has the period given above; System II, the rest of the planet, has an average period of 9 hours 55 minutes 41 seconds.

Even this is an over-simplification. Various features, such as the Red Spot, have their own rotation periods. The case of the old South Tropical Disturbance is of particular interest; its period was less than that of the Red Spot, so that at intervals it caught up and passed the Spot. Interactions were obvious. The Spot seemed to attract the Disturbance, the ends of the Disturbance being accelerated as they approached the Spot and returned after they had passed it, while the Spot itself was accelerated.

There have also been 'circulating currents' as was the case in the South Tropical Zone during 1919–20 and 1931–34, and altogether the study of these peculiarities of Jupiter's rotation period is an important branch of planetary astronomy.

TEMPERATURE. Jupiter was formerly believed to be self-luminous, at least in part, and to be extremely hot. Modern research has proved otherwise. The temperature is in fact about −138° C., which is about what would be expected for a non-luminous body revolving round the Sun at such a distance. The satellites must of course be equally cold, since they can draw no appreciable heat from Jupiter itself.

CONSTITUTION OF THE GLOBE. According to R. Wildt, Jupiter is made up of a central metallic core 37,000 miles in diameter, surrounded by an ice layer 17,000 miles thick which is in turn overlaid by a dense atmosphere 8,000 miles in depth. The outer atmosphere can be analysed, and has proved to be very rich in methane and ammonia. There is no mystery about the presence of hydrogen compounds of this sort, since Jupiter's great mass enabled it to retain all its original hydrogen.

A theory more favoured today is due to W. R. Ramsey, who considers that Jupiter is composed mainly of hydrogen, the pressure at great depths being so tremendous that the hydrogen starts to behave in the manner of a metal. The core of hydrogen would have a diameter of 76,000 miles. At the moment it is impossible to decide between these two theories, and both may prove to be wrong; but we can be certain that any form of life on Jupiter would have to be radically different from the terrestrial type if it exists at all.

JUPITER AS A RADIO SOURCE. In 1955, it was discovered that Jupiter is a 'radio star'. As a source of radio waves it is indeed quite powerful, the traces being sometimes stronger than those from the **Crab Nebula.** The discovery was more or less accidental, and was certainly unexpected. The exact mechanism is still a matter for debate, but we may be sure that Jupiter has a strong magnetic field and associated radiation belts.

SATELLITES. Jupiter is attended by twelve satellites. Four of these (Io, Europa, Ganymede and Callisto) are visible in any small telescope, and have even been seen without optical aid by exceptionally keen-sighted observers. The remaining eight are very faint.

Callisto is the largest of the four, but it is also the least massive, and must be constituted very differently from a body such as the Moon. Ganymede, which has a fairly high escape velocity, might possibly be expected to retain some trace of an atmosphere; but

JUPITER. The shadow of its satellite Ganymede can be seen slightly above and to the right of the Great Red Spot, which is especially prominent as this photograph was taken in blue light.
(*Mount Wilson – Palomar*)

none has been detected, and it now seems probable that it, like the other satellites, is devoid of any atmospheric mantle.

Physical data of the major satellites are of interest; the diameters are uncertain within fairly narrow limits.

The fifth satellite, known semi-officially as Amalthea, revolves round Jupiter at a mean distance of only 113,000 miles. It is a small world, only 150 miles in diameter. The distances of the four major satellites range from 262,000 miles (Io) to 1,170,000 miles (Callisto); their eclipses, occultations, transits and shadow transits may be followed with a small telescope, thus providing the observer with a never-ending source of interest. The remaining satellites are unnamed and lie at distances of between 7 million and 14 million miles from Jupiter; owing to perturbations by the Sun their orbits are not even

Satellite	Io	Europa	Ganymede	Callisto
Diameter (*miles*)	2,310	1,950	3,200	3,220
Mass (*Moon* = 1)	1·09	0·65	2·10	0·58
Density (*Water* = 1)	2·7	2·9	2·2	1·3
Escape Vel. (*Miles/sec.*)	1·5	1·3	1·8	0·9

approximately circular. All are minute (XII, discovered in 1951, has an estimated diameter of only 14 miles); VIII, IX, XI and XII have retrograde motion. It is possible that they are captured asteroids.

EXPEDITIONS TO JUPITER. From what has been said, it is clear that there can be no question of landing on Jupiter either now or in the future. The Giant Planet is an alien body; the nature of the surface, the intense cold, and the high gravity and escape velocity rule it out. It is not impossible that landings will eventually be made upon some of the satellites, but even these bodies are most inhospitable. At speeds at present achieved, a round trip to Jupiter would take many years. (P.M.)

K

K CORONA. The inner region of the solar corona. (See **Sun.**)

KEPLER'S LAWS. The three laws of planetary motion formulated by Johann Kepler early in the 17th century on the basis of his analysis of Tycho Brahe's observations. They are:

1. *The planets move in ellipses, with the Sun at one focus.* (See **Conic Sections.**)

2. THE LAW OF AREAS: *the line joining the Sun and a planet sweeps out equal areas in equal times.*

3. THE HARMONIC LAW: *the square of the time of revolution* (*in years*) *of any planet is equal to the cube of its mean distance from the Sun* (*in astronomical units*).

The Law of Areas enables changes in a planet's orbital speed to be calculated. From the Harmonic Law we can obtain either the distance or period of revolution, provided the other is known from observation.

The laws also apply to satellites, as the latter may be regarded as planets of their primary. It follows from the Harmonic Law that artificial Earth satellites whose mean distances are very small must have very short periods of revolution.

KINETIC ENERGY. The energy possessed by a moving body by virtue of its motion. (See **Energy.**)

KIRKWOOD GAPS. Owing to its great mass, Jupiter exerts strong perturbing forces on the **asteroid** orbits. A minor planet whose period is a simple fraction of that of Jupiter repeatedly suffers the same perturbations; the cumulative effect is that its orbit is changed so that its revolution period is no longer a simple fraction of Jupiter's. A graph showing the number of asteroid orbits for any given length of period displays marked minima called the Kirkwood Gaps for those simple-fraction periods.

L

LAGRANGIAN POINTS. These are five points in a binary system at which the combined centrifugal and gravitational forces are zero. They therefore represent a special case of the solution of the **three-body problem,** since a particle of extremely small mass could remain at such a point indefinitely. The most important of these points in astrophysics is the *inner Lagrangian point*, which lies between the two components of a binary system, on a line joining their centres. This may be simply described as the point at which the force of attraction by one star exactly cancels out the force of attraction of the companion star together with the centrifugal force at that point due to the rotation of the system as a whole about its centre of gravity. This division into two fields of attraction by the two stars gives rise to peculiar behaviour in the evolution of close binary systems. (See **Binary Star,** and picture on following page.)

LANDING TECHNIQUES. When a space craft is required to land on a planet or natural satellite, provision must be made to reduce the relative velocity at touch-down to a few feet per second, depending on the ruggedness of the space craft, whether it is required to be able to lift off again and other similar considerations. Where there is no atmosphere it is necessary to use thrust from a rocket motor to provide the necessary braking. The minimum change of velocity required will be the escape velocity appropriate to the body, and the maximum will depend on the trajectory flown from the Earth. For example, the return trip from Mars could result in an

 LANDING TECHNIQUES

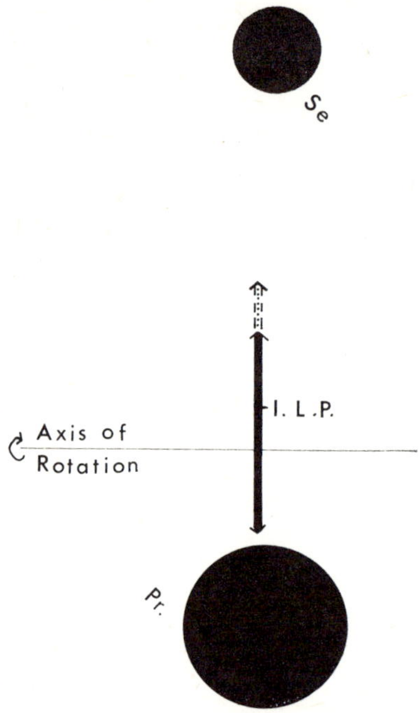

THE INNER LAGRANGIAN POINT. The gravitational attraction of the two stars at the inner Lagrangian point is indicated by the two solid *vector* arrows. The centrifugal force is indicated by the dotted vector arrow. The sum of the centrifugal force and the gravitational force of the secondary is exactly equal (but opposite in direction) to the gravitational force of the primary at the inner Lagrangian point. Therefore the force acting on a particle at this point is zero. To the left of the Lagrangian point matter is attracted towards the secondary by the resultant force, and, to the right of it, the attraction is towards the primary.

Earth approach velocity anywhere between 50,000 and 70,000 ft./sec. according to the initial launch date and the type of mission profile chosen. To provide this order of velocity change by rocket power alone would represent a considerable multiplication of initial launch weight, so if some way can be found of mitigating this problem it is worth considering. No alternative exists unless there is an atmosphere, but if even a very tenuous one exists (as on Mars) several methods may be employed to exploit the drag it can produce, i.e. the **air resistance.**

If the drag coefficient of the vehicle can be made high enough, and the atmosphere is dense enough, direct penetration at an accurately calculated angle is possible and reduces the velocity to a level where more or less conventional parachutes can be used, perhaps backed up by a rocket capable of slowing the vehicle the last few tens of feet per second. The main problem is in designing the vehicle to withstand the drag forces and heating rates involved, and tailoring the drag performance of the vehicle and the atmosphere entry angle to ensure that the accelerations imposed on the vehicle do not rise above tolerable levels. This approach has been demonstrated for Earth landings on return from the vicinity of the Moon, and could probably be applied also at velocities somewhat greater than the escape velocity of the Earth.

If the atmosphere is very much denser than the Earth's, another approach based on a number of passes through the outer atmosphere is needed. If the approach velocity is suitable, sufficient velocity can be lost by a single pass through the upper layers of the atmosphere where the aerodynamic forces and heating rates are tolerable to brake the vehicle into an eccentric orbit around the planet – a technique known as *aerocapture*. A relatively small expenditure of energy is then required to adjust the 'perigee' of the orbit to a suitable height. Successive passes through perigee (within the upper atmosphere) will slow the vehicle and reduce the eccentricity of the orbit until enough velocity has been lost in these **braking ellipses** for a direct descent to become feasible. At this stage quite a small rocket motor can be used to de-orbit the vehicle, or else the braking ellipses can be continued until the vehicle drops out of orbit. The small retro-rocket method would probably be preferable, as it offers much better control of the touch-down point. Essentially, though, when entering a dense atmosphere the braking ellipses are necessary to prevent the vehicle from being exposed to excessive forces and temperatures while its velocity is still high.

For a very thin atmosphere, the same basic method can be employed, but a large number of passes in very eccentric braking ellipses may be necessary. This could take an inconveniently long time, but reduces the weight of propellent needed for braking the descent.

LASER. A device using *L*ight *A*mplification by *S*timulated *E*mission of *R*adiation. Certain atoms have energy levels so spaced that the transition from one level to a lower level causes the emission of a **photon,** the energy of which is such that the radiation is at visible light or infra-red wavelengths. If the material is so excited that a sufficient number of atoms are at the higher energy level, a small stimulus will cause a large number of atoms to emit photons in phase with each other. The result is a *coherent* light source with rather unusual properties. The first of these is that a truly plane wave front is produced which can be focused to a geometrical point – or at least to a spot of the order of a wavelength in diameter. This means that enormous power densities become possible, and many industrial applications have been found using lasers to drill and cut otherwise difficult refractory materials. Certain types of laser can produce extremely short pulses of the order of nano-seconds and very high peak power, and use of the plane wave phenomenon (implying a parallel beam with no loss other than scattering from atmospheric particles) makes optical 'radar' systems possible which have remarkably long range, good security (only the target area receives any radiation) and very high range accuracy. The coherence of the radiation, if suitable modulators are used, offers a chance of exploiting the very large bandwidths potentially available in the optical region of the spectrum for communication.

The coherence of the light emitted by a laser makes lensless photography possible. The light scattered from an object on to a photographic emulsion forms a *hologram*, each part of which contains sufficient information to reconstitute an image of the original object when the hologram is illuminated with laser light. The hologram can be chopped up into little bits, and each piece is capable of producing the whole image. The hologram itself cannot be identified by ordinary light, and merely looks like a rather complicated interference pattern on the film – which is just what it is.

LATITUDE. The latitude of a point on the surface of a rotating sphere or ellipsoid is the angle at the centre of the body between the point and the equator.

LATITUDE, CELESTIAL. The angle in degrees between a point on the **celestial sphere** and the nearest point on the **ecliptic.** It is reckoned positive northwards from the ecliptic and negative southwards.

LAUNCH VEHICLE. A rocket of one or more stages used to accelerate a space craft to the velocity needed for its chosen mission. (See **Rocketry.**)

LAUNCH WINDOW. The interval of time within which a launch must occur if a space craft is to be able to carry out its specified task.

If it is desired to send a probe to another planet there will be a minimum energy trajectory or **transfer ellipse** which the probe must follow if only just sufficient rocket power is available. In order to follow this minimum energy trajectory, the probe must be launched when the Earth and the target planet are in just the right relative positions, and the probe's velocity must be exactly right on departure from the Earth. The exact launch times required can be predicted a long time ahead, but if any hitches occur and there is a delay in launch more energy will be required. To provide flexibility it is desirable to have sufficient rocket power to launch some time earlier or later than the optimum. For a given launch vehicle and planetary probe weight, the launch window can readily be calculated. In fact it will be a number of days each of which will have a short period in which launch could take place; a longer daily period will be available in the middle of the window, and the ends of the window will be when the possible launch time-interval has decreased to an unacceptable period. (See **Dawn Rocket.**)

For a rendezvous with a satellite already in orbit, much energy can be saved if it is possible to launch the second satellite into the same orbital plane as the first. Plane changes require a lot of extra velocity (see **Artificial Satellite**), so if too great a plane change is required rendezvous may not be possible. The concept of a launch window, and the performance compromise which has to be reached to provide such a window, has even more relevance to planetary missions than to Earth orbital operations. (See **Planetary Probe.**) (C.W.M.)

LENS. A disc of glass, or other transparent substance, with curved surfaces. The surfaces are convex or concave, or in some cases one of them may be flat.

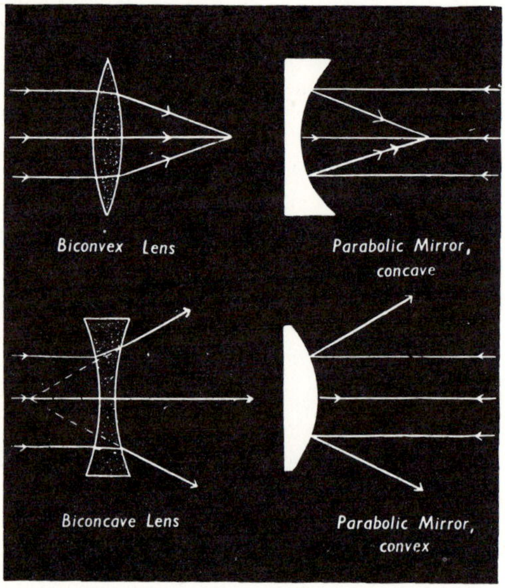

A *positive* lens causes light which enters it in a parallel beam, as from a distant star or planet, to converge upon a *focus*, where an image of the distant object is formed. The distance between the lens and the image is the *focal length*.

A *negative* lens does not bring parallel rays to a focus but causes them to diverge. Such lenses have been used in lieu of eyepieces in very crude telescopes, such as the early instruments and those sold as toys.

A single lens has various defects, including **chromatic aberration** and *spherical aberration*. By a tactful combination of lenses made of different types of glass it is possible to get rid of the most troublesome features of the simple lens, at least so far as the central part of the image is concerned.

LEONIDS. A **meteor** shower occurring annually about November 16; very intense returns were seen in 1799, 1833 and 1866, little in 1899, followed by the greatest meteor shower ever recorded, on November 17, 1966, when in certain regions the estimated number of meteors reached 300,000 per hour.

LIBRA, FIRST POINT OF. See **Celestial Sphere.**

LIBRATIONS OF MOON. **Tidal friction** has made the period of axial rotation of the Moon equal to its period of revolution round the Earth. This means that the same hemisphere of the Moon is always turned towards us.

However, we can at various times examine more than half the lunar surface; only 3/7 are concealed from Earth. Though the rate of axial spin is constant, the Moon has an orbit of appreciable eccentricity, and this results in a variable orbital speed. Consequently, the axial spin and position in orbit become periodically out of step, and the Moon appears to tilt very slightly to the East or West, exposing first one limb and then the other. This is called libration in longitude. There is also a libration in latitude, due to the fact that the Moon's orbit is slightly inclined to that of the Earth. (See **Moon.**)

LIFE. A substance may be said to be living if it can add to itself by chemical exchanges with its environment (growth and metabolism), can form replicas of itself (reproduce) and can react to stimuli (irritability). The constituents of living matter cannot be precisely defined if that definition is to include the simplest protozoa, algae, bacteria and a fertilized hen's egg. However, the simple characteristics used so far do not exclude the possibility that life could exist in forms which bear no resemblance to terrestrial organisms.

The only absolute prerequisites are the presence of carbon or silicon and a fair variety of other elements, and a temperature neither very cold nor very hot. A living substance must consist of extremely elaborate molecules, and only carbon and – to a much lesser extent – silicon atoms are reasonably small, common and capable of linking themselves in the long and branching chains and rings which provide the framework for such molecules. Extreme cold inhibits chemical activity, and heat much above the boiling point of water breaks up most complicated molecules; but a moderate temperature can support just that level of chemical interchanges which is most likely to result in the synthesis of living matter. Chance would play a part in this, but chance in the long run obeys inexorable laws; the evolution of higher organisms from simpler ones depends very largely on the systematic effect of great numbers of accidental events, and precisely similar principles can lead to the evolution of complex molecules from simpler matter.

Life on stars is impossible, because they are too hot. But life on bodies like the planets cannot be ruled out on the grounds that their atmospheres are 'unsuitable', that pressures are too great or too low, that there is no water or that 'poisonous' substances abound. So far as the planets are concerned, temperature alone is the factor that makes life in certain localities virtually impossible, and limits the width of the so-called *biothermal zone* around the primary of a planetary system. Where life does perhaps exist, it is most unlikely to resemble earthly forms even to an extent that would enable us to fit it into our plant or animal kingdoms. We already know viruses in the shape of crystals, spores like microscopic golf balls that can survive temperatures of $-220°$ C., fungi that do not breathe oxygen, plants without green chlorophyll, fish and plankton that can withstand pressures of thousands of pounds per square inch, highly-developed non-cellular animals that thrive in dark, boiling springs, bacteria that need no water; we know that living matter can look like jelly, like yeast, or like a sponge. It would be very rash to say where life is possible and where it is not.

It is another matter to try to assess the balance of probabilities. In the presence of the appropriate elements life may develop spontaneously from complex compounds in whose formation electromagnetic and ionizing disturbances from the primary, lightning, etc., may play a part. When simulated 'primeval brews' containing only inorganic compounds and carbon dioxide have been subjected more or less randomly to such influences in sealed containers in the laboratory, many complex molecules (including amino-acids) were formed which normally occur only in living matter.

The subsequent biochemical and biological evolution of the initial molecules obeys normal physical and chemical laws. Evidence for the ability of many molecular mixtures to aggregate spontaneously is common. The resulting coacervates can be shown to have a variety of structures (of cellular size) whose exact nature depends upon the components of the mixture from which they are formed.

It is theoretically possible for silicon to replace carbon as the fundamental chain-building atom. Many carbon compounds have silicon analogues, but the exact chemistry is different in detail. Similarly water can be replaced as the solvent in which biochemical reactions occur. Ammonia and hydroxylamine have been suggested as possible alternatives. It is considered, however, that the development of terrestrial-type life occurred not because carbon, hydrogen, oxygen and nitrogen are the most abundant elements, but because they are the most suitable – being the smallest atoms taking part in the appropriate reactions. Any search for extra-terrestrial life should therefore pay special attention to life processes similar to those which have succeeded in evolving on Earth.

There is strong evidence that such life is present on Mars. Spectroscopic observations have shown infra-red absorption lines in spectral positions characteristic of many of the molecular bonds found in carbon chemistry. The reflectivity of the surface material also shows properties similar to that of low forms of vegetative terrestrial life.

Alternative evidence suggesting the presence of Martian life comes from experiments in which terrestrial organisms have been exposed to conditions designed to simulate Martian surface conditions. It was found that there was a normal growth and reproduction of some terrestrial micro-organisms, and even of higher forms of life. The American Viking project is intended to land automatic life-searching devices upon the surface of Mars after 1971.

Evidence which suggests the presence of life elsewhere in the Universe is available from a study of the carbonaceous chondrites. The **meteorites** of this class contain up to 6% of their mass as carbon (less than 1 % is characteristic of aerolites). Microscopic examination of the surface of the meteorites reveals the presence in them of apparent organized structures of cellular dimensions. Whether these represent fossilized organisms or are the result of crystallization of inorganic salts on to an organic matrix is not known. However, the former view is favoured by a chemical analysis of the structures, which reveals a large organic content similar in composition and physical properties to that found in cells.

It has been estimated that of the 10^{20} stars in the observable Universe, one in a hundred million is theoretically capable of supporting terrestrial-type life in its biothermal zone. On this obviously rough estimate there are thus 10^{12} locations where such life had a chance, and it seems inconceivable that it should not have succeeded anywhere except here.

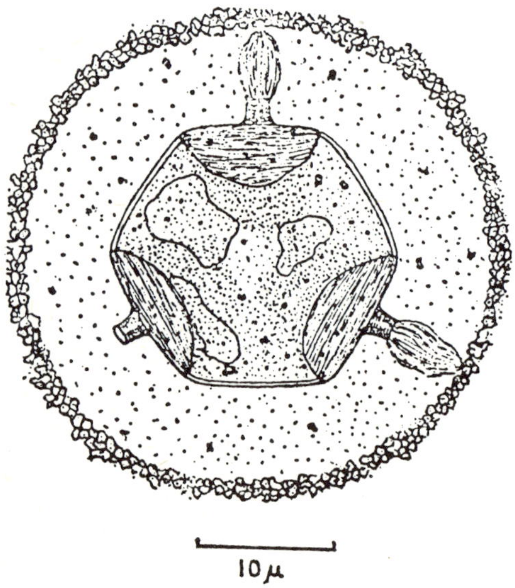

An organized element found in the Orgueil meteorite – a carbonaceous chondrite. Several structures have been found which have some similarity to fossilized unicellular algae.

COMMUNICATION WITH INTELLIGENT FORMS. If one accepts the above, it becomes reasonable to assume that in a number of places even within our own galaxy, there exist life-forms at least comparable in intelligence and technological achievement with mankind. The difficulties of establishing communications are truly daunting: (i) at both ends there must be the wish and the ability to watch for signals from no one in particular (at least as an incidental to other work) and to transmit a reply; (ii) a simple exchange of signals with our *nearest* neighbouring star system would necessarily take over 8 years; (iii) if such an exchange is ever to succeed, similar exchanges must almost certainly already have succeeded elsewhere. A signal from us is most likely to be picked up by those who are best at it, i.e. by those who are least likely to be interested in replying, particularly if our method of signalling strikes them as primitive.

Nevertheless, watch has been kept on Earth, signals have been sent out by responsible scientists, and for some brief but exciting hours the pulses of great regularity which were received from pulsars raised the question of signals in some very sober minds. The least of the problems is that of finding a language in which to communicate. A simple succession of groups of pulses such as

2, 2, 4; 2, 3, 6; 2, 4, 8, etc.
(factors and products)
or 2, 3, 5, 7, 11, 13 (primes)

would immediately establish the sender as a rational being. Groups which at first seem unintelligible to the recipient would presently be identified by him as meaning 'and', 'or', 'circle', 'line', etc., and with these as a basis a vocabulary can be built up and diagrams can be 'drawn'. A sequence of messages building up such a language could be transmitted without waiting for replies after the first one.

LIGHT. See **Electromagnetic Radiation, Electromagnetic Spectrum,** and **Spectroscopy** (section on Colour and Wavelength).

LIGHT YEAR. The distance travelled in one year by light, which covers 186,000 miles in one second. It is equal to 5,880,000,000,000 miles. (See table under **Distances.**)

LIMB. The edge of the visible disc of the Sun, the Moon or the planets.

LINE OF FORCE. See **Magnetism.**

LOGARITHMIC SCALE. A scale that is often used in graphs when quantities have to be represented that range from very small to very large values. The height scale in the diagram of the **Atmosphere of the Earth** is an example of its use in this book.

LONGITUDE. The longitude of a point on the surface of a rotating sphere is the angle at the pole between the **meridian** through the point and another meridian which marks the arbitrary zero of longitude (in the case of the Earth, the **Greenwich Meridian**).

LUNAR PROBE. A device designed to make observations of the Moon and its environment either by orbiting it or by landing instruments upon it. Some of the Russian *Luna* series have landed automatic vehicles on the Moon and returned rock samples to Earth—a considerable technological triumph.

LYMAN SERIES. A series of lines in the ultraviolet region of the spectrum of the hydrogen atom. (See **Spectroscopy.**)

β-LYRAE. An eclipsing **binary star** whose light variations, which have a period of nearly 13 days, can be followed by the naked eye. (See **Variable Star.**)

LYRIDS. A **meteor** shower occurring annually about April 21. It is now quite weak, but has been traced back to 15 B.C.

M

MACH NUMBER. The velocity of an aircraft or other vehicle moving in an atmosphere divided by the velocity of sound near the craft. For instance, at the surface of the Earth sound travels at about 750 m.p.h.; Mach 3 would therefore be $3 \times 750 = 2{,}250$ m.p.h. The speed of sound varies with temperature and therefore with altitude. The diagram illustrating **Atmosphere of the Earth** shows the exact form of the variation.

MAGELLANIC CLOUDS. Two irregular **galaxies,** the nest neighbours to our own stellar system. The naked eye sees them as faintly luminous patches. The *Large Magellanic Cloud* has a well-marked axis and is less irregular than the *Small Magellanic Cloud.*

Both clouds contain a remarkably high proportion of Cepheid **variable stars,** as well as O and B stars and some K type **supergiants.** The star of greatest intrinsic brightness known to us lies in the larger cloud; it is a very peculiar variable, *S Doradûs*, whose maximum absolute magnitude is −9.

The clouds are in motion around our own Galaxy. They both contain interstellar dust which affects the light of more distant galaxies seen through the clouds. **Globular clusters** and **novae** have been observed in them, together with the recently discovered **RR Lyrae** stars which furnish us with the most accurate distance estimates we have: both galaxies lie at about 50,000 parsecs from us.

MAGNETIC STORM. A period of violent fluctuations in the Earth's magnetic field caused by **solar flares.** The fluctuations induce electric currents in the **ionosphere** and in long-distance cables and severely interrupt radio and telephone communications.

MAGNETISM. The attribute of magnets. A magnet is a piece of metal (usually iron) which attracts other pieces of similar metal to itself.

The atoms of such metals are tiny natural magnets, but normally point in haphazard directions, so that they annul each other's effect. If some or all of these tiny magnets are aligned they reinforce each other's effect and the piece of metal containing them is magnetized. Such an alignment can be brought about by an already existing magnet, or by electric current passing through loops of wire around the metal.

Every magnet has two *poles.* The Earth itself acts as a magnet whose poles are near its geographic poles. A magnet which is suspended so that it is free to swing will set itself with its *north pole* pointing roughly north, and its *south pole* pointing south. Similar poles of two magnets repel each other, poles of opposite kind attract each other.

It is impossible to have a north pole without a south pole. Cutting a magnet simply results in two smaller magnets, each with two poles. The region round a magnet where its influence is appreciable is termed a *magnetic field.*

The pattern of a magnetic field is often represented by means of *lines of force.* A line of force is the path along which a free north pole would travel if such a thing could exist. At any point within a field, a compass needle will tend to lie along the line of force which passes through its centre, and the line can be traced by moving the compass needle farther and farther along it.

All lines of force run from one pole to the other, but they are not straight as a rule. The

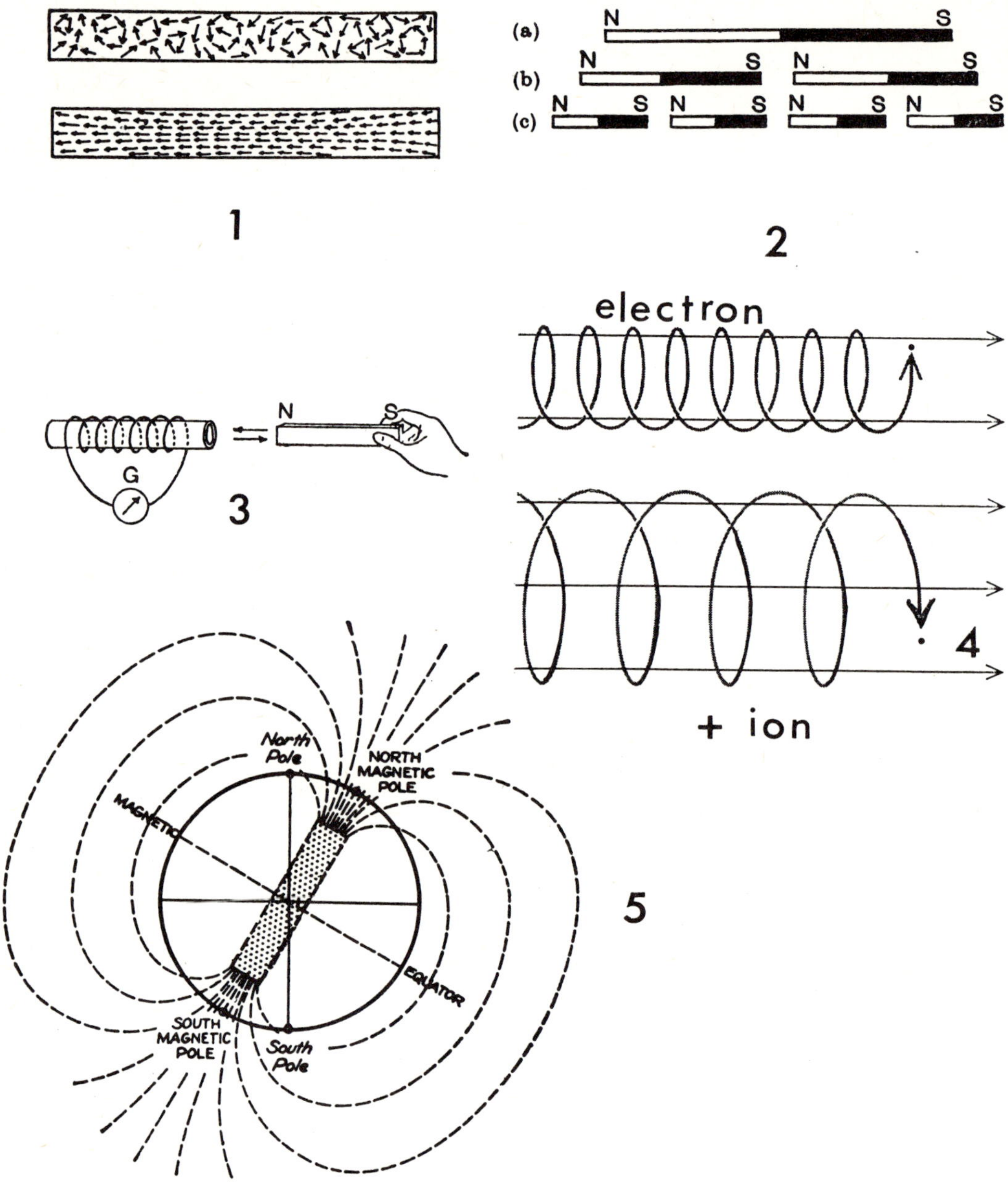

SOME BASIC FACTS OF MAGNETISM. (1) Each molecule in a piece of iron is a small magnet. When the molecules are aligned they act together, and the piece of iron becomes a magnet. (2) If a magnet is split in two, it yields two smaller, complete magnets. (3) If a conductor like a wire moves in a magnetic field (or if the field moves relative to the conductor) a current is made to flow through the conductor. (4) When an electrically charged particle enters a magnetic field, to the extent that it has a component of velocity perpendicular to the field it tends to spiral around the lines of force; to the extent that it has a parallel component it tends to move along the lines of force. (5) The Earth's magnetic field is shaped *as if* a gigantic bar magnet were buried within the Earth.

lines of force of the Earth's magnetic field all converge on a point which appears to lie within it below north-eastern Canada, and this point is one of the Earth's two magnetic poles, the other one being in the Antarctic. A compass needle suspended above the ground immediately over the poles assumes a vertical position. The magnetic poles are not stationary but move in small circles with a period of about 500 years. There has also been a less regular long-term movement; the direction of the Earth's field may once have been the reverse of what it is now, and during the last 1,000 million years the magnetic poles have wandered considerably both in relation to the land masses and to the Earth's axis of rotation. Some rocks which were deposited during this period still retain magnetism from which past patterns of the Earth's field can be deduced. Nothing definite is known about the reasons for these changes in the field.

The Sun and many other stars have magnetic fields whose detection is based on the **Zeeman effect.**

MAGNETOGRAPH. An ingenious instrument for mapping the magnetic field of the Sun. Its action depends on the **Zeeman effect.** It scans the Sun in narrowly spaced lines and measures the Zeeman splitting of the spectral lines continuously. The measurements are recorded in a trace which wanders above or below the scanning line according to the sign and strength of the magnetic field at each point along the line.

MAGNITUDE. The brightness of a star. *Apparent* magnitude indicates the brightness of a star as we see it; *absolute* magnitude is a measure of its intrinsic luminosity, independent of the star's distance.

APPARENT MAGNITUDE. Some of the brightest stars in the sky have brightnesses of the *first* magnitude. This is defined rather arbitrarily as the brightness of a candle flame at a distance of 1,300 feet. Stars of the *second* magnitude are fainter than the first by a factor of two and a half; those of the *third* magnitude are that much fainter again, and so on. A difference of five magnitudes means that one star is precisely a hundred times as brilliant as another. The faintest star visible

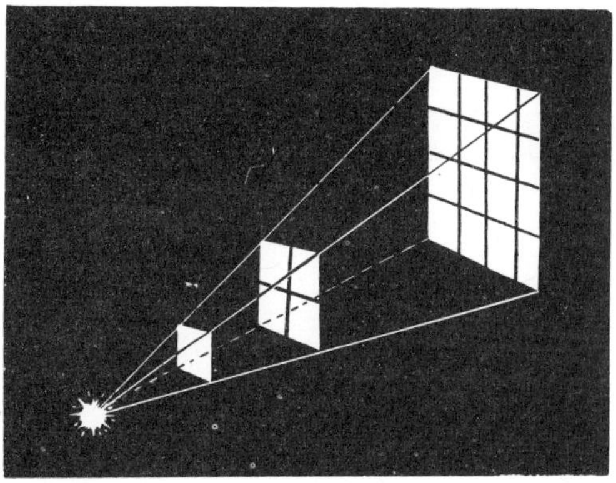

A certain area is illuminated by a source of light. At double the distance the same amount of light would be spread over an area four times as large, i.e. the intensity of illumination is reduced to a quarter. We say that the intensity of the light received from a source is *inversely proportional to the square of the distance.*

to the naked eye on a dark and clear night is of the sixth magnitude. The largest telescope in the world can photograph stars of the twenty-third magnitude.

It should be noticed that the *fainter* an object is, the *larger* its numerical magnitude.

A few stars are brighter than the first magnitude, and the scale is extended to zero and beyond into negative magnitudes. The Sun's magnitude is −27.

ABSOLUTE MAGNITUDE. The above takes no account of the distances of the objects concerned. The absolute magnitude of a star is the apparent magnitude it would have *if it were placed at a distance of* 10 *parsecs from us.* The Sun would there appear dimly visible to the naked eye as a star of magnitude 4·7; other stars, vastly more luminous, range up to −9.

PHOTOGRAPHIC AND VISUAL MAGNITUDES differ somewhat because the photographic plate is relatively more sensitive to blue light than the human eye, and so sees blue stars brighter and red stars fainter than the eye. Although this is in many ways a nuisance, it has been turned to good account and makes it possible to describe the colours of stars numerically. This point is dealt with under **Colour Index.**

BOLOMETRIC MAGNITUDE refers to the total brightness of a star, not only in the visible part of the spectrum, but at all wavelengths. It is discussed under its own heading.

DISTANCE MODULUS. This is defined as the apparent *minus* the absolute magnitude of a star. The apparent magnitude can always be measured by direct observation, and the absolute magnitude is often deduced from other evidence. Both together give the distance of the star.

Distance Modulus (*magnitudes*)	*Distance* (*parsecs*)
14	6,000
16	16,000
18	40,000
20	100,000
22	250,000
24	600,000

MAIN SEQUENCE. See **Hertzsprung–Russell Diagram.**

MARE CRISIUM. One of the smaller but most conspicuous of the lunar 'seas'. (See **Moon.**)

MARS is generally regarded as the most interesting of the planets. This is because it is the only world in the solar system, apart from the Earth, which has been regarded as capable of supporting life. Indeed, only a few decades ago the existence of 'Martians' was by no means ruled out. To-day we know better; Mars is a hostile planet, and even lowly life-forms are improbable but not impossible (see **Life**). Even so, it is less unfriendly than Venus, and there is a strong possibility that astronauts will go there within the next forty or fifty years.

ORBIT. Mars is the first of the *superior* planets, lying beyond the orbit of the Earth. The mean distance from the Sun is 227 940 000 kilometres, but the orbit is appreciably more eccentric than that of the Earth. The period of revolution is 687 days, and the orbital velocity is 24 km/sec, as against 29.8 km/sec for the Earth.

Mars comes to opposition at intervals of about 780 days, and detailed observations from Earth are therefore possible only for a few months every alternate year. When really close, as in October 1973, Mars may appear brighter than any other planet apart from Venus.

For obvious reasons, Mars can never appear as a half or a crescent; but when some way from opposition it is decidedly gibbous—an effect first noted by Galileo.

ROTATION. The rotation period of Mars, 24 hours 37 minutes 22·65 seconds, is known very accurately. The Martian day is therefore only about half an hour longer than ours, and the axial inclination is only slightly greater. As on Earth, the southern summer occurs near perihelion, so that the southern hemisphere experiences greater extremes of temperature than the northern.

DIMENSIONS AND MASS. Mars has a diameter of 6 790 km, roughly half that of the Earth. The density is 4 times that of water, corresponding to a mass 0·11 and a surface gravity 0·38 of that of our world. The escape velocity is 5 km/sec. Since there are no oceans on Mars, the land surface of the planet is roughly equal to that of the Earth.

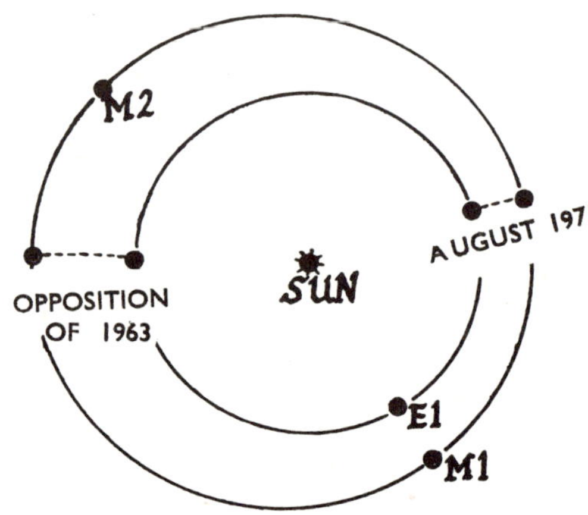

OPPOSITIONS OF MARS are not all equally favourable for observations of the planet owing to the eccentricity of the orbits. E 1 represents the Earth; M 1 the nearest and M 2 the farthest possible positions of Mars. (*P. Moore*)

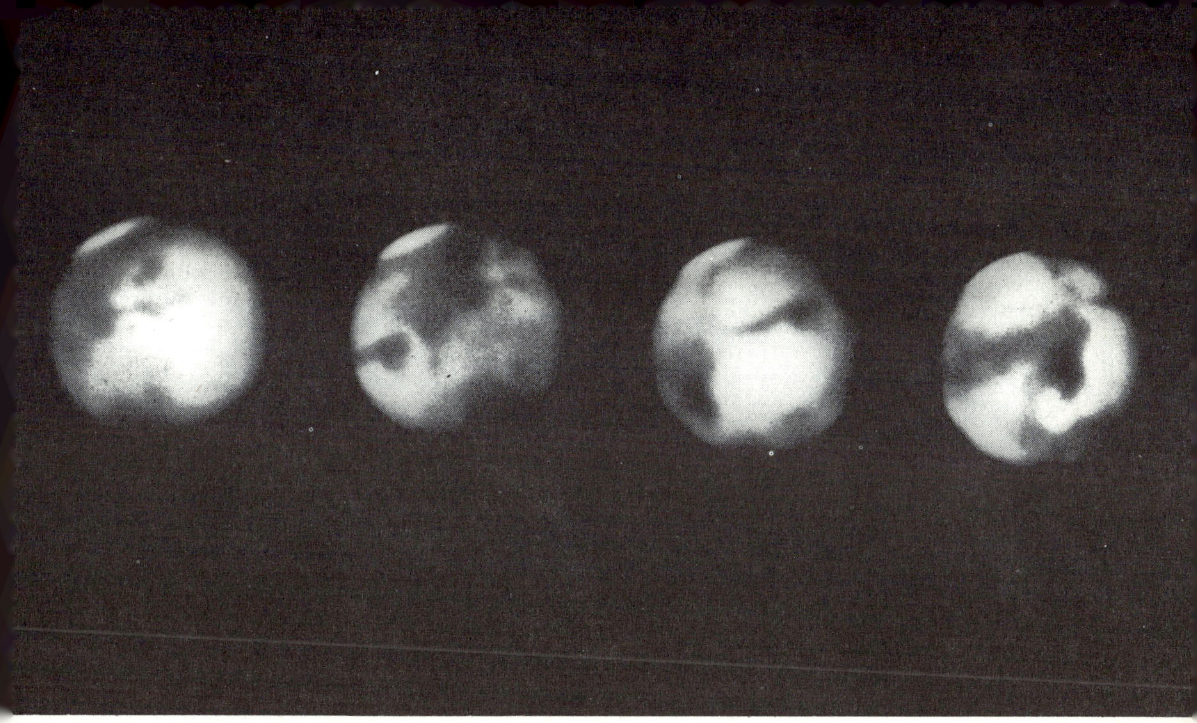

FOUR VIEWS OF MARS. The white area near the top is the South Polar Cap. The triangular dark area in the third and fourth photographs is the Syrtis Major.

TELESCOPIC APPEARANCE. When observed through an adequate telescope, under good conditions, Mars is a fascinating sight. The main colour of the disk is reddish-ochre, which led to our ancestors naming the planet in honour of the God of War; there are dark features, which are essentially permanent; and there are white areas covering the poles (though generally only one pole is visible, the other being turned away from the Earth). Originally it was believed that the bright areas were deserts, the dark patches seas, and the polar caps made up of ice or snow.

Undoubtedly there is a seasonal cycle on Mars. During winter, a polar cap is extensive and brilliant; with the arrival of spring, the cap begins to shrink, and by midsummer it has become very small. The southern cap has even been known to vanish completely. It has been claimed that when a cap shrinks (either by melting or by sublimation) the dark areas adjoining it harden and become more obvious, as though they were made up of plant areas being revived by the arrival of wet winds. For almost a century it has been known that there can be no oceans on Mars; the atmosphere is too thin and too dry, so that the vegetation theory for the dark areas seemed very plausible.

Before the age of space probes, astronomers were confident that they had drawn up a reliable picture of Mars. They agreed that the dark areas were covered with organic matter, and were probably depressed basins; the bright ochre regions were coated with a coloured mineral deposit, possibly felsite or limonite; and the polar caps were made up of a very thin layer of frost. The atmosphere was known to be very tenuous, with a ground pressure of around 85 millibars, and it was thought that the main constituent must be nitrogen, with a little carbon dioxide and practically no free oxygen.

In fact, most of these ideas are wrong. Far from being depressed regions, many of the dark patches on the planet have been found to be elevated above the general surface level; the polar caps are probably composed mainly of solid carbon dioxide, while the atmospheric pressure is much less than had been anticipated. Nowhere does it exceed 10 millibars. The new information has come from the various space-probes, beginning with *Mariner 4* in 1965. The most detailed pictures of the planet were sent back from *Mariner 9*, which was put into an orbit round Mars in late 1971 and remained active until well into the following year. (See also **Appendix.**)

THE MARTIAN SURFACE. *Mariner 4* sent back pictures showing that instead of being smooth and gently undulating, as had been expected, the Martian surface is rough, and is scarred with craters. This was confirmed by *Mariners 6* and *7* of 1969, and *Mariner 9* of 1971–2. (*Mariner 8*, the 'twin' of *Mariner 9*, failed immediately after being launched.) We now know that massive volcanoes exist, together with craters of all sizes and immense rift valleys, as well as narrower valleys which look remarkably like dried-up meandering riverbeds.

The most important volcanoes, such as Nix Olympica and Nodus Gordii, are similar in structure to our Hawaiian volcanoes, though they are much more massive. (The names are not new. The features were recorded by observers from Earth many years ago, though of course their nature was quite unknown, and they appeared only as tiny markings.) The Nix Olympica, for instance, towers to a much greater height above the mean surface of Mars than our Mount Everest does above sea-level, and it is crowned by a complex, multiple volcanic crater 65 km. in diameter. The total height of the volcano is estimated to be at least 25 km. It is not unique; another volcano, Nodus Gordii, is probably even higher. Extensive drainage systems can be traced from these great structures.

The craters, scattered widely over the planet, are shallower than those of the Moon, but are superficially similar to the lunar structures. Even more interesting are the immense valleys, such as that of the Tithonius Lacus. Here we have a system extending over a total length of 5 500 km, and with a maximum breadth of 400 km. The valley gives the impression of having been cut by water—but no liquid water can exist on Mars now. Note also that there is no obvious evidence of erosion, indicating that the valleys cannot be very old on the geological (more accurately, *areological*) time-scale. It has therefore been suggested that some 50 000 years or so ago Mars had a denser atmosphere than it possesses now.

TOPOGRAPHY. When seen from Earth, the dark features of Mars are well-defined. They were first seen by Christiaan Huygens as long ago as the year 1659, and their outlines are essentially permanent. There are also some bright features, such as the circular, whitish plain Hellas—which can at times look almost as brilliant as the polar cap. It was natural to assume that the dark features were sunken, while the bright Hellas was taken for a high plateau.

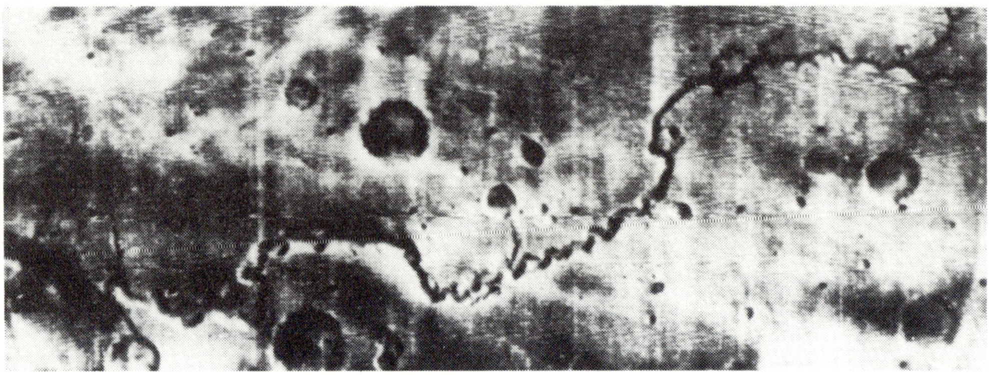

MARTIAN RIVER BED? This meandering, 400 km long valley and other, similar Martian features are difficult to explain as anything other than dry river beds, but we do not know of enough water on Mars to account for them.

Yet this is not confirmed by the *Mariner* results. There is no basic difference in topography between the bright and the dark areas, and the most obvious of the dark patches, the Syrtis Major (once called the Hourglass Sea) proves to be a high plateau, sloping off quite steeply. It seems, in fact, that the dark regions are 'albedo features' only, so that they cannot possibly be old sea-beds. On the other hand, Hellas proves to be a depressed basin, devoid of interior detail.

ATMOSPHERE AND CLOUDS. The atmospheric density has been measured from all the Mars *Mariners*. The procedure is to send the probe behind Mars; just before and just after the actual occultation, radio signals from the vehicle come to us by way of the Martian atmosphere, and the manner in which these signals are distorted gives us information about the density and composition of the atmosphere itself.

The density is very low, and at the mean ground level the pressure is only about 5 millibars. This rarefied atmosphere means that the temperatures are low; at noon on the equator a thermometer might rise to 12° C, but it would drop to below −60°C at midnight.

There can no longer be any doubt that the main constituent of the Martian atmosphere is carbon dioxide. There are only traces of other gases, and there is very little water vapour. Whether the atmosphere is adequate as a screen against harmful short-wave radiations from space seems rather doubtful.

High-altitude clouds are quite common, and may be due to ice crystals or solid carbon dioxide crystals. Much more dramatic are the great dust-storms, which may at times cover much or all of the planet. Such a storm was in progress when *Mariner 9* went into orbit round Mars, in late 1971; for some weeks virtually nothing could be seen apart from the top of the dust-layer, together with a few of the lofty volcanoes (such as Nodus Gordii) which were high enough to be visible. The cause of these storms is still a matter for debate. It has been assumed that they are due to material whipped up from the 'deserts' by winds of the cyclone variety; but the thin atmosphere means that even a high-velocity wind would have little force, and it seems strange that particles could remain suspended for a period of months. Active vulcanism has also been proposed, though the *Mariners* have shown no evidence of it.

THE CANALS. No description of Mars would be complete without reference to the 'canals'—straight, artificial-looking lines running across the surface, first described in detail by G. V. Schiaparelli in 1877. The American astronomer Percival Lowell believed them to be true waterways, constructed by an intelligent civilization to form a planet-wide irrigation system. Between 1895 and his death in 1916 he made numerous drawings of Mars, and showed a highly complicated network. However, the idea of intelligent Martians has long since been discarded, and there is nothing at all artificial in the appearance of Lowell's 'canals'. Some of them are lines of craters, while others merely mark the boundaries between light and dark areas.

LIFE ON MARS? When we consider the question of Martian life, we must be cautious. The low temperatures and the thin atmosphere, together with the absence of surface water, indicate that no advanced life-forms of terrestrial type can exist there. Even the existence of vegetation is improbable. However, we must not be dogmatic, and it may be hoped that the problem will be solved in 1976, when the Americans land a probe upon the surface (probably in the ochre region known as Chryse) and obtain direct information. The Russians have already attempted this, and in 1972 their probe *Mars 3* deposited a capsule on the planet between the regions of Electris and Eridania; unfortunately signals

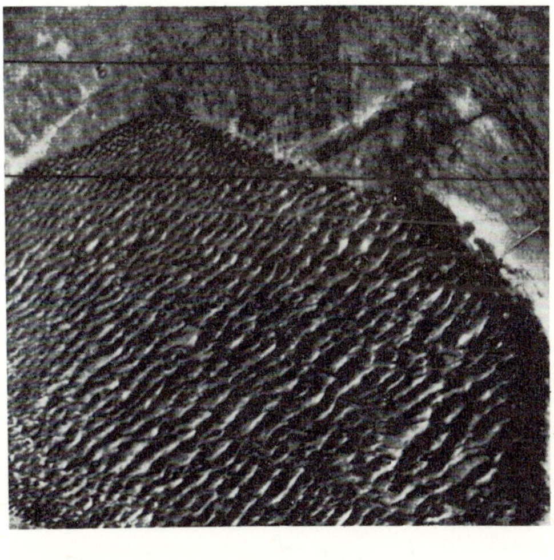

A MARTIAN DUNE FIELD, shaped by relatively strong winds blowing from the southwest over loose material inside a 160 km wide crater in the Hellespontus region.

from the capsule failed after only 20 seconds. Four more Soviet probes were sent up in the summer of 1973, and all will reach the neighbourhood of Mars in 1974. It is quite likely that further soft landings will be attempted.

Meanwhile, we can say only that the brilliant-brained Martians must be relegated to the realm of fiction, and that the existence or absence of lowly organic matter remains to be proved.

SATELLITES. The two dwarf attendants of Mars, Phobos and Deimos, were discovered by Asaph Hall in 1877, using the large refractor at Washington in the United States. Statistical details are as follows:

	Phobos	*Deimos*
Mean distance from centre of Mars, km ...	9 270	23 400
Sidereal period ...	7h 39m	30h 18m
Orbital eccentricity ...	0.02	0.003
Orbital inclination to the Martian equator	1°·1	1°·8
Mean opposition magnitude	11·6	12·8
Maximum diameter, km	22	12

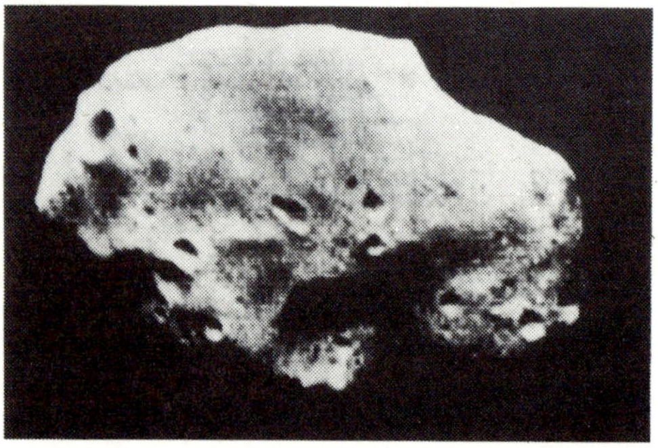

PHOBOS, the larger of the two satellites of Mars, in a computer-enhanced photograph taken by *Mariner 9*. The profusion of small craters suggests that it is very old and has considerable structural strength.

The satellites are not only small, but are very close to the planet; Phobos has a revolution period which is less than the rotation of its primary—a case unique in the Solar System so far as natural satellites are concerned. Neither can be of much use in providing illumination during the Martian nights, and their tide-raising powers would be negligible even if Mars had any oceans.

Both were photographed from close range by *Mariner 9*. They are irregular in shape, and are pitted with craters which may be either blowhole structures or due to impact. It seems likely that they are captured asteroids rather than genuine satellites.

TRAVEL TO MARS. Automatic probes have been to Mars, and more will follow. Technically, a manned flight should be possible before the end of the twentieth century—though whether a journey there will be attempted remains to be seen. At any rate Mars, hostile though it undoubtedly is, will be the first target for human astronauts after the colonization of the Moon. (P.M.)
(See also **Appendix.**)

MASS is defined under **Gravitation.**

MASS RATIO.

$$\textit{Mass ratio} = \frac{\textit{take-off mass of rocket}}{\textit{all-burnt mass of rocket}}$$

The mass ratio determines the final velocity attainable by the rocket in terms of its **exhaust velocity.** If the mass ratio is 2·7 to 1 (i.e. if the propellent accounts for just under two-thirds of the full weight of the rocket), the final velocity, ignoring **gravitational loss** and other minor factors, is equal to the exhaust velocity; if it is 7·4 to 1, the rocket can travel twice as fast as the exhaust velocity, while a ratio of 20 to 1 would give it three times the exhaust velocity. The first figure was already bettered in the V–2 rocket. The second might conceivably be achieved in the future. The third would require so light a structure – not to mention any **payload** – that it could not possibly support the strain of high accelerations.

MEAN SUN. In its elliptical course round the Sun, the Earth accelerates as it moves towards the Sun and slows down as it recedes from it. This causes the Sun's apparent motion in the sky to vary throughout the year. The *Mean Sun* is an imaginary celestial body which moves through the sky with a constant speed equal to the average speed of the true Sun. The interval by which the true Sun is ahead of or behind the Mean Sun is called the *Equation of Time*. It never exceeds seventeen minutes, and four times a year it is zero. It is tabulated in almanacs.

MERCATOR'S PROJECTION is used to portray the curved surfaces of planets, including the Earth, on a flat sheet of paper. A Mercator projection of a hollow transparent sphere could be made by placing a lamp at the centre and a paper cylinder round the sphere. The features on the sphere are drawn on the paper cylinder, which is then cut along a meridian and opened out flat. A great circle becomes a straight line on Mercator's projection. The scale of a Mercator map varies according to latitude, and great distortions arise near the poles.

MERCURY is the closest to the Sun of the nine major planets. It is never a conspicuous object as seen with the naked eye, owing to its relatively small size and to the fact that it can never be seen in a dark sky, far from the Sun. At favourable **elongations** it can, however, be seen shining like a moderately bright star low in the West after sunset or low in the East before sunrise.

ORBIT. Mercury revolves round the Sun at an average distance of 36 million miles. The orbital eccentricity is 0·2056, greater than that of any other major planet apart from Pluto, so that the distance varies from 28·5 million miles (perihelion) to 43·5 million miles (aphelion). The corresponding orbital velocities are 35 and 23 miles per second. The inclination of the orbit is also unusually high, 7°00′14″.0, so that transits of the planet across the disc of the Sun are infrequent. A photograph of a transit appears under **Transit.** A small discrepancy between the true orbit and that calculated by Newtonian mechanics led to one of the earliest practical confirmations of the theory of **Relativity.**

ROTATION. It used to be thought that the rotation period of Mercury was the same as its revolution period, i.e. 88 days. This would have meant that tidal friction had slowed its rotation until it presented the same side to the Sun at all times, and this may yet be its ultimate fate. Recent studies have shown the true period to be two-thirds of the old value, i.e. $58\frac{2}{3}$ days. It follows that every part of Mercury is in sunlight at some time or another, and that three Mercurian days cover the seasonal changes of two years.

DIMENSIONS AND MASS. Mercury has a diameter of 3,100 miles, so that its volume is 0·06 and its surface area 0·15 of that of the Earth. It is the smallest major planet in the solar system, and is actually inferior in size (though not in mass) to several of the satellites. As it is moonless, its own mass is not easy to determine with accuracy. One recent estimate gives 0·063 of that of the Earth, in which case the density would be 5·8 times that of water, but others prefer a density value of about 3·8. The escape velocity is 2·6 miles per second, and only a very tenuous atmosphere is to be expected.

SURFACE TEMPERATURE. Because of its nearness to the Sun, Mercury can become extremely hot, with a temperature of perhaps 400° C. The very long nights must, however, be bitterly cold, since the atmosphere is too thin to carry any appreciable heat round to the dark side of the planet by convection, or to provide an effective blanket against loss of heat from the surface by radiation.

ATMOSPHERE. Until recent years, Mercury was believed to be devoid of any trace of gaseous mantle. In 1950, however, a slight atmosphere with a ground density of about 3/1000 of that of our own was detected. Nothing is definitely known about its composition.

OBSCURATIONS. Antoniadi reported frequent hazy appearances on Mercury, which he described as being commoner and more obliterating than those on Mars. These hazes have not been confirmed by recent studies. Such appearances would indeed be hard to account for; the hypothesis of dust-storms of volcanic origin has little to recommend it, since extensive vulcanism is hardly likely to occur upon a world such as Mercury.

An imaginary view of Mercury's half-molten surface. It *could* look like that, but it could also be entirely different.

SURFACE FEATURES. Owing to its small size and considerable distance, Mercury is a difficult object to map. The best chart is probably that of Katterfeld. The dark patches are undoubtedly permanent, and the most prominent of them, such as the Solitudo Hermae Trismegisti, may be seen with moderate instruments. Mercury may be mountainous, but as yet there is no definite proof one way or the other. Nor do we know anything about the composition of the surface layers, but it is worth noting that the reflecting power or albedo, 0·07, is exceptionally low.

VISUAL OBSERVATIONS. Like Venus, Mercury shows phases which can be seen with a small telescope (see **Solar System**). Otherwise the planet is of little interest to the amateur astronomer or to the average professional, since the examination of surface detail requires a very large instrument. Mercury is best observed in full daylight, when it is invisible to the naked eye and so cannot be found easily without the aid of a telescope equipped with setting circles.

POSSIBILITY OF LIFE. So far as we can tell, Mercury must be a dead world. It is almost without atmosphere, subject to extremes of temperature, and water is completely lacking, so that no form of life known to us could survive upon the surface. Even if it can be reached by space-explorers of the future, it seems unlikely that any permanent base will be established there. (P.M.)

MERCURY PROJECT. The first U.S. manned spaceflight programme. Four manned orbital missions were flown, and there were two manned sub-orbital flights. The two sub-orbital flights used Redstone boosters, and the orbital missions used a version of the *Atlas* ICBM. The single-seater capsule, which could charitably be described as cramped, was provided with an attitude control system but apart from the retro-rocket for de-orbiting purposes had no manoeuvre capability, and the astronauts were little more than passengers.

The programme provided a basis for **Space Medicine** on which to assess the likely capabilities of men in space, and served also to build up experience of the problems associated with manned launches and recoveries. The parachute descent (after retro-rocket initiated re-entry) and ocean recovery pioneered methods subsequently used in the U.S. Gemini and Apollo programmes. The final Mercury mission, Mercury–Atlas 9, had a 22-orbit flight which splashed down a mere 7,000 yards from the recovery aircraft-carrier.

MERIDIAN. Any **great circle** passing through the poles of a rotating sphere.

The meridian of a place on the surface of the Earth is the great circle on the celestial sphere passing through the North and South points of the observer's horizon, and through his zenith.

MERIDIAN PASSAGE. The meridian passage or *transit* of a celestial object occurs when the latter crosses the observer's **meridian.** As a body describes its daily circle about the celestial pole it will cross the meridian twice; one transit occurs as it rises to its greatest altitude and is called culmination; the other usually takes place after the object has sunk below the horizon and is therefore unobservable.

MESSIER NUMBER. The number assigned to nebulous objects in the catalogue compiled by the French astronomer Messier. Some of these objects are star clusters, some true nebulae; the catalogue lists 103 objects, and was intended to prevent future confusion between these and comets. The Messier numbers of some well-known objects are:

M 13 Globular Cluster in Hercules.
M 31 Andromeda Galaxy.
M 42 Orion Nebula.
M 57 Ring Nebula in Lyra.

METEOR. Originally any atmospheric phenomenon, such as lightning, but now referring solely to a shooting star. The word is used somewhat loosely, and may connote either the luminous streak of light which can be seen and photographed, the trail of ionized gases which are formed in the process, or the actual solid particle which causes them in its swift passage through the atmosphere. The small particles which enter the atmosphere are now also called *meteoroids*, and they may range in size from the very smallest objects detected by radar methods to the more spectacular bodies which are visible even in daylight. The majority of meteors are completely destroyed in the course of their flight through the atmosphere, but the few that manage to penetrate and to fall on the surface of the Earth are called **meteorites.**

Most meteors are vaporized in less than a second, and this makes visual observation difficult. The larger and brighter meteors may endure for a longer time and have path lengths of hundreds of miles. The brightest of these bodies emit so much light as to cast shadows, and may be visible in full sunlight; such *fireballs* often present a very complex appearance, throwing off sparks and leaving a luminous train which persists after the meteor itself has vanished. The trains are of value in the study of the upper atmosphere, durations of over an hour being not uncommon. Fireballs are sometimes accompanied by noise like thunder – presumably the same effect which accompanies the passage of an aeroplane through the 'sound barrier'. Fireballs which actually explode in flight are termed *bolides*.

The number of meteors to be seen on a moonless night may reach 6 or 8 in the course of an hour's watch; the hourly rate, however, varies with the time of night and the time of year; it is always greater after midnight than before, and in northern latitudes is greater in the autumn and winter months. At certain seasons showers of meteors may reach quite astonishing proportions, with an hourly rate of many thousands. Such great storms of meteors have occurred in 1833, 1866, 1872, 1885, 1933 and 1966. The meteors which are seen by one observer during a shower may be in any part of the sky, but their lines of flight, when produced backwards, intersect in a small area known as the *radiant* of the shower. Radiants are named from the constellation in which they appear; thus the Leonid meteors of November appear to radiate from Leo, the August Perseids from Perseus, etc. In a few cases the name is taken from an associated **comet;** thus the Andromedids are often called Bielids (after **Biela's Comet**) and the October Draconids are the Giacobinids.

The astronomical significance of meteors was not appreciated until the great Leonid shower of 1833, November 12–13. On that night America witnessed a storm of meteors lasting nine hours, during which the shooting stars fell as thickly as snowflakes, reaching at one period an estimated hourly rate of 35,000. With so many meteors visible at one time, the fact that they all appeared to radiate from a point in Leo was obvious. The effect was shown to be one of perspective, the meteors actually travelling in parallel paths which appear to vanish at a point, as in the more familiar case of railway lines. During the next few years it became clear that there were

THE LEONID RADIANT. The thick white lines represent the observed trails of meteors drawn on a star map. The thin white lines produce these trails backwards; all but a few intersect with the others in a small region, the *radiant* of the Leonid meteor shower. The remaining trails are those of sporadic meteors. Very compact meteor streams have sharply defined radiants.

well-defined showers on the same date each year, and this led to the theory that the meteoric dust was distributed in orbits intersecting the Earth's. Each year a shower is seen as the Earth passes the point of intersection. In many cases the dust is not uniformly distributed in the orbit.

Research showed that the Leonids had been recorded as far back as A.D. 902, and tended to recur at intervals of 33 years. Another shower was therefore predicted for 1866. In that year a fine shower of Leonids was seen mainly in Europe, although it was not so spectacular as that of 1833. There was a repetition in 1966, but not in 1899.

The close connection between comets and meteors was first revealed by the similarity of the orbits of the Perseid meteors and the comet of 1862. This was quickly followed by the identification of the Leonids with Tempel's Comet of 1866, of the Lyrids with the comet of 1861, and of the Andromedids with Biela's Comet.

PHOTOGRAPHIC METHODS. The chance passage of a bright meteor across the field of view of an astronomical camera has occasionally resulted in a record of interest, and a valuable collection of plates is preserved at Harvard. Systematic attempts to photograph meteors began at Yale in 1891, a number of cameras being attached to equatorial mountings at each of two stations, the exposure being interrupted by a rotating sector placed in front of the lens. The base line was far too short (about two miles) to give accurate results, but the method is one on which all subsequent work has been based. In 1932 further experiments began at Harvard and by 1936 Whipple was using cameras there and at Oak Ridge (Mass.), the base line in this case being 24 miles. The more rapid emulsions which were then available gave better results, but the number of successful duplicate plates was still disappointingly small. Only the very brightest meteors gave any impression at all, and on the average only one meteor in 100 hours' exposure time could be expected. In 1946 plans were made for a more efficient camera, and these led to the Super-Schmidt cameras which are now in use. (See **Schmidt Camera.**)

Exposures of ten minutes are used, and the cameras are capable of photographing meteors almost to the limit of naked-eye brightness. A rotating shutter, which causes breaks in the trail of the meteor at intervals of 1/60 of a second, is placed just in front of the photographic film. It is estimated that the length of path can be determined to within a few feet and the velocity to one part in a thousand, while the actual retardation of the meteor in the atmosphere can be measured on some of the longer trails at several points in the path. The results obtained cannot compete in numbers with those from radar methods (see **Radio Astronomy**), but they afford the most accurate means yet devised for height and orbit determination.

HEIGHTS. In general, meteors appear and disappear at heights which lie in the range 70 to 120 km. above the Earth's surface, but the individual values vary. The faster meteors begin and end at the greater heights.

NUMBERS AND MASSES. The general hourly rate of 6 or 8 is due partly to a number of minor showers, but in the main to sporadic meteors. Although this rate appears small, it is merely the number seen by an observer with a limited range of vision over a very small fraction of the Earth's surface. The total number of meteors that enter the atmosphere in the course of one day is estimated at about 100 millions. Much larger values are obtained when the fainter telescopic or radio-echo meteors are included.

In spite of such large numbers, the actual mass of meteoric material entering the Earth's atmosphere each day is not considered to be greater than a few tons. The individual meteor is *very small*, and the ordinary shooting star may be regarded as a grain of dust often weighing as little as a thousandth of a gram. The brighter meteors and fireballs may reach 100 to 500 grams, but there is no upper limit, since masses of many tons are known to have fallen as meteorites. There is, however, a lower limit, since the very smallest particles, like the very largest, cannot be vaporized and will fall to the Earth as **micrometeorites.** The average density of the meteoric dust in the solar system is probably of the same order as the density of matter in interstellar space.

Calculations indicate that meteoric matter is extremely light, porous and friable, and is so weak that the smallest force is enough to shatter it. The smaller bodies are thus destroyed at once, or scattered into hundreds of minute particles which are instantly rendered luminous. The result is of great interest in connection with the formation of interplanetary dust.

MAGNITUDES. In general, the brighter meteors begin at a greater height and end at a lower height than the faint ones. The following values are averages for groups of Perseids:

Effect of Height on Magnitude

Mag.	−3·4	−1·1	+1·8	+3·4
Begin	116	115	112	109 km.
End	87	95	99	98 km.

SPECTRA. There are two distinct classes of meteor spectra, which differ mainly in the strength of the calcium lines, and since the spectrum consists of lines (superposed on a continuous background) it must originate from material in the vapour state. The lines of sodium and magnesium are often seen, and when magnesium is present the light of the meteor is predominantly green. Other elements detected include nitrogen (almost certainly atmospheric), hydrogen, oxygen, aluminium, silicon, chromium, manganese, iron and nickel – all found in meteorites.

TRUE AND APPARENT RADIANTS. The marked variation in the number of meteors seen at different times depends on the fact that the observed radiant is not the actual

Part of a meteor trail photographed through a rotating shutter which cuts off the light at regular intervals. The number of breaks along the whole trail can be used to calculate the time taken by the meteor to traverse it; changes in the meteor's speed show up as changes in the length of successive 'dashes'; the heating effects due to friction with the atmosphere lead to fluctuations of brightness. A pair of such photographs taken from two separate stations yields information not only about the meteor itself and its path, but about the temperature and pressure of the atmospheric layers in which it has burnt itself up. (*Harvard University Observatory*)

direction in space from which the meteors are travelling, but is only the *apparent* direction resulting from the combination of the motions of the meteors and of the Earth. The effect is exactly that of rain falling on the window of a moving train. Even if the rain is falling vertically, its trace on the window is slanting, the direction from which it appears to come being displaced in the direction of motion of the train, whose speed determines the amount of change. Similarly, the apparent meteor radiant is displaced towards the *apex of the Earth's way*, the point on the celestial sphere towards which the Earth is for the moment travelling. The apex comes into view on the eastern horizon of an observer at local midnight, and is due South at about 6 a.m., and this accounts for the fact that more meteors are to be seen during these hours. Moreover, since the apex lies on the **ecliptic,** there will be more meteors visible on autumn mornings, when the ecliptic rises to its greatest altitude above the eastern horizon (in northern latitudes).

VELOCITIES. The velocities of meteors range from about 12 to 72 km./sec., but until recent years such figures depended solely on theoretical values. In the photographic method, the breaks in the trail caused by the rotating shutter represent time intervals of 1/60 of a second; if the height of the meteor is known (from duplicate photographs) the velocity may be determined with great accuracy.

ORBITS. The position of the true radiant in the sky gives the actual direction in space from which the meteor is coming; this direction is a tangent to the **orbit** at the time of the observation. If the velocity of the meteor is also known it is possible to determine all six elements of the orbit. The calculation of the orbit of a meteor stream is somewhat complicated, but much of the work can now be done on electronic computing machines. The results that have been obtained for the major showers agree well with the orbits of comets known to be associated with them, but work on minor showers has provided a number of surprises.

SHOWERS. Details of the major *meteor showers* are given in the table, which includes the names of the comets which are associated with these meteor streams. In the past, and in all cases in which a determination of velocity is not practicable, these associations have been based on four important criteria: (1) the orbits must, of course, be similar. (2) The meteor shower may give periodic displays, as in the case of the Leonids and Giacobinids. (3) If the meteor stream is of sufficient width, the display may last for several days, but in such cases the radiant shows a definite motion; thus the Perseid radiant moves from 27°, +53° to 54°, +58° in the period July 27 to August 17. (4) Planetary perturbations advance or retard the date of the shower.

None of the major showers gives complete agreement on all four points, but the balance of evidence is in favour of these associations. In other cases, some discrimination must be used. For example, the δ-Aquarids and Orionids may be associated with Halley's Comet, since there is some resemblance between the orbits, but Halley's Comet does not come closer to the Earth than 0·15 astronomical units. This does not preclude the comet from having a common origin with the two showers in the remote past, since perturbations over a long period may have caused serious changes in the orbit. Such changes must once have taken place in the orbit of Encke's Comet, and the ejection of matter from the head of the comet, particularly at two epochs 1,500 and 4,700 years ago, would account satifactorily for the occurrence of the Taurids and ζ-Perseids of today.

METEOR SHOWERS

Name	*Duration*	*Maximum*	*Radiant*	*Hourly Rate*	*Velocity km./sec.*	*Associated comet*
NIGHT-TIME SHOWERS:						
Quadrantids . . .	Jan. 3	Jan. 3	231° +50°	40	41	
Lyrids	Apr. 20–22	Apr. 21	270° +33°	8	48	1861 I
δ-Aquarids	May 1–11	May 5	337° − 1°	12	66	Halley ?
June Draconids . .	June 28	June 28	208° +54°	12		Pons-Winnecke
η-Aquarids	July 24–Aug. 6	July 30	340° −15°	20	41	
Perseids.	July 27–Aug. 17	Aug. 12	46° +58°	50	61	1862 III
October Draconids .	Oct. 9	Oct. 9	262° +54°		23	Giacobini-Zinner
Orionids	Oct. 15–25	Oct. 20	94° +16°	16	66	Halley ?
Taurids	Oct. 26–Nov. 16	Oct. 31	52° +21°	6	31	Encke
Andromedids . . .	Nov. 14	Nov. 14	23° +44°	?	16	Biela
Leonids	Nov. 15–20	Nov. 16	152° +22°	6	72	1866 I
Geminids	Dec. 9–13	Dec. 12	113° +32°	60	35	
Ursids	Dec. 21–22	Dec. 22	206° +80°	12	35	1939 X
DAYTIME SHOWERS:						
o-Cetids.	May 13–23	May 15	30° − 3°	15	37	
ζ-Perseids	June 1–16	June 8	59° +22°	40	29	Encke
Arietids	May 30–June 18	June 8	44° +23°	60	38	
β-Taurids	June 25–July 7	June 29	85° +17°	24	31	Encke

SPORADIC METEORS. Much of the early work on meteors was concerned with mere numbers, and particularly with the variation of hourly rate. Statistical analysis pointed to a high proportion of hyperbolic orbits, but today all approaches lead to the opposite conclusion – that there is no evidence whatever for a preponderance of hyperbolic velocities among meteors. In 300 photographs, reduced in duplicate, Whipple has found no case of hyperbolic motion, all the measured velocities being below the parabolic limit. The radio-echo results are similar, although less accurate; here a small percentage of meteors appear to have velocities slightly in excess of parabolic. It is reasonable to suppose that this represents merely the spread of the errors of measurement.

STRUCTURE OF METEOR STREAMS. The width of a stream of meteoric dust is always very small. For instance, the Perseid shower, which may last for about 18 days, has a width of only 0·075 astronomical units (6,800,000 miles).

Within this stream the particles must be widely separated. Even in a great shower such as the 1966 return of the Leonids, when the hourly rate was about 4,000 times the normal, the particles must have been at distances of the order of 15 km. Thus the dust particles will each have a separate existence, and the stream will consist of a tangled array of individual paths. The old idea of a uniform structure – the 'bicycle tube' theory – is wrong; planetary perturbations alone would disrupt such a structure, and other forces are present which also have a disturbing influence on small particles.

Perturbing forces, however, must last for some time to be effective, and the persistence of such showers as the Perseids, Leonids and Lyrids over many centuries must be due to the fact that they have escaped prolonged disturbance. The Lyrids have an orbit which is inclined at 80° to the ecliptic, while the Perseids and Leonids travel in retrograde orbits, so that they move in the opposite direction to Jupiter, whose influence is therefore short-lived.

Planetary perturbations affect each particle individually, and so the orbits in which the particles travel can never be identical with that of the parent comet, and there are, in fact, difficulties in accepting the view that all meteors are the débris of comets. The Perseids and Leonids have been visible for centuries, yet the comets with which they are associated were discovered only in the 1860s. Although comets can come within 0·05 astronomical units of the Earth, and 60 within 0·1 units, the only accordances with meteor showers are the eight given in the table. If these can give showers, it is surprising that the others do not.

Meteoric dust travelling in a comet's orbit does not prove the dust to be the débris of the comet. In the most recently formed shower, that of the Giacobinids, the material of the shower of 1926 was *in front* of the comet. But families of comets are known which have similar orbits. This would seem to be a parallel case, and it appears unwise to depart from the simple view that *meteors and comets have a common origin.*

METEORS AND THE UPPER ATMOSPHERE. The presence of high wind-speeds has been shown repeatedly in the past by the behaviour of visual meteor trails, which frequently become greatly distorted in the course of a few minutes. Radio-echo methods show the presence, in the 80–100 km. range, of a prevailing wind, of about 30 m.p.h., blowing to the East in summer and winter, and to the West in spring and autumn; in addition there is a wind of 40 m.p.h. whose direction rotates clockwise twice during the day. Similar results have been found in the southern hemisphere, but the rotation is anti-clockwise. These directions appear to be reversed at lower levels, and ionospheric measurements at higher levels show a similar semi-diurnal change, but this occurs about 3 hours earlier. The winds are, in fact, highly stratified and variable.

The onset and fluctuation in the incandescence of meteors have been used to assess atmospheric temperatures at different heights, and the results are in agreement with corresponding data from rockets and artificial satellites.

METEORS AND THE LUNAR SURFACE. The enormous size of some of the lunar craters makes it obvious that each must have originated in some form of violent explosion. The suggestion of a meteoric origin is not new, but has received considerable support in recent years, especially among those who are not experienced in observing the Moon in a large telescope. It is assumed that in the remote past the Earth and Moon were subject to intense bombardment from large meteors. Craters formed in this way on the Moon would be larger than those on the Earth, which has a protective atmosphere; and the lunar craters, moreover, would not be exposed to atmospheric erosion. The impact of a large body on the Moon's surface would be so great that the energy liberated would be comparable with that of a hydrogen bomb. On this theory, the craters are just like bomb-craters, and the flat floors of the craters are merely settled dust, caused by the crumbling of the rocks as they are alternately baked in the heat of the Sun and frozen at night.

The older volcanic theory, in spite of its weaknesses, still receives support from observers. They point out that the craters are not distributed at random, as they would be on the meteoric theory, but occur in pairs, lines and overlapping chains, and even some of the great walled plains occur in lines. This might happen along lines of weakness of the crust. The level floors are easily explained as solidified lava, and a crater such as *Wargentin*, filled to the brim, which is many hundreds of feet above the surrounding plain, appears to have been formed by the flow of lava underneath. Summit craterlets are found on the tops of mountains, of isolated hills and the central peaks of craters. These small formations occur in considerable numbers, and the probability that meteors could make precisely central impacts on such a number of peaks is almost zero. The subject continues to form a fruitful source of speculation, but it will very soon be possible to decide between the rival theories. (See also **Moon.**)

After a century of very slow progress, it seems likely that meteoric astronomy is at last assuming its true importance, as a probe of the upper atmosphere and of interplanetary space. (J.G.P.)

An aerial view of the meteorite crater in Arizona. Its size may be judged from the roads leading up to it. (*Fairchild Aerial Surveys*)

THE GREAT METEOR SHOWER OF 1833.

METEORITES are solid bodies which have entered the Earth's atmosphere from outer space and have struck the Earth's surface, either as a single mass or as a shower of smaller pieces – the latter very often being the result of the body fragmenting during its flight through the atmosphere. Meteorites are too large to be vaporized completely in the air like meteoroids. They fall into three

Upper left: the 'Ahnighito' Meteorite in the Hayden Planetarium.

Upper right: an etched section of the Knowles iron meteorite, showing the characteristic Widmanstätten figures.

Lower left: a large iron meteorite. Very few meteorites weighing more than a ton are stony.

broad physical classes: (*a*) the *irons*, which are always associated with a certain amount of nickel, and which on being etched and polished show the curious Widmanstätten figures; (*b*) the *stones*, which are mainly silicates; and (*c*) the *stony-irons*, which as the name implies are a mixture of the previous two. The greater proportion of meteorites found on the Earth's surface are irons while stones account for only one-quarter. However, observed falls comprise 80 % stones, but only 6·5 % irons. The reason for this apparent difference is that the stony meteorites resemble certain terrestrial rocks, and because of their greater tendency to weather, they escape detection much more easily if they are not recovered soon after the fall. The blocks of nickel-iron, on the other hand, resist weathering and are generally so conspicuous that they can be easily recognized.

The largest meteorite ever seen to fall weighed 820 lbs. (Arkansas, U.S.A., 1930); the largest one known weighs 60 tons, and it remains, buried with its top on a level with the surface, at Hoba West in South Africa. This, like most large meteorites, is an iron. Larger falls than this have certainly occurred, but it would appear that in such cases the body disintegrated. In the Arizona Desert, about 35 miles east of Flagstaff, there stands a single hill in what is otherwise a flat plain. At the centre of this hill is a crater three-quarters of a mile across and 600 feet deep, its rim being 140 feet above the surrounding plain. This is *Meteor Crater*, and is believed to have been caused by the fall of a meteorite some thousands of years ago.

Actual falls of this order of size have been reported during the present century, the two most important being in Siberia. On June 30, 1908, a great collection of stones estimated to weigh some thousands of tons landed in the desolate region of the river Yenesei. Expeditions found great areas of the forest laid waste for twenty miles around the scene, the trees lying with their tops all pointing outwards from the centre, and completely bare, as if they had been burned. At the time of the fall a column of fire appeared over the forest, while the noise of the impact was heard as a crash with several thunderclaps, more than 600 miles away.

A more recent event was the fall of irons in a dense forest in eastern U.S.S.R. on February 12, 1947. On this occasion the actual fireball was seen in flight. Using all the resources of modern expeditions (aeroplanes, mine-detectors, etc.) various parties were able to find many craters, and succeeded in collecting 30 tons of nickel-iron meteorites.

The largest single crater which has been associated with such a fall is Chubb Crater in Ungava; this is about 3 miles across and almost perfectly round, with walls at an angle of 45°.

It is likely that a number of large meteorites hit the Earth's surface every day, but go unseen since many fall in the oceans or on uninhabited parts. Many of the smaller meteorites produce impact effects which can hardly be noticed and so are missed. There are numerous reports about the impressive and frightening sounds which often accompany even the smallest fall and resemble the booming of cannon, thunderclaps and the now familiar noises which accompany the breaking of the sound barrier by jet aircraft. Horses may bolt and dogs cringe or hide away. Hissing noises like the wind singing through telegraph wires and which are probably electromagnetically induced are also heard – but very often a meteorite or its fragments are never found. The probability of a person being struck by a meteorite is quite small, and although there are reports of animals killed by meteorites and buildings damaged, there are no confirmed cases of a human being killed by one. It has been calculated, for example, that in the United States a human will be struck by a meteorite once every 9,300 years.

Meteorites are generally given the name of the geographical locality in which they were recovered such as the *Barwell Meteorite* 1965. This latter fall, which occurred in the village of Barwell in Leicestershire, England, on Christmas Eve 1965, was well observed and many detailed accounts have subsequently appeared. Although fragments totalling over 50 kilos were recovered over an area of about one square mile, it is thought that the original fragment weighed perhaps twice that amount, but a great deal of it may have been dissipated into fine dust when it first broke up in the upper atmosphere.

It is not always easy for the inexperienced to recognize a meteorite, and only by using several criteria is one able to distinguish them from terrestrial rocks and artefacts. In many private collections there are objects classified as meteorites which originated in blast furnaces or as pyrite nodules in the Earth's crust! Generally, all meteorites have thin, black fusion crusts, either dull or lustrous, covering the object completely (or only partly if fragmentation has occurred), which is the result of melting as the meteorite passes through the atmosphere. Indications of shallow pits or regmaglypts – also caused during the flight through the atmosphere – are good evidence and can indicate the size of the parent body.

Characteristic of all iron meteorites is their nickel content which ranges from 5 % to 20 %. Terrestrial native iron also may contain nickel, but in smaller or larger amounts, less than 3 % or greater than 35 %. If there are doubts that a stony or metallic object may be indeed a meteorite the fact should be brought to the notice of authority.

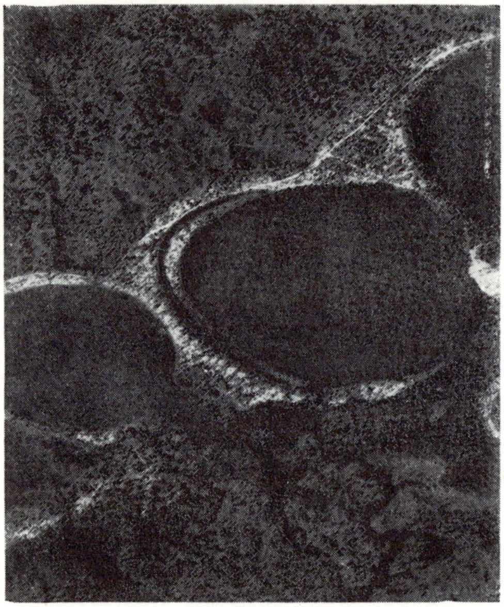

METEORITE CRATERS that have become filled by erosion. These giant scars, some measuring 10,000 feet across, were caused by a shower of meteors, and pit an area 80 miles wide from Virginia to Georgia, U.S.A.

(*Fairchild Aerial Surveys*)

THE BARWELL METEORITE fell in the afternoon of December 24, 1965, in the village of Barwell, Leicestershire, England. The largest of the fragments shown here is about one foot long.

THE ORIGIN OF METEORITES would appear to be different from that of meteoroids. Whereas the latter are known to be genetically associated in most cases with **comets,** meteorites show a much closer relationship to the **asteroids.** First conclusive evidence for this idea was provided by the trajectory of the Pribram (Czechoslovakia) stony meteorite which fell in 1959. It was photographed by two meteor stations 40 kilometres apart, and from the photographic record it has been possible to calculate with great accuracy the orbit of the meteorite prior to its encounter with the Earth. This brilliant fireball of magnitude -19 passed through zenith, and after its descent seventeen fragments were recovered. This is the only case on world record where we know reliably both the orbit of the meteorite in the solar system and the chemical nature of the body itself. Its age was determined as 4×10^9 years, and its orbit indicated that at perihelion it came close to the orbit of Venus and at aphelion near the orbit of Jupiter. Whether the meteorites have, in fact, originated in the disruption of one or more of the asteroids (or Olbers' hypothetical planet) or whether they represent original condensed primordial material is still open to debate.

Meteorites can be dated using various analytical techniques. A technique using the potassium–argon method, which utilizes the decay of radioactive isotopes, was applied to the stony Barwell Meteorite referred to previously. Age estimates round $4{\cdot}2 \times 10^9$ years were indicated since the crystalline structure was formed, and this compares fairly closely with other meteorites of the same kind and also approximately to the Earth's age which is believed to be about 5×10^9 years.

CARBONACEOUS METEORITES are among the most interesting and controversial meteorites that have been studied by modern analytical techniques. They are comparatively rare and only about twenty examples are known, all of which were collected within a short time of their falling to Earth. It is generally assumed that many more specimens have fallen, but have gone unrecognized owing to their very friable texture and the presence of water-soluble compounds which would cause them to disintegrate within a few days of landing if left exposed.

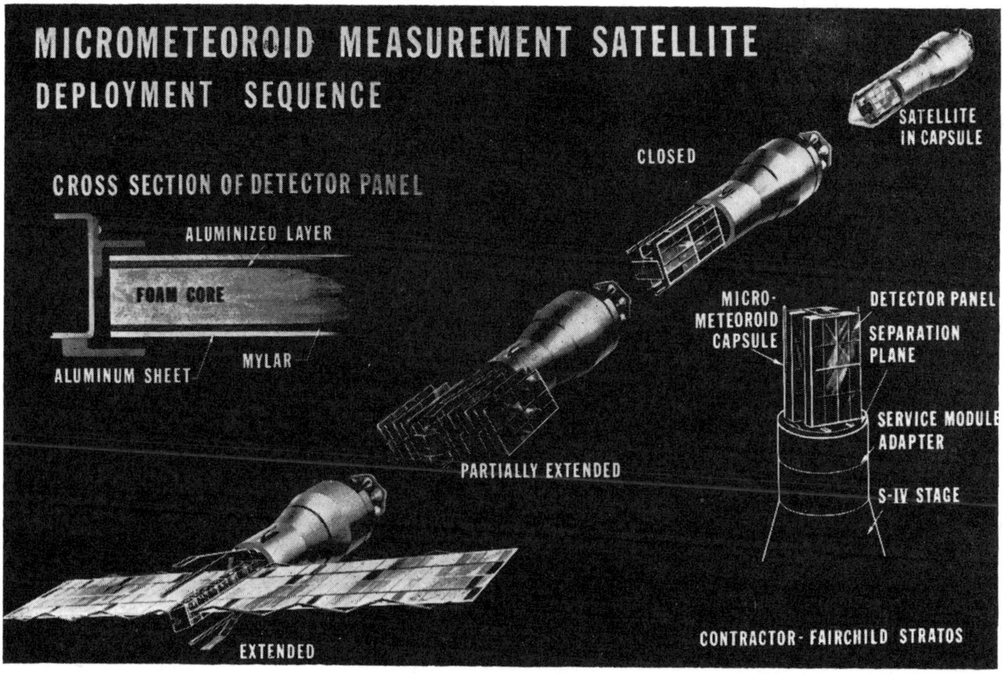

PEGASUS SATELLITE DEPLOYMENT. Three such instruments were launched by *Saturn* rockets as secondary loads to *Apollo* command modules under test. When a micrometeorite strikes one of the electrically charged panels on the 96-foot wide pair of wings it is instantly vaporized, allowing the panel to discharge; it is recharged by solar batteries within 0·003 sec. At intervals, on command from ground stations, *Pegasus* telemetered times of impact together with its attitude, location of struck panel, and 160 other items of information concerning itself and its measurements. In less than one year over a thousand impacts were recorded.

They differ from ordinary stony meteorites to which they belong (as three sub-groups), in that they contain hydrocarbons, water-soluble salts and a multitude of organic compounds. They are generally found in small pieces weighing only a few kilograms. Some specimens contain veins of magnesium sulphate, indicating that they might possibly have been deposited from a water solution – so providing some evidence pointing to liquid water in the parent meteorite body.

A number of theories have been proposed as to their origin. One idea is that they have originated from the lunar surface as a result of explosive impact events. It has also been proposed that comet nuclei may consist of such carbonaceous matter. Another idea is that they represent accretions of the first 'dusty' material that condensed in the solar system. Their very friable nature – or ginger-bread texture – is consistent with formation in a low gravitational field where little binding energy would be available – as might well be expected if primitive matter does, in fact, coalesce in space. (P.L.B.)

METHANE (CH_4) or 'marsh gas' occurs in the atmospheres of the giant planets and in comets. Its molecule contains one atom of carbon and four of hydrogen. Methane condenses to a liquid at $-160°$ C. at atmospheric pressure.

MICROMETEORITES. In recent years the study of these smallest cosmic particles has attracted considerable attention owing to the possible hazards they may incur if they are encountered by manned space vehicles. These small particles are only a few microns in diameter and can enter the atmosphere and fall to the Earth's surface undamaged, since generally whatever cosmic velocity was originally possessed is reduced to almost zero in the higher atmosphere above the ablation zone. They fall freely under the influence of gravity and reach only modest terminal velocities. However, if sufficiently small, they can be retained in the upper atmosphere and drift about for indefinite periods. Dust collected from the atmosphere and from the Arctic and Antarctic ice caps is found to contain microscopic black spherules which are magnetic and range in size from 5–30 microns, but some of these particles may have originated from terrestrial sources such as industrial furnaces rather than from outer space.

The most reliable identifications have been the result of collections made by rocket and space probes at high altitudes above the Earth's surface. These specimens are more often found to have loose or fluffy consistencies and are very fragile rather than homogeneous 'chips' of material. Perhaps the denser particles are all results of melting processes. Although some recovered micrometeorites show a crystalline structure, many do not. This could be due to cosmic radiation damage, or because the material has been formed in an environment where there was insufficient energy available to align atoms into a definite crystal lattice. In some literature 'micrometeorite' is the name now generally given to small meteoroids which have partially melted, but have not been significantly ablated during their passage through the Earth's atmosphere. *Micrometeoroids* is a term restricted to describing small, solid, unmelted particles which have not lost a significant fraction of their mass in passing through the Earth's atmosphere above. (P.L.B.)

MINOR PLANETS. See **Asteroids.**

MIRROR. In reflecting **telescopes,** the image is formed by a concave mirror of parabolic cross-section. These mirrors are silvered, or coated with aluminium, on the front surface, in contrast to ordinary looking-glasses. The light does not therefore have to enter the material of the mirror at all, and glass of the highest optical quality, essential for lenses, is unnecessary. Although glass is usual, opaque materials could be used, since they act only as supports for the real mirror which is the silver or aluminium film. A special alloy, speculum metal, was formerly employed without any coating, but it tarnished and had to be re-polished periodically – a very delicate task, since the surface has to be brought to the correct curve or *figure* anew each time. With a glass-based mirror, the metal film can be replaced as often as necessary without prejudice to the figure of the glass.

MISSILE. An unmanned rocket vehicle that can be sent to a definite target or location. (See **Rocketry** and **Guided Missile.**)

MIZAR (*ζ-Ursae Majoris*) is one of the seven bright stars in the Plough. Very close to it is **Alcor.** A small telescope reveals that Mizar has two components, one of which is itself a spectroscopic **binary star.**

MOON. The Earth and Moon are companions in space. They affect each other strongly, and each bears traces which help to elucidate the history of the other within the solar system. The surface of the Moon, unprotected by any atmosphere and unaffected by erosive processes, presents an uninterrupted and unmodified record of all the events that have happened in this part of the solar system since its formation 4 600 million years ago. Although the Earth was formed at about the same time as the Moon, its oldest known extant surface rocks are no more than 3 800 million years old. There is no remaining record on the Earth of the first 1 000 million years of its history. Therefore the Moon rocks and structures provide invaluable evidence which spans these 'dark aeons' in the history of the Earth–Moon system and is of unique interest to astronomers and Earth scientists.

In prehistoric times the waxing and waning of lunar phases provided the first astronomical basis of the reckoning of the time. Whenever we go sufficiently far back in the history of almost any primitive civilization, we find it invariably dependent on the lunar, rather than the solar, calendar. The month became a unit of time long before the concept of the year emerged from accumulating observations.

FULL MOON, North at the bottom to agree with map on pages 184–185. The new convention is to place North at the top.

The Moon aged 27 days, a day before New Moon. (*Lick Observatory*)

ORBIT. The Moon moves round the Earth in approximately an ellipse of mean eccentricity 0·055 which is slightly distorted by the attraction of the Sun and by the uneven attraction of the non-spherical Earth. The plane of the Moon's orbit is inclined to the ecliptic (the plane in which the Earth revolves around the Sun) by a little more than 5° 09′, so that **eclipses** do not occur at every revolution. The mean period of revolution of the Moon around the Earth is equal to 27 days, 7 hours, 43 minutes and 11·5 seconds. This is the so-called *sidereal month*, the time it takes for the Moon to return to the same place with respect to the Earth. However, because of the apparent motion of the Sun it takes 29 days, 12 hours, 44 minutes and 3·0 seconds for the Moon to attain the same phase. This period is termed the *synodic month.*

How far away is the Moon? Astronomers have determined this by triangulation and, more recently, from the measured time-lag of radar and laser echoes reflected from the Moon. The results show that the distance from the Earth to the Moon varies in the course of each month from ca. 356 000 to 407 000 km. The mean distance in the course of a month is 384 400 km, or 60·267 times the Earth's equatorial radius. In other words, it is 0·002 57 times the mean distance separating the Earth from the Sun, and less than one per cent of the distance separating us from our nearest other celestial neighbours, the planets Venus and Mars, at the time of their closest approach.

The mean velocity of the Moon's motion is 3 681 kilometres per hour, or about 1 023 metres per second, which corresponds to a mean angular velocity, as viewed from Earth, of about 33 minutes of arc per hour. This is just a little greater than the apparent diameter of the Moon itself.

PHASES. Since the Moon has no inherent light, it exhibits regular phases from new to full. During the crescent stage, the darkened half of the Moon is often faintly visible. The first correct explanation of this phenomenon was given by Leonardo da Vinci, who realized that it is due to light reflected from the Earth (see **Earthshine**).

The boundary between the daylit and night hemispheres, known as the *terminator*, naturally appears rough and broken, since an elevation is bound to catch the sunlight in preference to an adjacent valley.

DIMENSIONS AND MASS. The Moon has a diameter of 3476 km, roughly one-quarter that of the Earth. Compared with its primary, the Moon is far larger than any other known satellite, and in some ways it is better to regard the Earth–Moon system as a double planet (see **Satellites**). The density of the globe is 0·606 of that of the Earth, the mass being 0·0123 and the surface gravity 0·16. Like the Earth, the Moon is not quite a spheroid flattened at the poles, but has depressions which give it a pear-like shape in the sense of the diagram under **Earth.**

ROTATION. **Tidal friction** in past ages has resulted in the rotation of the Moon being 'captured', or made equal to the period of rotation round the Earth. Consequently, the same hemisphere is always presented, and part of the surface we can never see from the Earth. We can, however, see more than half the total area, since the varying velocity of the Moon in its orbit results in a slight tilting from side to side, known as libration in longitude, while there is also a libration in latitude due to the Moon's orbital inclination (see **Librations of Moon**). Only 0·411 of the total surface is permanently averted.

TELESCOPE APPEARANCE. Even in a small telescope, the Moon is a superb sight. In addition to the lofty mountains and the broad dark plains or *maria*, there are numerous walled circular formations (*craters*), peaks, valleys, clefts (*rilles*), domes and many finer features. Before the dawn of the Space Age, the best lunar maps were those compiled by amateur astronomers; nowadays, of course, *Apollo* and *Surveyor* missions have sent back photographs showing the whole of the lunar surface in close-range detail, so that Moon-mapping from terrestrial observatories has become obsolescent.

Formerly, lunar maps were drawn with south at the top and east to the left. The astronautical authorities have reversed this, putting north at the top and east to the right.

The relatively low gravitational acceleration prevailing over the lunar surface means that much less effort is required there to lift weights or throw stones. However, it also means that a man's own body-weight works less effectively when he wishes to compress anything, as when driving a shovel into the ground by stepping upon it.

ATMOSPHERE. The smallness of the mass of the Moon and the relatively low value of the surface **escape velocity** have several important consequences. One is the complete absence of any atmosphere or protective gaseous mantle above the Moon's surface. Under temperatures ranging between some 130°C at noon in the lunar tropics and −200°C in the later part of the lunar night, any hydrogen or helium atmosphere would dissipate from the Moon in a matter of days. Water vapour would escape more slowly, and at noon temperatures oxygen and nitrogen as well as carbon dioxide would completely escape in a relatively short time. In fact, the

MOUNT MERU CRATER, TANGANYIKA. An unusually well-rounded volcanic crater, but still very different from lunar craters. – An aerial view of a meteor crater is given in the article on Meteorites.

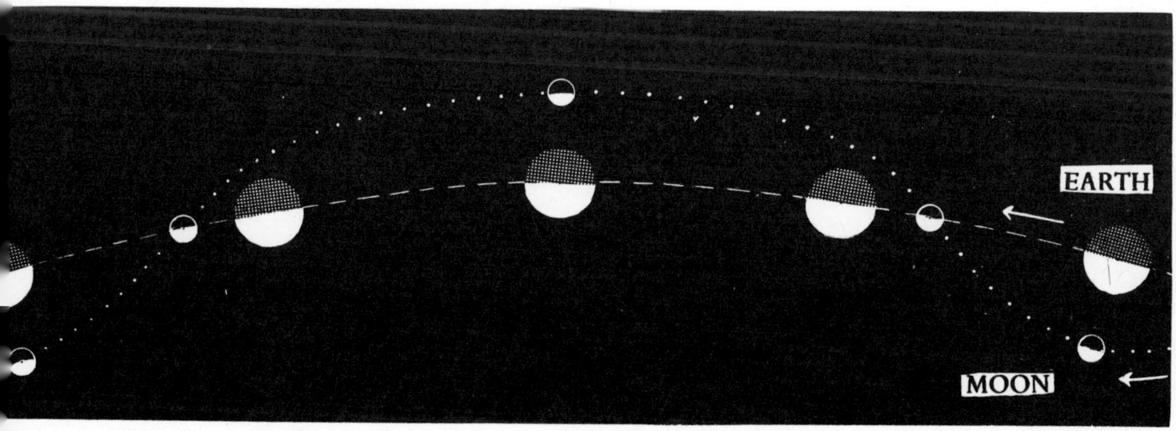

THE MOON'S ORBIT. While circling the Earth, the Moon also follows the Earth in its path round the Sun, weaving in and out of the Earth's orbit. The diagram greatly exaggerates the effect, and the Moon's orbit is in fact convex everywhere.

MOON

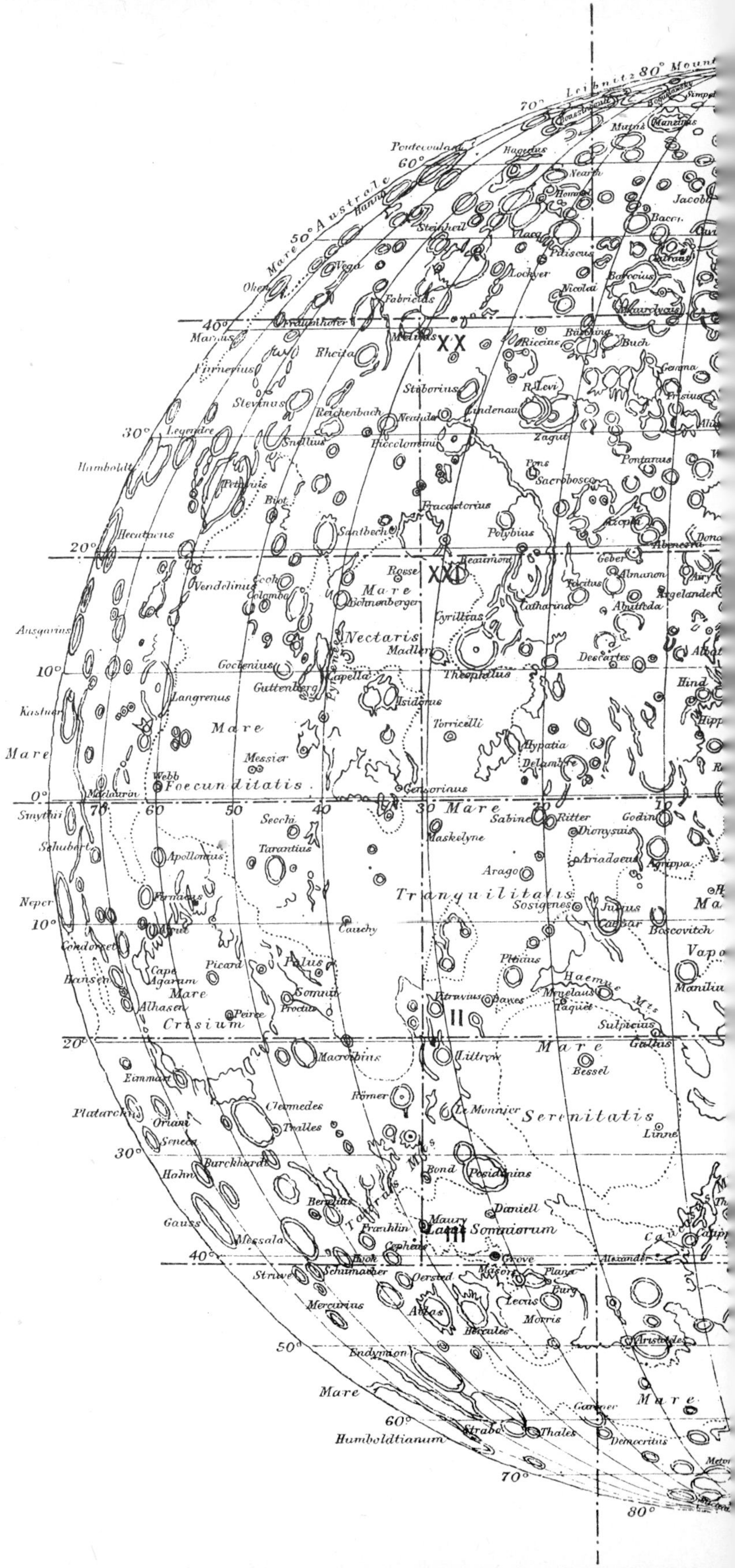

Leibnitz
80°
70°
60°
50°
40°
30°
20°
10°
0°
Mare Australe
Hagecius
Mutus
Manzinus
Nearch
Jacobi
Steinheil
Pitiscus
Vega
Lockyer
Nicolai
Barocius
Oken
Fabricius
Marinus
Metius
XX
Riccius
Büsching
Buch
Rheita
Furnerius
Stiborius
R. Levi
Stevinus
Reichenbach
Neander
Lindenau
Legendre
Snellius
Piccolomini
Zagut
Humboldt
Pons
Sacrobosco
Pontanus
Petavius
Biot
Fracastorius
Hecataeus
Santbech
Polybius
Beaumont
Rosse
XXI
Geber
Almanon
Airy
Vendelinus
Cook
Colombo
Mare
Nectaris
Bohnenberger
Cyrillus
Catharina
Abulfeda
Argelander
Ausgarius
Madler
Descartes
Gocleniu
Capella
Theophilus
Guttenberg
Pyrenees
Langrenus
Isidorus
Hind
Kastner
Torricelli
Mare
Hypatia
Delambre
Messier
Webb
Foecunditatis
Maclaurin
Censorinus
Smythii
Secchi
Sabine
Ritter
Godin
Dionysius
Schubert
Tarantius
Maskelyne
Apollonius
Ariadaeus
Agrippa
Arago
Tranquilitatis
Neper
Sosigenes
Julius
Caesar
Boscovitch
Condorcet
Cauchy
Plinius
Picard
Palus
Somnii
Proclus
Bansen
Cape Agarum
Mare
Crisium
Alhasen
Peirce
Vitruvius
Dawes
Menelaus
Taquet
Sulpicius
Gallus
II
Macrobius
Littrow
Mare
Bessel
Eimmart
Römer
Le Monnier
Serenitatis
Linne
Cleomedes
Plutarch
Oriani
Seneca
Tralles
Burckhardt
Hahn
Bond
Posidonius
Taurus Mts
Berzelius
Daniell
Gauss
Franklin
Maury
Lacus Somniorum
III
Messala
Cepheus
Grove
Alexander
Struve
Schumacher
Oersted
Plana
Bürg
Mercurius
Atlas
Lacus
Morris
Hercules
Aristoteles
Endymion
Mare
Humboldtianum
Strabo
Thales
Gartner
Democritus
Mare

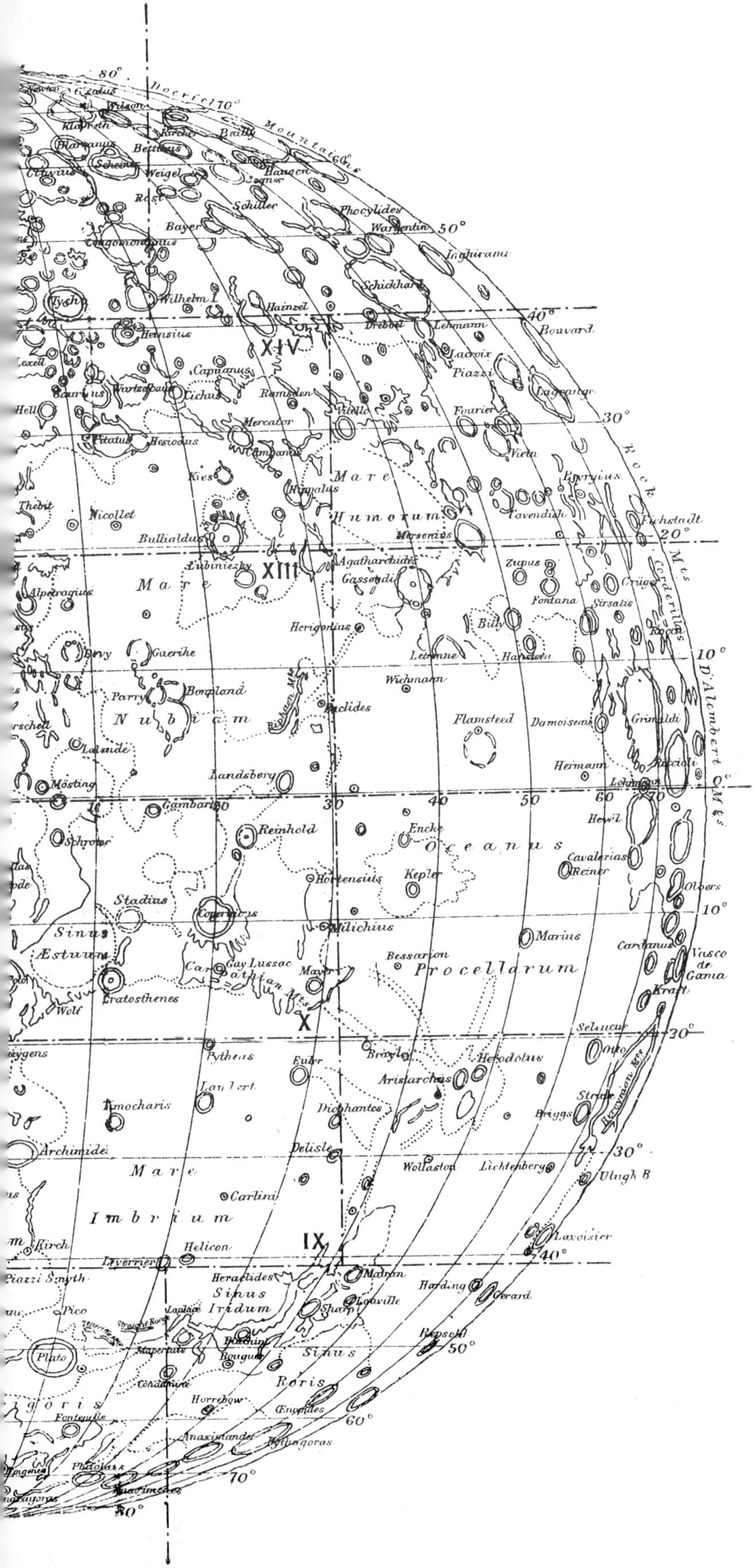
80°
70°
Wilson
Klaproth
Blancanus
Bettinus
Scheiner
Kircher
Bailly
Weigel
Hansen
Rost
Schiller
Phocylides
Bayer
Wargentin
50°
Longomontanus
Inghirami
Schickhard
Tycho
Wilhelm I.
Hainzel
Heinsius
Drebbel
Lehmann
40°
Bouvard
XIV
Lexell
Lacroix
Capuanus
Piazzi
Saussurus
Wurtzelbauer
Cichus
Ramsden
Hell
Mercator
Vitello
Fourier
30°
Pitatus
Hesiodus
Campanus
Vieta
Kies
Mare
Rook
Thebit
Nicollet
Hippalus
Humorum
Cavendish
Eichstadt
Mersenius
20°
Bullialdus
Lubiniezky
XIII
Agatharchides
Gassendi
Zupus
Mare
Alpetragius
Fontana
Sirsalis
Cordillera
Herigonius
Billy
Rocca
Gaerike
Hansteen
10°
Letronne
Wichmann
Parry
Bonpland
Nubium
Euclides
Flamsteed
Damoiseau
Grimaldi
Lalande
Hermann
Riccioli
Mösting
Landsberg
Lohrmann
0°
Gambart
30
40
50
60
70
Hevel
Reinhold
Encke
Schröter
Oceanus
Cavalerius
Reiner
Hortensius
Kepler
Olbers
Stadius
Copernicus
10°
Sinus
Milichius
Marius
Æstuum
Cardanus
Carpathian Mts
Gay Lussac
Bessarion
Procellarum
Vasco de Gama
Mayer
Eratosthenes
Kraft
Wolf
X
Seleucus
30°
Huygens
Pytheas
Euler
Brayley
Herodotus
Otto
Lambert
Aristarchus
Timocharis
Diophantes
Briggs
Archimides
Delisle
30°
Mare
Wollaston
Lichtenberg
Ulugh B
Carlini
Imbrium
Lavoisier
Kirch
Helicon
IX
40°
Leverrier
Piazzi Smyth
Heraclides
Malran
Harding
Gerard
Sinus Iridum
Louville
Pico
Laplace
Sharp
Straight Range
Repsold
50°
Plato
Maupertuis
Bouguer
Sinus
Condamine
Roris
Horrebow
Fontenelle
Œnopides
60°
Anaximander
Pythagoras
Philolaus
70°
Anaximenes
80°

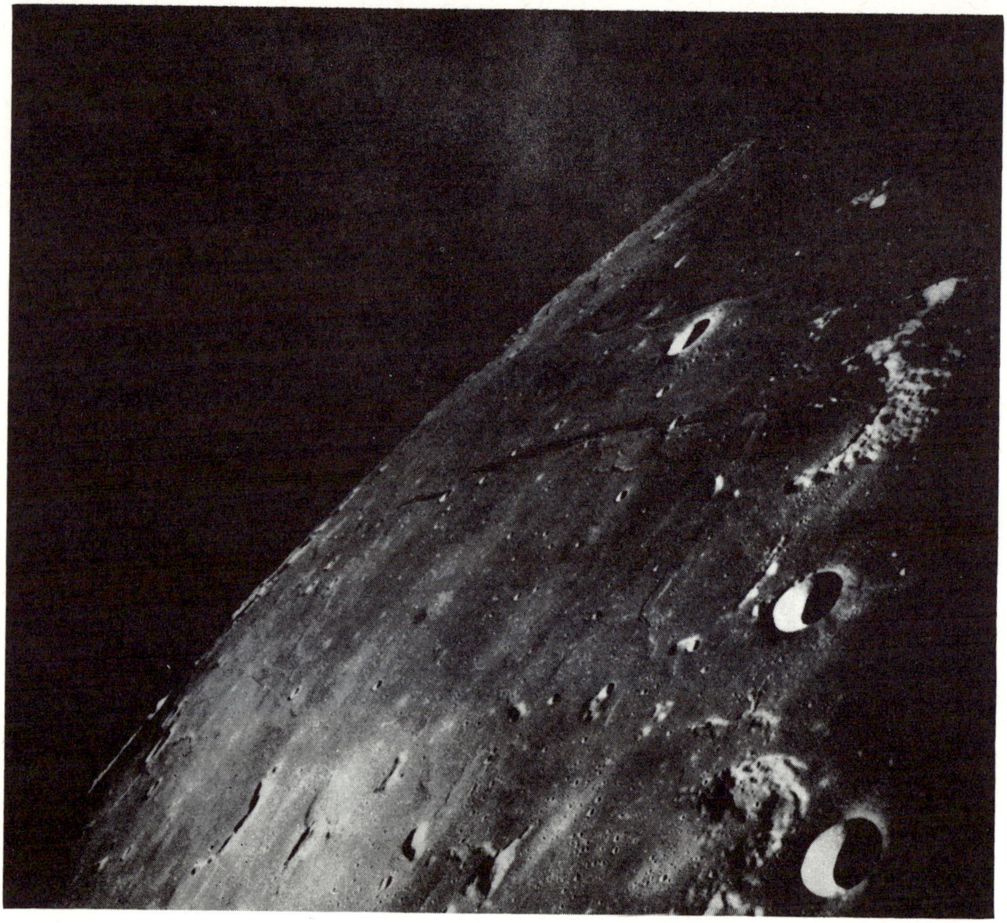

THE SEA OF TRANQUILITY is partly shown in this oblique photograph from *Apollo 8*. The view is towards the north-west, and includes the Cauchy Scarp in the foreground and rilles in the background, with the Cauchy Crater between them.

rate of dissipation into space of all but the heaviest gases, which are cosmically very scarce, is so high on the Moon that we do not expect any appreciable permanent atmosphere around it. All observational evidence available to us supports this conclusion.

If, therefore, the Moon possesses no detectable atmosphere because of its low gravitational field and relatively high daytime temperature, it cannot maintain any liquid on its surface. The surface of the Moon is thus bone-dry, and has been so from time immemorial. No feature visible on the surface of the Moon could have been formed or modified by the effects of running water or by the effect of repeated freezing and melting of water entrapped in surface rocks. One of the most important terrestrial agents of erosion therefore is, and always has been, absent from the Moon, and cannot be invoked to explain any of the Moon's surface features.

THE INTERIOR. The Moon's surface is controlled by internal thermal and stress processes, but is also an 'impact counter' of external events which the Moon has experienced since the days of its formation.

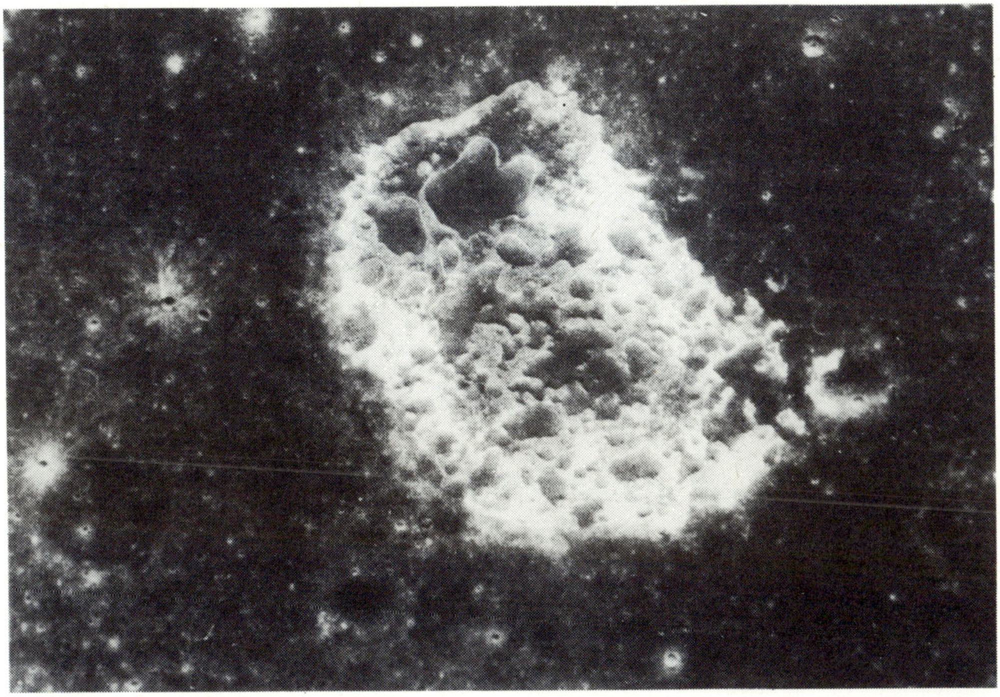

A LUNAR VOLCANO is believed to have created this caldera-like basin between the Sea of Serenity and the Sea of Rains. This unique feature, 3 km wide, is strewn with white material and dark domes whose surfaces are smooth, and presumably, therefore, the domes are young formations. The photograph was taken from the *Apollo 15* command module in lunar orbit.

If we wish to examine the essential properties of the Moon's interior, the primary clues are already in our hands, namely its observed mass and its size, combining as they do in a mean density of $3{\cdot}34 \times 10^3$ kg m^{-3}. What is the pressure prevailing inside this mass? Even in so small a celestial body as the Moon, existing under the influence of its own self-attraction, the internal pressure is essentially hydrostatic throughout most of the interior. In other words, the strength of the Moon material is unable to withstand its own weight anywhere except, possibly, in the very outer part of its crust. The internal pressure in *any* globe of the lunar mass and size cannot exceed some 50 000 atmospheres even at its centre—a pressure exceeded at a mere 150 km below the Earth's surface.

The internal temperature of the Moon is much lower than that of our Earth. The mean electrical conductivity of the lunar globe (as inferred from its interaction—or, rather, lack of it—with the **solar wind** of charged particles emanating from the Sun) is that of common silicate rocks at a temperature of less than 1 000°K—well below the melting point of silicate rocks; therefore the Moon should behave as a solid body throughout most of its mass.

That it actually does so is now well attested. An analysis of the lunar gravitational field (deduced from the motion of lunar artificial satellites since 1968) disclosed the presence, at shallow depths, of localized mass concentrations (*mascons*). The locations of such mascons are usually found to coincide with the maria (in particular, circular maria), of which Mare Imbrium is the largest; and it is very probable that their formation is due to cosmic impacts which produced the respective mare.

HALF-EARTH FROM MOON, photographed by the *Apollo 8* astronauts on their first orbit around the Moon, Christmas Eve 1968. The lunar horizon is about 780 km. from the spacecraft. The sunset terminator bisects Africa on the Earth.

The excess loads exerted by the mascons would not be tolerated by the Earth's crust for any (astronomical) length of time. That the Moon has tolerated them for more than 3 000 million years discloses that its crust is at least 1 000 times more rigid than the terrestrial one —consistent with the fact that the visible face of the Moon exhibits no evidence whatever of lateral shifts (folding); all mountain chains we see on the Moon are nothing but partly preserved ramparts of impact formations ranging from craters to maria.

SEISMICITY. Seismometers deposited on the Moon by *Apollo 11–17*, and operative almost continuously since 1969, have shown that tremors and quakes on the Moon dissipate very little energy. Frequent but minor moonquakes have been registered by the Apollo seismographs—due partly to external (meteoritic impact) and partly to internal causes. The most interesting of the latter have emanated repeatedly from the same epicentre (of lunar coordinates $\lambda = 28°W$, $\beta = 21°S$) at a depth of 800 ± 40 km below the surface of the Moon (i.e. almost halfway between the centre and the surface). A number of additional centres of moonquakes have already been detected, and some of them found to give rise to whole swarms of such quakes in rapid succession. Their activity was, moreover, found to fluctuate with varying distance from the Earth in the period of exactly one month. The frequency of seismic events seems to be maximal when the Moon is closest to the Earth—a fact which suggests that bodily tides may trigger deep-seated instabilities of as yet unspecified nature. However, the magnitude of such events is really tiny in comparison with current terrestrial standards. The energy of individual moonquakes is, on the average, less than that liberated by the explosion of one kilogram of TNT. But, while terrestrial disturbances are damped out in a few minutes, lunar tremors persist for 60 to 100 minutes. Even the impact of an abandoned lunar module ascent stage set the Moon 'ringing like a gong' for an appreciable time.

THE SURFACE. The 'seas', or MARIA, can be detected with the naked eye. Of course, they are not seas, and have never been water-filled; they are vast plains, smoother than the bright uplands although still very rough. The

MOON

THE MOON NEAR 'LAST QUARTER' photographed with the 100-inch telescope. The large isolated crater near the middle is Copernicus; below it and to the left are the Apennine Mountains. The two photographs were made at different dates, and show clearly the effects of libration. (*Mount Wilson – Palomar*)

Oceanus Procellarum (Ocean of Storms) is larger than the Mediterranean. Some of the maria are basically circular, and it has been suggested that they are similar in nature to the large craters. The Moon's far side (i.e. the side always turned away from Earth) is conspicuously deficient in maria—a fact which is undoubtedly of great significance; but its real meaning still eludes us.

The MOUNTAINS of the Moon are not at all similar to those of Earth, since most of the 'ranges' constitute only the borders of regular maria (for instance, the Apennines and the Alps form part of the border of the 870 000 km^2 Mare Imbrium, or Sea of Rains). Some of the heights exceed 6 500 metres above the mean level of the Moon, so that the mountains are relatively much higher than those of the Earth. Separate peaks are also common, and hills and hummocks almost innumerable.

The Moon is dominated by the walled circular formations known generally as CRATERS. They are everywhere; they cluster thickly on the bright uplands, ruining and distorting one another, and they are found also on the maria, on the floors and walls of larger craters, and on ridges and peak-crests. No part of the Moon is free from them. The number of craters with diameters of more than 1 km exceeds 300 000 on the visible side of the Moon alone, and there are many more on its far side. Smaller craters are vastly more numerous.

When seen near the terminator, a crater gives the impression of great depth and precipitous walls. This is somewhat misleading. It is true that the depths are considerable, but in relation to its diameter a crater is comparatively shallow. Clavius, for instance, is 5 500 m deep and 230 km in diameter, so that in form it is more like a saucer than a well. A typical crater has a wall rising to only a moderate height above the surrounding country, but to several hundreds of metres above the depressed interior. An observer standing inside a large crater has no impression of being shut in by towering walls, particularly when the sharp curvature of the lunar surface is taken into account. Many of the craters possess central mountains rising to considerable heights, though never to altitudes equal to the surrounding ramparts. Others have clusters of central peaks, while central craters are also fairly common.

The distribution of the craters is not random; from the largest plains to the smallest craterlets they tend to arrange themselves in chains, pairs and groups, while when one crater breaks into another the larger formation is almost always the sufferer. One object 90 km across, Wargentin, has a raised floor, and is evidently filled to the brim with lava.

The problem of the origin of the craters has caused much argument in the past. Most authorities now attribute the majority to meteor impacts, and there was and is enough debris in interplanetary space (ranging from microscopic specks of dust to major meteorites, asteroids or comets) to satisfy this theory. Others maintain that the non-random distribution of the walled features implies that at least some craters are of volcanic origin, but they are limited in number and relatively small in size (e.g. Regiomontanus A).

Certainly there are other structures, like the domes or wrinkle ridges, which are undoubtedly of internal origin, but the specific mechanism of their formation remains obscure. On the other hand, there are no mountain chains on the Moon that could have been formed by folding; nor, for that matter, are there any features to indicate lateral movement of the Moon's crust. The structures in the mountain chains of the Moon—the lunar Alps, Apennines and Carpathians surrounding Mare Imbrium or the Cordilleras bordering on Mare Orientale—originated, so far as we can tell, by the same impact processes that produced craters; there is no indication of folding. Thus, one of the most important terrestrial mountain-forming processes seems to be completely lacking on the Moon, and its absence suggests that the rigidity of the Moon's 'crust' is much greater than that of the terrestrial mantle.

Some of the craters, notably Tycho and Copernicus on the Earth-turned hemisphere, are the centres of systems of bright rays extending for great distances across the Moon. These rays constitute only shallow surface deposits. Other minor features include long, narrow depressions known as clefts or rilles; domes; and shallow, rimless pits.

The Moon's surface experiences great variations of temperature, ranging, at the equator, from about 130°C down to almost −190°C at night—a value much lower than anything experienced on Earth. However, the temperature not far below the surface layer is much more constant, indicating that the surface materials are excellent insulators.

It is no longer thought that the Moon is entirely inert; indications of mild activity (possibly volcanic) have been seen in and near some craters, notably Alphonsus and Aristarchus. The existence of these so-called Transient Lunar Phenomena (TLP) is admitted, but their precise origin is still a matter for debate.

LUNAR ROCK. A large number of samples of rock from the surface or immediately below it have been brought back by the Apollo missions. The earlier samples were given a preliminary examination in quarantine, but no traces of living or fossil organic matter were found. They were then released for over 130 research projects in many different countries, and many results are still awaited.

By atomic contents, the most abundant constituents of the material are—as in the Earth's crust—oxygen (about half by weight), silicon, and aluminium; followed by magnesium, titanium, and others in diminishing amounts. It is not until we come to less abundant elements that differences between the Moon and the Earth's crust become apparent. Thus titanium or chromium are more abundant on the Moon than on the Earth (or in the Sun); whereas nickel or the alkali metals (sodium, potassium, rubidium etc.) or rare earths like europium are very much less abundant. The ratio of iron to nickel appears to be larger in the Moon than in any other sample of cosmic matter available for analysis; and carbon or nitrogen are conspicuously deficient on the Moon.

What kind of molecules and minerals do these elements constitute? About 40% of lunar surface material by weight appears to be silica, followed by iron oxide (around 20%), and aluminium and calcium oxides (10%). Numerous minerals well known from the Earth—such as olivine, pyroxene,

CRATERS ON THE FAR SIDE OF THE MOON, in an oblique photograph taken from *Apollo 8* near 160° West, 10° South – nine years after *Lunik III* transmitted the first vague glimpses of the averted side which can never be observed directly from the Earth.

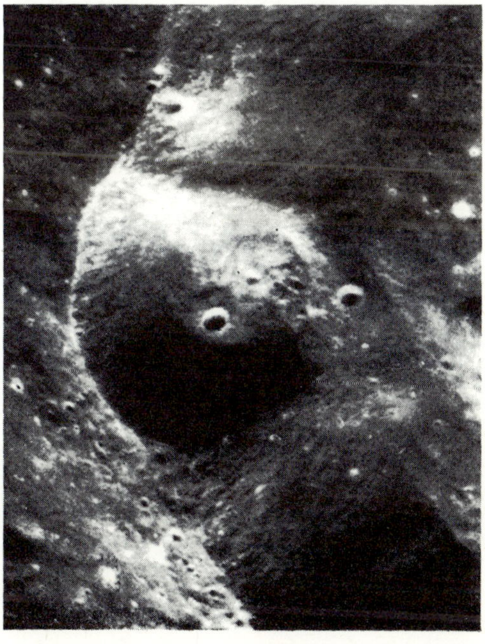

The smaller picture on the right shows a telephoto detail of the area indicated by the black arrow, photographed nearly vertically a few moments later. It is here purposely placed 'upside down' to indicate how the eye can mistake high relief for low relief in such pictures. The small twin craters in the centre facilitate exact comparison.

This unusually defined crater is 500 feet in diameter. The double wall is due to a continuous landslide around the circumference. The photograph was made by *Lunar Orbiter III*, and the smallest craters revealed in it are no more than 2–3 feet across. Another of its photographs (*below*) shows *Surveyor I*, which soft-landed earlier, casting a 30-foot shadow.

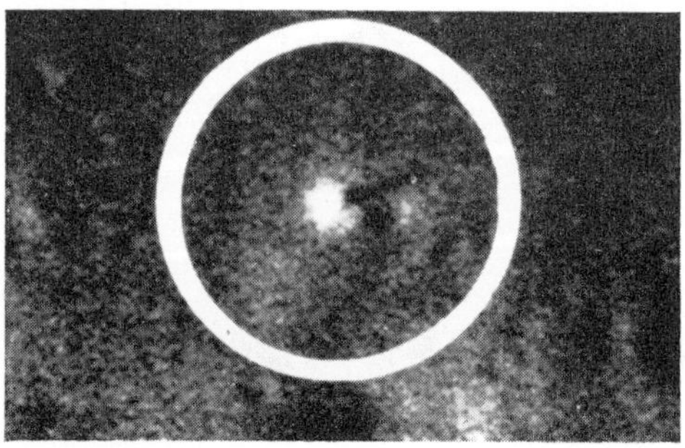

plagioclase, feldspar and ilmenite—have also been found on the Moon. Only three minerals not previously known on the Earth have been identified in lunar samples so far. About half the total volume of the material forming the topmost layer of the surface is empty space; and this accounts for the relatively low bearing strength encountered on the surface.

Many rocks brought back from the Moon suffered from evident effects of shock metamorphism produced, in all probability, by meteoritic impacts. All rocks are igneous (no trace of sedimentary material has been found anywhere on the Moon), and generally of basaltic composition. Their crystalline structure indicates, moreover, that lunar rocks solidified at temperatures between 1 000° and 1 200° C under highly reducing conditions, in which the partial pressure of free oxygen was less than 10^{-13} of an atmosphere; for all oxides present correspond to the lowest states of oxidation.

The sources of heat which produced melting and thermal differentiation were undoubtedly local rather than global in nature. Global melting of the Moon at any time in the past is inconsistent with the shape (and gravitational field) of our satellite. A completely molten Moon would not have had time to solidify in 4 600 million years; and if at least its surface has been all molten at any time, it would have solidified to a form approximating a state of hydrodynamical equilibrium much more closely than it actually does. More localized melting could be the result of impacts, and local pockets of lava may have been produced in this way; but 'volcanoes' as we understand the term in terrestrial geology probably played a much smaller role in shaping the face of the Moon and its composition than they did on the Earth.

The greatest single piece of information furnished by man-made spacecraft in recent years has been a determination, by chemical methods, of the age of the lunar rocks from different localities on the lunar surface—or, more specifically, of the time which has elapsed since these solidified. Radiometric methods were applied to determine this age of crystalline rocks, and we can now say that the principal pattern of lunar maria as we see them to-day was formed between 3 300 and 4 000 million years ago, as a result of an apparently unrelated series of distinct events that occurred within this time.

PRESIDENT JOHNSON being briefed by Dr. Pickering on the results emerging from photographs taken by *Ranger 7* before its impact on the lunar surface.

MOON HOUSE. Stabilizing blocks being placed under an inflatable shelter designed to house astronauts during prolonged stays on the Moon.

However, perhaps the most interesting result which has emerged from radioactive dating of lunar rocks so far is the fact that, on each landing site, the smaller-size debris (the *fines*) were found to be of substantially greater age than the lumps of rock. The time which has elapsed since the solidifications of the fines proved, in fact, to cluster around 4 600 million years in all localities visited so far; and this age is virtually identical with that of the oldest known meteoritic material intercepted by the Earth. These fines provide an irrefutable testimony that solid matter existed on the lunar surface as far back as 4 600 million years ago and that it has not been re-melted since that time.

Nothing of comparable age has been found anywhere on the Earth, where the oldest rocks preserved in rare localities are not more than 3 800 million years old (all older strata having been 'consumed' by the interior).

TIDAL EFFECTS. The oceanic tides raised on the Earth by the Moon are, of course, a well-known phenomenon. Less well known, because their effects are not immediately visible, are the tidal motions produced in the solid Earth. On the Moon, where there are no oceans, the tides raised by the attraction of the Earth are produced entirely in the solid body. What, then, are the effects of these tides on the dynamics of the Earth–Moon system?

BROKEN ROCK on the surface of a relatively new crater.

The solid parts of the Earth and the Moon are not perfectly elastic, and the liquid part of the Earth is viscous. The movement of masses attracted by the gravitational force of the Moon therefore lags behind the direction of this force; the flow of these masses generates tidal friction, and the rotation of the Earth displaces the tidal bulge so that it does not lie directly under the Moon. As a consequence of the resulting asymmetries, the Earth continuously loses some of its rotational energy, part of which is dissipated as heat and part transferred into altering the orbital motion of the Moon to conserve the angular momentum of the Earth–Moon system.

For how long has the rotation of the Earth been decreasing? One of the effects of a decreasing rotation is, of course, an increase in the length of the day. Astronomical observations over the past two centuries show that tidal friction is lengthening the day by two thousandths of a second per century. That this figure is also the average value for the past 370 million years has recently been confirmed from palaeontological evidence. The daily growth rings in corals show both monthly and yearly variations due to seasonal changes. It is thus possible to determine the number of days in a year by counting the number of growth rings in corals of various ages. Since tidal friction does not appreciably affect the length of the year, the ring counts lead directly to the length of the day. Recent corals show about 360 rings a year, whereas fossil corals from the Middle Devonian have about 400 a year. Thus about 370 million years ago the day was only 21·9 hours long.

Calculations based on the figures for the lengthening of the terrestrial day show that the semi-major axis of the Moon's elliptic orbit is currently increasing at the rate of 3·2 cm per year and that the orbit is becoming even more eccentric. It is also possible to extrapolate these calculations backward in time in an attempt to determine what the orbit of the Moon looked like in the remote past. The results of such an extrapolation should, of course, be regarded with some reserve because they are inevitably based upon the effects of tidal friction as they are observed *now*. The internal structures of the Earth and Moon could have been very different from what they are to-day—for example, the Earth may not have had its present metallic core.

CLOSE-UP OF A LUNAR RILLE running north-west and east from the crater Hyginus. The crater is 6½ miles wide and 2,600 feet deep, nearly in the centre of the earthward side of the Moon.

ORIGIN OF THE MOON. This topic, like the origin of the solar system as a whole, is still a rich field for speculation. From the above it is clear that since the separation of the Earth and Moon is increasing, the Moon must have been much closer to the Earth in the past. It may be that it originated from the same mass that gave rise to the Earth itself; it may have been an independent body captured by the Earth during the early history of the solar system; or it may have formed in close proximity to the Earth. But the theory that it originated outside the gravitational field dominated by the Earth and was captured by our planet may still be regarded as the most likely hypothesis. If there was capture, it must have been virtually simultaneous with the formation of the Earth, and at that time moon-like objects may have been more numerous in interplanetary space than they have become since then. But whatever uncertainty still surrounds the genesis of the Moon, its future is not in doubt: the continuous operation of tidal friction will make it recede from the Earth until its distance from us has 1·6 times the present value and its orbital period is approximately 60 days of our present time. The same cause will have slowed the rotation of the Earth to the point where a terrestrial day is as long as a month, and the Earth and the Moon will henceforward continue to show each other the same face. Several thousand million years are likely to elapse before this comes to pass; but it is improbable that the topography of the Moon will undergo any important changes in the meantime.

We now take it as certain that the solid crust of the Moon (constituting perhaps the major part of the lunar continents) was already formed as far back as 4 600 million years ago, and that for the main part its surface features were sculptured during the first few hundred million years of its existence; since then it has probably been disturbed only by incidental impacts. For reasons which we do not yet fully comprehend, the greatest events which disfigured the Moon—namely, the formation of the maria—occurred between 500 to 1 500 million years after the Moon was formed; and since that time things have become increasingly quiet. (Z.K.)

(See also *Apollo Summary* in the **Appendix.**)

Above:
VARIED FEATURES appear in this oblique view, looking north, taken from the *Apollo 15* command module. The foreground shows Hadley Rille and the Apennine Mountains. The two craters at upper left are Autolycus and Aristillus, each of ca. 50 km diameter. At upper right are the Caucasus Mountains and part of the Sea of Serenity; the Marsh of Decay is at lower left. The white arrowhead marks the *Apollo 15* landing site.

Below:
The landing site itself, with the lunar module and the rover.

MOUNT PALOMAR OBSERVATORY.

MOON PROBE. A rocket with instruments designed to collect and transmit information on the Moon and the space surrounding it. (See **Artificial Satellite** and **Moon.**)

MOONS. Satellites of planets, a moon being in the same relation to its planet as the planet is to its primary. Of the nine planets in the solar system, six have a total of thirty-two moons; their origin is not known, nor has it been possible to determine whether stars other than the Sun possess planets and moons.

MOUNT PALOMAR OBSERVATORY in southern California, U.S.A., has the largest optical telescope in existence, with a mirror 200 inches in diameter. The instrument is almost invariably used photographically. The mirror is so large that no great obstruction is caused by the observer (who must be constantly vigilant during an exposure to keep the telescope correctly aligned) sitting in a 'cage' in the upper end of the telescope tube. The observer is carried round with the telescope, which turns on an ingenious type of mounting: the northern bearing is shaped like a horseshoe 46 feet across, and the telescope can look between its arms to observe objects near the north celestial pole. The telescope and moving parts of the mounting together weigh 500 tons. A sectional drawing of it will be found under **Telescope.**

Palomar also has the largest Schmidt telescope in the world, with an aperture of 48 inches. It takes photographic plates 14 inches square, and was occupied for over seven years by the 'Sky Survey' whose result is a monumental atlas containing nearly 2,000 photographs covering all the sky accessible from Palomar.

MOUNT WILSON OBSERVATORY is situated on a mountain top 5,960 feet above sea-level, in California, U.S.A. It possessed for several decades the largest telescope in the world, a 100-inch reflector.

This instrument and a 60-inch have done sterling work since the beginning of the century, especially in the photography of faint nebulae beyond the reach of lesser telescopes. Mount Wilson also has two 'tower' telescopes used for observing the Sun. In such instruments the image is reflected into a subterranean observing room by a moving mirror on top of a high steel tower, and the worst effects of atmospheric turbulence caused by ground heating are avoided.

MULTIPLE STAR. See **Binary Star.**

THE OBSERVER'S CAGE at the prime focus of the 200-inch telescope; an astronomer is riding in the cage.

MOUNT WILSON OBSERVATORY: the dome housing the 100-inch telescope – a view taken from the 150-foot tower; the tower itself; and a sectional view of the dome.

N

NADIR. The point on the celestial sphere opposite the zenith and therefore vertically beneath the observer.

NAUTICAL ALMANAC. A book which gives the **ephemeris** or time-tables of predicted movements of celestial bodies for a whole year, in so far as they can be of value in navigation.

NEBULAE. See **Galactic Nebulae,** and **Galaxies.**

NEPTUNE. The outermost of the four giant planets. It was first recognized in 1846, by Galle and d'Arrest at the Berlin Observatory, but the discovery was not due to pure chance. Uranus had been wandering from its predicted orbit, and Le Verrier, a French mathematician, had decided that an unknown planet must be responsible. By careful calculation he had worked out just where the new body must lie, and all that Galle and d'Arrest had to do was to search in the position indicated. Actually, a Cambridge graduate, John Couch Adams, had reached the same solution some time before, but the search in England was not carried out with the same energy as on the Continent. It is now agreed that the honour for the tracking down of Neptune must be shared equally between Le Verrier and Adams.

ORBIT. Neptune's perihelic, mean and aphelic distances from the Sun are 2,769, 2,793 and 2,817 million miles respectively. The orbital eccentricity is 0·009, lower than that of any other planet except Venus; the inclination is 1°.8, and the mean orbital velocity 3·4 miles per second. Neptune has a sidereal period of 164·8 years, so that it has completed just three-quarters of a revolution since its discovery.

DIMENSIONS AND MASS. Neptune has a diameter of 31,000 miles, and is thus rather larger than Uranus. It has, however, a greater mass, 17 times that of the Earth; and the density is 1·6 times that of water, appreciably greater than for the other giants. The surface gravity (1·4 times that of the Earth) is higher than for any other planet except Jupiter, and the escape velocity is 14 miles per second.

ROTATION. The curious axial tilt of Uranus is not shared by Neptune (see **Uranus**). The inclination of the equator to the orbit is only 29°. The axial rotation period is not easy to determine, but appears to be about 14 hours.

SURFACE FEATURES. Owing to its remoteness, Neptune shows little or no surface detail. The colour is bluish, and differs from the green hue of Uranus.

TEMPERATURE AND COMPOSITION OF THE GLOBE. The temperature of Neptune seems to be about −200° C. There is no reason to suppose that the globe is built in a manner radically different from that of the other giants (see **Jupiter**). On Wildt's model, the rocky core is 12,000 miles in diameter, while Ramsey considers that the composition of Neptune is almost identical with that of Uranus.

SATELLITES. Less than three weeks after Neptune itself had been discovered, it was found to possess a large satellite that has been named *Triton.* It appears to be almost the equal of *Titan* (see **Saturn**), with a diameter of over 3,000 miles; it lies 220,000 miles from Neptune, and has a sidereal period of 5 days 21 hours. The orbit has low eccentricity, but high inclination, and the motion is retrograde. Despite its size, Triton would not be a brilliant object in Neptunian skies, as it would shed only 1/400 the light of our Full Moon.

Triton must have an escape velocity of at least 2 miles per second, and should therefore be capable of retaining an atmosphere. Indications of a methane mantle have been suspected, but definite proof is still lacking.

Nereid, the second satellite, was discovered in 1949. It is only about 200 miles in diameter, and has a remarkable orbit, the eccentricity being 0·760. The sidereal period is 359 days, and the motion direct.

SUMMARY. Neptune and Uranus may be considered as true twins, and are far more similar than Venus and the Earth. The main difference is that Neptune is slightly larger, slightly denser, slightly more massive and appreciably colder.

So far as is known at present, Neptune is the last member of the group of giant planets. If another exists, it must be so faint that its discovery will be a matter of extreme difficulty. (P.M.)

NEUTRAL POINT. The point on the line joining Earth and Moon at which the gravitational fields of the two bodies just cancel each other. Neutral points, of course, exist between other pairs of neighbouring bodies.

NEUTRINO. An elementary particle which has zero rest mass and zero electric charge. Its existence was predicted by Pauli in 1930 but it was not observed until 1956. The detection of neutrinos is extremely difficult because they interact very rarely with matter; a neutrino of moderate energy could pass through a slab of lead sixty light years thick without colliding with an atom. The very small interaction rate between neutrinos and matter means that once energy has been converted into neutrinos it is very seldom returned to matter. In this sense, neutrinos can be said to be a *sink* of energy.

NEUTRINO ASTRONOMY. **Neutrinos** are produced in some of the nuclear reactions which occur at the centres of stars (see **Stellar Energy**). Because neutrinos interact very rarely with matter they travel right out of the star, so if they can be detected on Earth they will give direct information about the physical conditions at the centres of stars. Neutrinos are also produced by the interaction of **cosmic rays** with **interstellar matter** and with the atmosphere of the Earth. With the present detection equipment, which includes **scintillation counters** and **Cerenkov detectors,** the Sun is the only star close enough to give a detectable neutrino flux. The detectors are placed deep underground, usually in mines a mile or two deep, in order to separate the solar neutrinos from other elementary particles. The observed solar neutrino flux is much less than the theoretically predicted flux; the discrepancy may be due to experimental difficulty, but the theory may need to be modified.

Neutrinos, originating at the centres of stars, can remove a large amount of energy, which may have important consequences in the later stages of **stellar evolution.**

Some of the *big bang* theories of **cosmology** predict the existence of a cosmic neutrino background. These theories say that the Universe is bathed in a 'sea' of very low energy neutrinos which were formed in the early stages of the initial explosion. (J.A.J.W.)

NEUTRON. A particle that exists in the nuclei of all **atoms** except those of normal hydrogen. The neutron's mass is nearly equal to the mass of a hydrogen atom, but unlike the proton and the electron it possesses no electrical charge, and this accounts for its relatively late discovery (1930).

NEUTRON STARS. Stars composed of the neutral elementary particles, **neutrons,** tightly packed together and behaving as **degenerate matter.** They are formed by the evolution of **white dwarf** stars.

When white dwarf stars collapse under the attraction of their powerful gravitational forces, the electrons recombine directly with the protons of the nuclei to form neutrons. This results in the destruction of the nuclei themselves and eventually, at least at the centre of these stars, the matter consists only of neutrons. The existence of neutron stars is predicted by present theories of physics. Whether such stars have been observed is still uncertain. (See **Pulsars.**)

Stars must collapse so much before the pressure at the centre becomes high enough for neutrons to be formed that a neutron star's radius is only about 10 km. This is about a hundred thousand times smaller than the present radius of the Sun. Because they still have a mass of about that of the Sun concentrated within this radius, a lump of sugar would weigh rather more than 100,000 tons at the surface of such a star.

N.G.C. Abbreviation for the *New General Catalogue* of nebulae and star clusters.

NOCTILUCENT CLOUDS. Peculiar clouds at heights of about 50 miles in the Earth's atmosphere. They can be seen at night as, owing to their altitude and to the bending of sunlight by atmospheric refraction, they are still illuminated by the Sun long after nightfall on the Earth's surface.

NODE. One of the two points in which the line of intersection of any two planes cuts the **celestial sphere.** In the solar system, it is one of the points in which the Moon's or a planet's orbit cuts the ecliptic.

When the Moon or a planet crosses the plane of the Earth's orbit from South to North it passes through the *ascending node*; when it crosses from North to South it passes through the *descending node*.

A body may not cross the ecliptic in the same point every time. This causes its nodes to move slowly forwards or backwards along the ecliptic, a phenomenon which is called *progression* or *regression of the nodes*. The nodes of the Moon regress so that each makes a complete circuit of the ecliptic in about 19 years.

NOMENCLATURE. Most of the brightest stars have names of their own as well as systematic designations. For easy reference, the sky is divided into 88 **constellations,** of differing sizes, bounded by lines of Right Ascension and Declination for the epoch 1875. In each constellation the brightest stars are given Greek letter prefixes, in approximate order of brightness, followed by the name of the constellation. When the Greek alphabet is exhausted, small English letters are used, and finally English capitals.

All stars down to about the sixth magnitude also have numbers within each constellation, and these, called Flamsteed Numbers after the astronomer who allotted them, are now generally used in preference to English letters.

Stars down to almost the tenth magnitude have been catalogued in the monumental **Bonn Durchmusterung.** This takes no account of constellations, stars being named by their declination and a number. Thus B.D. $+59°$ 1376 is star number 1376 in the zone between declinations $+59°$ and $+60°$. The brightest star in the northern hemisphere is Vega or α-Lyrae; in the Flamsteed catalogue it is 3 Lyrae, and in the Durchmusterung it is B.D. $+38°$ 3238.

Variable stars have their own designations, described under their own heading.

There exist three main catalogues for nebulous objects, i.e. galaxies, nebulae and star clusters: the *Messier* Catalogue, the *New General Catalogue* (N.G.C.) and the *Index Catalogue* (I.C.). The I.C. catalogue is an extension of the N.G.C. catalogue. In all three, the objects referred to are merely given a number in the particular list; hence, M 82 and NGC 188 refer to the eighty-second entry in Messier's catalogue and the one hundred and eighty-eighth entry in the N.G.C. catalogue.

Since the discovery of radio sources in large numbers, radio astronomers have been preparing catalogues of increasing scope and accuracy. These include:

(1) The MSH catalogue compiled by Mills, Slee and Hill. The list refers to the Southern Hemisphere and the numbering system describes the position of the source in the sky.

(2) The various catalogues prepared at Cambridge, called 1C, 2C, 3C, 4C and 5C. The 1C, 2C and 3C catalogues have been superseded by the 3CR (3C revised) catalogue, which is much more accurate. The 3CR catalogue sources are numbered consecutively in order of increasing **right ascension,** whereas the sources in the 4C catalogue have been arranged in lists in order of increasing right ascension, where each list contains all the sources observed in the $1°$ range of **declination.**

(3) The PKS catalogue compiled at the Parkes Radio Observatory. The numbering system, like the MSH catalogue, describes the position in the sky.

(4) The NRAO list, prepared at the National Radio Astronomy Observatory.

(5) The CTA or CTD list prepared at the California Institute of Technology.

(6) Lists of sources discovered with the Arecibo radio telescope, which are prefixed AO and use the same numbering system as the PKS catalogue.

NOVA (*literally* 'new one'): a star which undergoes a sudden and enormous increase in brightness. (See also **Supernova.**)

About 25 novae appear every year in our Galaxy. Of these, most are only discovered long after their outbursts during investigation of photographic plates; but a few, which are relatively close to us, reach naked-eye brightness and are discovered almost immediately. These latter are the ones which we are able to study at the most interesting time, near the maximum.

Faint stars have been found recorded on plates taken before a nova outburst in the position of the nova. The only two known pre-nova spectra are those of a normal A-star and G-type giant. The outburst is characterized by an extremely rapid rise of brightness through about 12 magnitudes (representing a 60-thousand-fold gain in brilliance) in something like two days. Two novae this century, those in Perseus in 1901 and Aquila in 1918, have reached zero apparent magnitude. The absolute magnitude of a nova at maximum is around −8. Typical behaviour after maximum is an immediate decline, rapid at first but soon becoming slower and having superimposed on it rapid and erratic fluctuations. These short-term oscillations die away after a few months and the star gradually sinks to an unsteady magnitude around its pre-nova level.

During the outburst the star's surface is expanding – we see it hurtling towards us at one or two thousand kilometres per second – and at maximum light the surface bursts and material is thrown off in many directions.

While all novae have their own peculiarities, the behaviour of T Aurigae 1891 is significant because it was repeated by DQ Herculis in 1934. The latter reached nearly first magnitude, declined irregularly and rather slowly to fifth and then suddenly and without warning fell to the thirteenth magnitude. It immediately recovered almost to sixth magnitude and continued its decline. Some other novae develop remarkably slowly: the extreme example is RT Serpentis which exploded in 1909.

Some novae have exploded more than once; for instance T Pyxidis has exploded four times in seventy years. These novae have a smaller range of luminosity than normal novae. The resemblance between the initial and final states of a nova, and the possibility of recurrent outbursts, strongly suggests that the explosion is quite superficial and has little effect on the star as a whole: it is probably some form of safety-valve mechanism for the star. It may well be that all novae are recurrent, the period of ordinary novae being comparable with a million years.

A few **variable stars** with very peculiar spectra seem to be related to novae. One of these, η Carinae, after a highly irregular rise of brilliance through several decades, reached −1 magnitude in 1843, when it ranked second only to Sirius in brightness; thereafter it slowly declined, reaching seventh magnitude in 1880. It is now between eighth and ninth magnitude.

It has also been suggested that the class of variable stars known as U Geminorum stars are related to novae. Indeed, they are frequently termed *dwarf novae*, because of the resemblance of their behaviour, on a minor scale, to that of novae. Very precise observations with the world's largest telescopes have recently shown that the stars in which both nova and U Geminorum type outbursts occur are in some cases, and possibly all, close **binary** systems. The stars which are orbiting about one another in such very close proximity and with very short periods of rotation have always been found to be of very different spectral type. The system is always a blue dwarf star with a cooler, redder giant or main-sequence companion. Furthermore, the red component is thought to have evolved to such an extent as to be in contact with the inner **Lagrangian point** of the binary system.

The cause of these outbursts of novae is still unknown. It is being increasingly suggested that the binary nature itself may play a critical role in inducing the instability of one of the stars in the system. But how this could come about, and whether it is the hotter or cooler star which explodes is still a question of some controversy amongst astronomers.

NUCLEAR PROPULSION. The energy released by the processes described under **Nuclear Reactions** can be used to propel a rocket. Nuclear *fusion* is by far the most powerful source of energy, but the problems of controlling and containing a process which involves temperatures of millions of degrees Centigrade will not be overcome for several years. Nuclear *fission* on the other hand can be used in two ways: the first converts the heat into electric power for the systems described under **Electric Propulsion**; the second uses the heat directly to expand a propellent gas and so drive it at high velocity through a nozzle. This is the principle of the *nuclear thermal rocket*.

The heat can be transferred to the gas by letting it flow through and around the solid core elements of a reactor using rods of, for example, uranium-235 and zirconium hydride, or through the core of a reactor in

SNAP 10A is one of a series of successfully tested Systems for Nuclear Auxiliary Power. The reactor (*left*) contains 37 fuel rods composed of a mixture of uranium 235 and zirconium hydride. It is mounted on top of the conical structure (*right*) which itself can surmount a conventional space vehicle. The heat of the nuclear reaction is converted into electrical power, which may be used either to supply the instruments and equipment in the vehicle, or to accelerate a stream of particles and so impart further speed to the vehicle, gradually but continually for long periods.

which the fissionable material is maintained in the gaseous state.

Sufficient progress has been made with solid core systems to show that they can provide the propulsion for missions to the nearer planets. Thrusts in the 50,000 lb. force class have been demonstrated, with **specific impulse** in the region of 800 seconds, with the U.S. KIWI/NERVA ground test systems. Such a rocket, used in the third stage of the *Saturn 5* launch vehicle, could increase the payload landed on the Moon in an Apollo-type mission by around 60%. Realistic estimates of the effect of using appropriately staged and clustered nuclear rockets based on this technology suggest that the weight required in Earth orbit for a 500-day round trip to Mars might be around 2·1 million pounds, compared with about 6·5 million pounds for the most optimistic chemical rocket. What is more, the weight required in Earth orbit per pound of weight returned to Earth after the Mars round trip is proportionately greater for chemical rockets than for nuclear. In other words, for a given total of payload left at Mars plus payload returned to Earth, a chemical rocket is much more sensitive to the proportion returned to Earth. Thus use of nuclear rockets, quite apart from their intrinsic attraction on performance grounds, would confer much greater mission flexibility, and a lot of money could be saved by not having to develop too much specialized hardware for particular missions and launch dates (which affect the energy requirements appreciably).

If the vehicles using nuclear rockets are of suitable configuration, with the crew compartment a long way away from the reactor cores, the shielding problem is probably no more serious than that arising from the need to provide protection against high-energy particles from solar flares and from cosmic rays. Quite obviously, however, fission rockets would be unsuitable for the initial launch into Earth orbit, even if their thrust/weight ratio approached that of chemical rockets, because of the atmospheric contamination which would result.

So far no suggestions have been published which would make the gas-core approach feasible, but in principle the temperatures available could be a great deal higher. The difficulty would be to prevent the (gaseous) fissile fuel from blowing away in the blast of propellent gas. Dilution of the fissile gas by the propellent might also be a problem, but if fuel containment in the reactor zone could be achieved, or the fuel loss limited to economic levels, specific impulses of 2,000 or more might be attainable. (C.W.M.)

NUCLEAR REACTIONS. When an **atom** is bombarded with particles such as neutrons, the effect in most cases is not very catastrophic: the nucleus of the bombarded atom may for instance absorb a neutron and perhaps emit some other particle. In the case of the isotope of uranium of atomic weight 235, however, such bombardment with neutrons can split the atomic nucleus into two more or less equal parts, with the liberation of several more neutrons and a great deal of energy. This process is *nuclear fission*. In a large lump of uranium the newly emitted neutrons cause the fission of other atoms before they can escape from the lump, and a chain reaction is initiated which builds up from small beginnings like an avalanche within less than one second. This is the principle of the atomic bomb; it has been modified in nuclear reactors, where instead of one large lump being used several smaller pieces of fissionable material are placed at such distances from each other that the chain reaction maintains its own level and does not grow out of control.

A few elements besides uranium have fissionable isotopes.

At the other end of the atomic weight scale, several lighter nuclei can give up energy by *nuclear fusion*: joining together to give a heavier nucleus. This process can take place only if a tremendous amount of energy is first invested to trigger it off. Hence, hydrogen bombs are detonated by exploding a fission bomb. A great deal of research is being carried out to harness the power of fusion processes by stabilizing the reaction, e.g. by restricting it inside a powerful ring-shaped magnetic field as in the British Zeta apparatus. Temperatures must be produced comparable to those in the interior of the Sun and other stars, where fusion processes occur naturally. (See **Stellar Energy.**)

NUCLEOGENESIS. The origin of the elements. Before discussing this it is necessary to discover the relative abundances of the elements in the Universe. These can be investigated in several ways.

The **Sun,** stars, gaseous **galactic nebulae** and **interstellar matter** emit light. The lines in their spectra (see **Spectroscopy**) tell us the quantities of the various elements in the region from which the light comes. Samples of the Earth, **meteorites** and **cosmic rays** can be obtained and analysed directly. Light elements such as hydrogen and helium can escape from solid objects such as the Earth and meteorites but, provided that allowance has been made for this, all abundance determinations yield similar results and it is possible to construct a *cosmic abundance* table giving the number of atoms of an element per million million hydrogen atoms. The greatest difference between results is that those stars which we believe to be oldest, and hence formed first, have a much smaller abundance of elements other than hydrogen and helium

RELATIVE ABUNDANCES OF SOME ELEMENTS IN THE SOLAR PHOTOSPHERE. (These values are typical of all but the oldest stars.)

Element	*Relative number of atoms*
Hydrogen	1,000,000,000,000
Helium	90,000,000,000
Carbon	350,000,000
Nitrogen	85,000,000
Oxygen	590,000,000
Sodium	1,500,000
Magnesium	30,000,000
Aluminium	2,500,000
Silicon	35,000,000
Phosphorus	270,000
Sulphur	16,000,000
Potassium	110,000
Calcium	2,100,000
Chromium	300,000
Iron	3,200,000
Nickel	120,000

than the younger stars. The relative abundances of these heavy elements vary very little with the age of the star. By far the most abundant elements are hydrogen and helium; all the others comprise no more than 2% by mass of the Galaxy. The observed stellar abundances are believed to be those of the medium from which the stars were formed and they therefore show that there has been hydrogen and helium in the Galaxy since shortly after it was formed at the latest, but that other elements have been synthesized during the lifetime of the Galaxy.

COSMOLOGICAL HELIUM. The production of helium in the initial stages of a big bang is described in the article on **cosmology.** The amount produced is just that required to explain the observations.

ASTROPHYSICAL NUCLEOSYNTHESIS. All other element formation is thought to be due to **nuclear reactions,** mainly in stars.

The *carbon-nitrogen cycle* and the *proton-proton chain* convert hydrogen into helium in stellar interiors and provide the main source of **stellar energy.** If the Galaxy has been about as luminous throughout its lifetime as it is now these reactions could have produced only about one-tenth of the observed helium. The *steady state* theory produces no helium cosmologically and objects with the mass of thousands of suns existing shortly after the formation of a galaxy have been suggested to be capable of producing the observed abundance of helium.

In the later stages of a star's lifetime the temperature in the centre rises and helium is converted into carbon and oxygen. A complicated series of reactions can then follow in which elements such as neon and magnesium are formed. All the reactions up to this point release energy (they are *exothermic*) but when the temperature reaches about four thousand million degrees absolute, energy-absorbing, or *endothermic*, reactions can also occur and equilibrium is set up between all reactions, both endo- and exothermic. The result is the formation of mainly iron and nickel and a few other metals, known collectively as the *iron group*. On balance there is a net energy release.

All these reactions are believed to provide energy sources at various stages of **stellar evolution.** They produce most elements lighter than iron and the members of the iron group, but heavier elements, such as bismuth and lead, cannot be built up by exothermic reactions. It is suggested that they are formed by **neutron** capture. The capture of more than one or two neutrons makes an atomic nucleus unstable and it radioactively decays into a stable nucleus. If the capture of neutrons is rapid (the *r-process*) many are captured before decay occurs and a nucleus relatively rich in neutrons is formed. Slow capture (the *s-process*) allows decay to occur after fewer neutrons have been captured and a nucleus with a higher proportion of protons results. The neutrons come from reactions of the light elements present; just which ones depends on the physical conditions. Neutron capture also works with elements lighter than the iron group and it is thought that some of these are formed in the same way.

All these processes, and a few minor ones in addition, together can produce, with the appropriate abundances, all observed elements.

Determining the responsible nuclear reactions is only part of the study of nucleogenesis; objects must be found in which the reactions can take place. Detailed studies of stellar evolution have not yet been taken to the stages when the important nuclear reactions take place and while some nucleosynthesis certainly occurs in stars, it is still not clear that ordinary stars are the main site of the synthesis of all the elements.

It is also necessary that the elements produced in the centres of stars be expelled into space so that they can be incorporated into a later generation of stars. Steady mass loss from the surfaces of some stars is observed, but for this to be effective it must occur at a late stage of evolution, after elements have been formed, and these elements must be brought to the surface by convection. The study of mass loss and convection is very difficult and although it appears that steady mass loss must play an important part in the distribution of the lighter elements such as helium, carbon, nitrogen, oxygen and neon, it is not yet possible to say whether it is important for heavier elements. The outbursts of **supernovae** are thought to be sufficiently violent to produce heavy elements and to expel them into interstellar space, but difficulties are met when attempting to account for the total quantity of heavy elements and for details of the abundances.

Thus, although astronomers think that they know which reactions produce the observed elements, they are not at all sure where they occur and how their products are expelled into space. (P.R.O.)

NUTATION. A small oscillation or 'wobble' in the Earth's axis in addition to precession, with a cycle of 19 years. It is caused by the Moon's tendency to pull the Earth's equatorial bulge into line with its gravitational attraction. (See **Precession.**)

O

OCCULTATION. The disappearance of a star behind the disc of the Moon or of a planet, or of a satellite behind its primary. An occultation is basically an **eclipse.** Occultations by the Moon are of great value in the accurate determination of its orbit. They also confirm that the Moon can have none except the most tenuous atmosphere, since the occultation of a star is always very sudden and without traces of atmospheric refraction at the Moon's surface.

OCTOBER DRACONIDS. A **meteor** shower which gave spectacular displays in 1926, 1933 and 1946 when the Earth passed through the orbit of the Giacobini-Zinner comet shortly before or after the comet itself.

OPPOSITION. See **Conjunction.**

ORBIT. The path in which a celestial body moves about the **centre of gravity** of the system to which it belongs. Every orbit is basically in the shape of a **conic section** with the centre of gravity at one focus.

The Sun is so massive in comparison with the planets that the centre of gravity of the solar system is always in or very near to it. All other members of the system therefore move in more or less elliptical orbits around the Sun.

An orbit is completely described if six of its characteristics or *elements* are stated. The six elements usually given for planets are:

1. ECCENTRICITY, which determines the shape of the orbit, whether it is hyperbolic or elliptical, elongated or nearly circular.

2. SEMI-MAJOR AXIS, which fixes the size.

The next three elements describe the orbit's orientation in space:

3. INCLINATION is the angle between the plane of the orbit of the planet and the plane of the Earth's orbit, i.e. the ecliptic.

4. LONGITUDE OF ASCENDING NODE tells us the direction of the point in which the planet crosses the ecliptic from South to North.

5. LONGITUDE OF PERIHELION states how far around the orbit the point on it nearest to the Sun lies as measured from the ascending node.

Now the shape, size and orientation of the orbit are fixed. To find how a planet moves in this orbit we only need to know –

6. TIME OF PERIHELION PASSAGE on any one occasion. This provides a starting point for reckoning based on **Kepler's third law,** from which we can calculate the position and orbital speed of the planet at any other time, *provided* we already know the mass of the Sun (or, more strictly, the mass of the entire solar system). Failing this, a seventh element is required; viz.

7. PERIOD or time taken to complete one circuit of the orbit.

The orbit of each member of the solar system is perturbed by the other members; Jupiter's influence is particularly powerful, and has profoundly affected many asteroid, comet and meteor orbits. Some of the last two have hyperbolic orbits: they approach the Sun once and recede again, never to return. All others move in elliptical paths. A parabolic orbit cannot in practice be followed, because the slightest perturbation converts it into either a hyperbolic or an elliptical one.

Outside the solar system the ecliptic becomes irrelevant as a plane of reference,

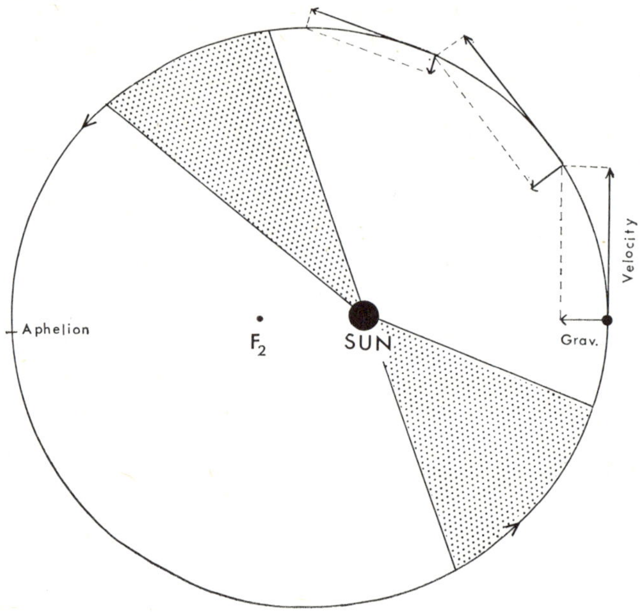

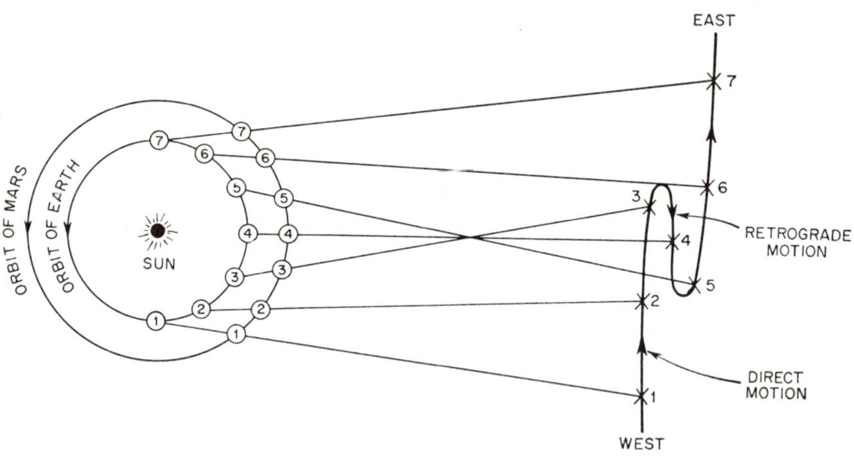

PLANETARY ORBITS. The upper diagram shows an ellipse of eccentricity $0 \cdot 2$, with the Sun at one focus and nothing at the other, F_2. The planet moves under the combined forces due to gravity and to its momentum, the latter depending on its velocity. The result is movement in a closed curve, with velocity and gravitational attraction decreasing towards aphelion and increasing away from it. The shaded areas illustrate **Kepler's Law** of Areas.

The lower diagram explains why the motion of a planet as seen from the Earth is occasionally *retrograde*, i.e. backwards or looped.

and inclination is usually measured relative to the plane at right angles to the line of sight.

There is a close connection between the shape and size of an orbit and *orbital velocity.* The Earth, for instance, travels around the Sun with an average speed of $18\frac{1}{2}$ miles per second, in a slightly elliptical path. If somehow it were to slow down a little when it is at perihelion, its orbit would become more nearly circular, and if it continued to slow down it would gradually spiral towards and into the Sun. If, on the other hand, it increased its speed, its orbit would elongate, and for a particular speed called *parabolic velocity* the orbit would 'split' at the far end of the ellipse, so that instead of returning the Earth would escape entirely from the solar system. The same principle applies to all orbits, including those of artificial satellites and space ships, and explains why parabolic velocity is often called **escape velocity.** The least speed for maintaining an orbit *at a given distance* from the centre of attraction and the least speed for escaping from it altogether are connected by the equation:

$$\textit{Parabolic vel.} = \sqrt{2} \times \textit{circular velocity.}$$

ORBITING SOLAR OBSERVATORY OSO-C being checked at the Goddard Space Flight Center.

ORBITING ASTRONOMICAL OBSERVATORY. Often referred to as an OAO, an orbiting astronomical observatory is an artificial satellite containing experimental equipment designed to record information concerning radiation emitted by astronomical objects. The chief advantages of such a system are: (1) the radiation received is not affected by absorption in the **atmosphere of the Earth,** as is the case for ground-based observatories; this is particularly valuable in the X-ray and ultraviolet range. (2) An OAO is not affected by troubles of **seeing,** and it is this and not size which limit the effective magnification of which terrestrial telescopes are capable. One may therefore expect much better resolution of detail from OAO photographs. (3) Large structures such as the paraboloid 'dishes' of radio telescopes, and mirrors like the 200 in., are extremely heavy and subject to many distortions through their own weight and temperature changes. It is therefore very expensive to build, balance and protect them. The weightless conditions in an OAO remove some of the design problems for heavy components, and make it possible to have quite large and occasionally inflatable structures whose components are relatively flimsy but do not have to withstand great stresses. Current techniques allow instruments to be stabilized and trained with an accuracy of 0·1 second of arc.

The performance of an OAO launched from the U.S. on December 7, 1968, may be taken as an example. It carried 11 telescopes to study the extremely hot, young stars that emit most of their energy in the ultraviolet, and in 30 days collected 20 times more relevant information from a 485-mile high orbit than sounding rockets had obtained in 15 years.

Orbiting geophysical observatories (OGO) are conducting some of the studies mentioned under **Artificial Satellite.** For details of a manned observatory, see **Skylab.**

ORION. Perhaps the most striking of all constellations. It is prominent in the winter sky in northern latitudes. Orion was a hunter in ancient mythology. The three stars in a line are said to represent his belt, and the line below it, including the **Orion Nebula,** contains the stars that are the jewels on his sword.

ORION NEBULA. The brightest of the gaseous nebulae in our Galaxy (see **Galactic Nebulae** for photograph). The naked eye sees it as a haze around the interesting multiple star υ Orionis.

ORIONIDS. A meteor shower with multiple radiants; it is most active about October 20, and may be associated with **Halley's Comet.**

ORRERY. A mechanical model which reproduces the motions of the planets in the **solar system.** Named after the Earl of Orrery, for whom one was made.

A LARGE ORRERY or clockwork model of the solar system, made in 1733. The spheres representing the planets travel in grooves about the Sun at the centre, being raised and lowered by their supporting rods according to the inclinations of their orbits. The brass rings represent the ecliptic (resting on the horses' heads), the celestial equator, the Tropic of Cancer and the Arctic Circle; the two vertical hoops cross at the celestial North Pole. The orrery can keep time with the motions of the planets, and shows their configurations at a glance.

OXYGEN. A chemical element, whose atomic nuclei contain eight protons. Oxygen forms about one-fifth of the Earth's atmosphere, and is essential to almost all forms of terrestrial life. It can be liquefied at normal pressure at a temperature of $-183°$ C.; liquid oxygen is often used as an oxidant in rocket motors.

There is little oxygen in the atmospheres of the other planets, and its abundance on the Earth may be due to its release by green plants.

P

PALLAS. The second largest of the minor planets, discovered by Olbers in 1802. It has a diameter of 304 miles, inferior only to **Ceres.** The orbital period is 4·61 years; the orbit has the exceptionally high inclination of 34° 44′. At its brightest, Pallas shines as a star of magnitude 8. (See **Asteroids.**)

PARABOLA. See **Conic Sections.**

PARABOLIC VELOCITY. The velocity which a particle must possess in order to describe a parabolic **orbit** in the gravitational field of another body. This is a boundary case: a slightly lower velocity causes the particle to move in a closed ellipse, while a slightly greater speed gives it a hyperbolic orbit in which it never returns to the same vicinity. Parabolic velocity is the minimum escape velocity. (See **Orbit.**)

PARALLAX. The difference between the directions in which an object is seen from two places at the same time.

The separation of the viewpoints is called the *baseline*. For example, the separation of our eyes is constantly being used as a baseline, and from the parallax we judge how far away things are. Baselines of thousands of miles between observatories in different parts of the world serve for determinations of the parallax of the Moon, from which can be found the size of its orbit.

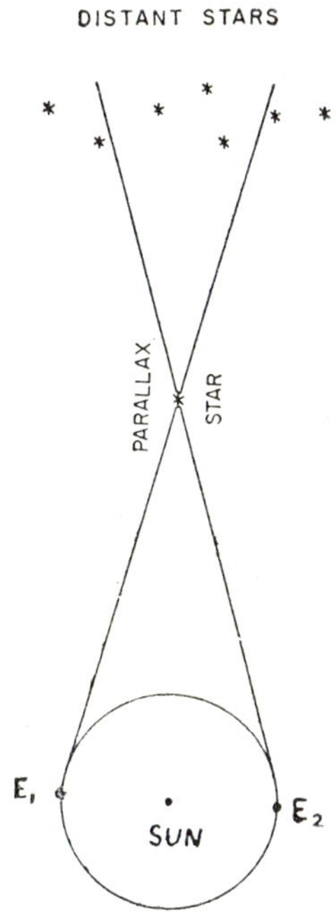

WHILE THE EARTH MOVES THROUGH HALF OF ITS ORBIT, nearby stars appear to shift against the background of the more distant stars. This parallactic shift allows the distances of the nearer stars to be calculated by triangulation.

The solar parallax is never measured directly for a variety of reasons; the parallax of a reasonably close member of the solar system is obtained instead (Venus, Eros and other asteroids have been used) and the solar parallax derived from this. The latter must be accurately known because on it depends the value of the **astronomical unit,** the mean distance of the Earth from the Sun, which is fundamental in many other measurements.

The stars are so far removed that it is necessary to use a larger baseline: the diameter of the Earth's orbit. As the Earth makes its annual journey round the Sun, the nearest stars appear to describe tiny circles against the background of stars which are too far away to show any appreciable parallax. Refined methods permit quite exact distances to be found in this way. (See also **Parsec.**)

PARSEC. A unit of distance equal to 19,150,000,000,000 miles. The word *parsec* is a contraction of *parallax second,* i.e. the distance at which the mean radius of the Earth's orbit would subtend an angle of one second of arc. If a supersonic aircraft had taken off in the days of Julius Caesar it would by now have covered less than one-thousandth of a parsec. Our Galaxy has a diameter of 30,000 parsecs approximately.

The parsec is compared with other units in the table under **Distance.**

PASCHEN SERIES. A sequence of lines in the infra-red region of the hydrogen spectrum. (See **Spectroscopy.**)

PAYLOAD. The weight of everything in a rocket or missile that can be described as 'useful cargo', such as scientific instruments, passengers, supplies or, in the case of weapons, the warhead. The payload is usually less than a tenth of the total weight of the missile with full propellent tanks.

In a step-rocket, the payload of each step except the last consists of the succeeding steps.

PECULIAR A-STARS. These are stars of **spectral type** A with peculiar abundances of metals in their atmospheres when compared with normal A-type stars. Thus the intensity of certain lines in their spectra is not that normally found in A-type stars, but is frequently so much more intense as to indicate an over-abundance of these metals by more than a hundredfold. They generally rotate more slowly than normal A-stars, and some are thought to possess strong magnetic fields. Nothing comparable to these peculiar A-stars is found in stars of other spectral types.

PENUMBRA. See **Umbra.**

PERIGEE. The point in the orbit of the Moon or of an artificial satellite which is nearest to the Earth. Opposite of *Apogee.*

PERIHELION. The point closest to the Sun in the orbit of any member of the solar system.

PERSEIDS. A **meteor** shower with a very regular maximum on August 12.

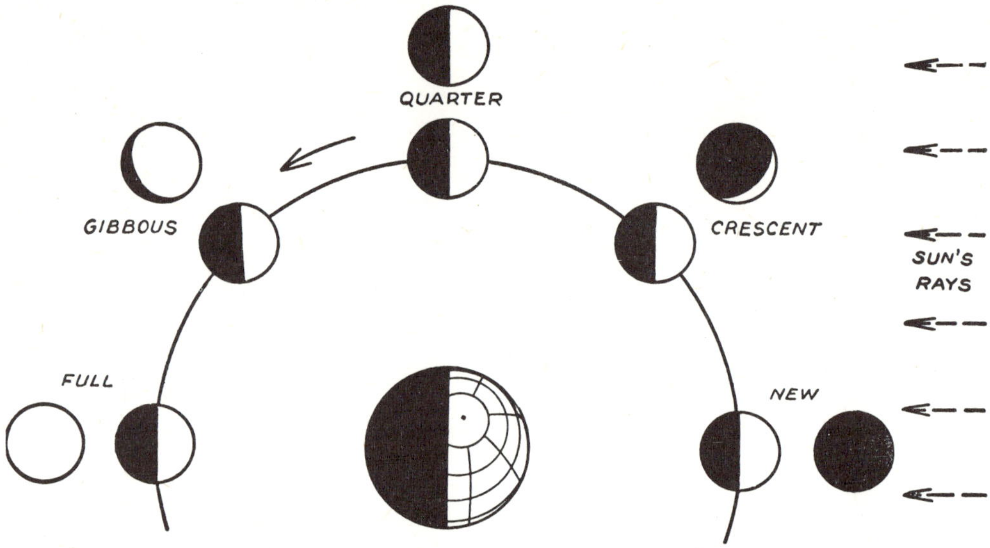

PHASES OF THE MOON; their appearance as seen from the Earth is given by the outer ring of the diagram.

PERTURBATION. The effect of the gravitational pull of one body upon the orbit of another. It leads to small regular or irregular departures from what would otherwise be a smooth orbit having the shape of a **conic section.** Thus the attraction of the Moon causes the Earth to 'weave' somewhat in its elliptical path about the Sun; Jupiter's satellites perturb each other as well as Jupiter itself; and comets may be so perturbed by planets near their course that the latter is entirely changed.

The most accurate determinations of the masses of planets are made by noting the perturbations caused by their satellites or by other planets. Irregularities in the orbit of Uranus led to the prediction of the existence of a planet which perturbed Uranus, and so to the discovery of **Neptune** near the predicted position.

PHASES. The varying apparent shapes of the Moon and the inner planets.

The planets in the solar system, their satellites and our Moon shine by reflected light from the Sun, and consequently only half the surface of each is illuminated at any one time. Unless we happen to see just this half, the object does not appear to be circular in outline but exhibits a *phase*. It may look crescent, semicircular or gibbous. The planets beyond Mars do not show phases because the Earth is so close to the Sun that we are always looking from virtually the direction from which the solar rays illuminate them. Mercury and Venus, like the Moon, can show any phase from New to Full.

The horns of the crescent phase always point away from the Sun. A little consideration will show that the Full Moon must rise roughly when the Sun sets, and that the crescent phase can never be above the horizon in the middle of the night.

PHOTOMETRY. The measurement of the intensity of light. It is of particular importance in the study of **variable stars.**

PHOTON. The 'packet' of light or other **electromagnetic radiation** which possesses one **quantum** of energy. It is owing to the emission of radiation in discrete photons that light sometimes behaves like a stream of corpuscles.

PHOTOSPHERE. The layer of the **Sun's** surface that we normally see.

PHYSICAL UNITS. Some of the physical units and constants that have been used repeatedly in this book are brought together here for convenience, as well as certain basic definitions. Some units are being replaced by SI (Système International) units.

THE METRIC SYSTEM.

1 kilometre (km.)	= 1,000 metres or 0·621 miles.
1 metre (m.)	= 100 centimetres or 39·37 inches.
1 centimetre (cm.)	= 10 millimetres.
1 millimetre (mm.)	= 1,000 microns.
1 micron (μ)	= 10,000 Ångström (Å).

1 GRAM (gm. or g.) is the mass of one cubic centimetre of water at a temperature of 4 degrees Centigrade. (See **Gravitation** for the distinction between mass and weight.)

DENSITY of a substance is its mass per unit volume.

FORCE is that which tends to alter the uniform motion (or rest) of a body. Force is stated in terms of the weight that exerts an equal force, i.e. a force of *a* lbs. weight or of *b* gm. weight.

FORCE can also be measured by the acceleration it can produce: a force of 1 *dyne* will accelerate a mass of 1 gram by 1 cm. per sec. per sec.

WORK is done when a force moves its point of application. It is expressed as the product of the force and the distance through which it has moved its point of application without diminishing or getting stronger, i.e. as *a* ft.lbs. or *b* gram-centimetres, or in ergs.

1 *erg* = 1 *dyne-centimetre.*

ENERGY is the ability to do work. It is measured in the same units as work.

POWER is the rate of doing work. It is measured in units of work per unit time, e.g. in gm.cm./sec. or ft.lbs./sec.

SPEED is the distance covered in any direction per unit time. It is expressed in cm./sec. or miles per hour (m.p.h.) or any equivalent combination, and can never be negative.

VELOCITY is speed *in a given direction.* It is measured in the same units as speed, but a motion in a direction opposite to the given direction has negative velocity.

ACCELERATION is rate of increase of velocity. It is stated as the change of velocity per unit time, e.g. as (centimetres per second) per second. The expression in brackets here represents the velocity change that accrues during one second. A more usual form of writing the units is as *a* ft./sec.^{2} or *b* cm. sec.$^{-2}$. A negative acceleration is a *retardation.*

HEAT is measured in *calories.* One calorie is the amount of heat required to raise the temperature of one gram of water by one degree Centigrade (from 14·5° to 15·5°). One degree Centigrade is one-hundredth of the difference of temperature between melting ice and boiling water (at normal pressure).

Temperature = *heat per unit mass.*

PRESSURE is stated as force per unit area, e.g. lbs. per square inch. It is also given in *millibars* (1,000 dynes per square centimetre) or as the height of a column of mercury which exerts such a pressure. *Normal atmospheric pressure* equals 760 mm. of mercury or 14·79 lbs. per square inch or 1,013·4 millibars.

PLANET. A satellite of a star. The only planets known are the nine belonging to our Sun. Their origin is uncertain, but it is reasonable to suppose that the Sun is not unique among stars in possessing them. As yet, no methods are available by which planets of other stars could be detected.

None of the planets of the solar system give out any light of their own; they are rendered visible by the reflection of sunlight. Strictly speaking, they revolve in orbits not around the Sun but around the common centre of gravity of the whole system. (See **Solar System** and the entries for the individual planets.)

PLANET, ARTIFICIAL. See **Artificial Satellite.**

PLANETARIUM. A theatre with a domed ceiling on to which an intricate optical instrument projects an image of the sky as it

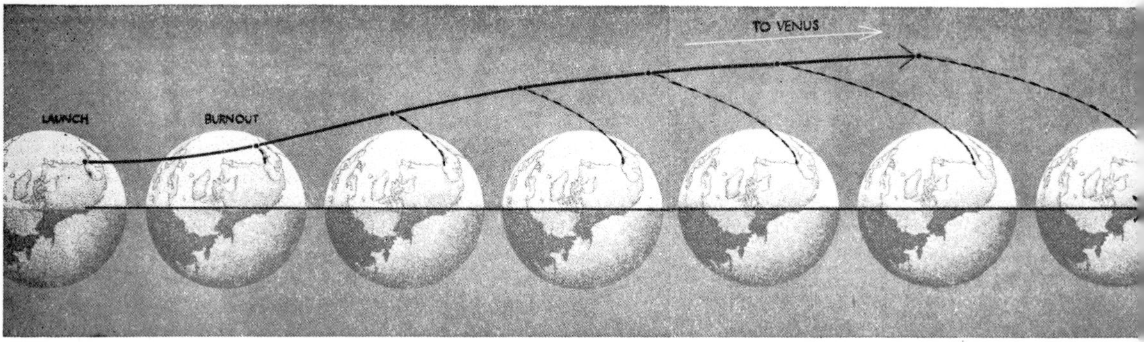

INITIAL TRAJECTORY OF A VENUS PROBE. The Earth is shown viewed from high above the North Pole at intervals of eight minutes. The solid line indicates the path of the probe as seen from this position. To an observer at the launching point it would appear as shown by the broken line. The direction of the Sun is towards the top of the page.

appears at different times from various parts of the globe. The instrument can usually also reproduce the apparent motion of the planets in accelerated form.

PLANETARY PROBE. An unmanned space craft designed to acquire information about a planet by passing near it, orbiting it or landing upon it and transmitting the information back to Earth. This information is not only of academic interest, but also an essential prerequisite for further exploration. The following topics are usually of special interest:

1. *Atmospheric depth, density, temperature and chemical composition.* Without some detailed knowledge of these, soft-landing of instruments cannot be attempted.

2. *Radiation* near the planet. This affects radio communications and the protection of equipment and future space crews.

3. *Nature of planetary surface.* Venus is permanently shrouded in dense cloud; no definite surface features have been observed on Mercury; the markings on Jupiter, including the Red Spot, may well have their origin in the atmosphere; problems of seeing limit the detail that can be discerned on Mars, and the existence on it of craters not unlike the lunar craters in shape and distribution was unsuspected until *Mariner 4* transmitted 20 photographs from its fly-by at 6,118 miles. No surface features have been observed on any satellites of the planets.

4. *Magnetic fields.*

5. Refinement of values for *rotation*, *inclination*, *density*, etc., of the planets, particularly for Venus and Mercury.

6. Conditions in different regions of interplanetary space.

A planetary probe is best launched as a **dawn rocket** into a heliocentric orbit of slightly greater eccentricity than the planet's orbit, i.e. into a **transfer ellipse.** This launching must be timed so that the probe arrives at a neighbourhood of the planet's orbit at the same time as the planet itself. Since the planet and probe reach this area on nearly parallel courses, their relative velocities are moderate. This increases the time available for close observation, and reduces re-entry problems, but greatly lengthens the distance and the time taken for the journey. With more powerful rockets than are at present available, it will be possible to send probes 'straight across' from the Earth to the planet when the two are near **conjunction.**

Mariner 4 was launched in November 1964 and was the first really successful planetary probe. It passed close to Mars eight months

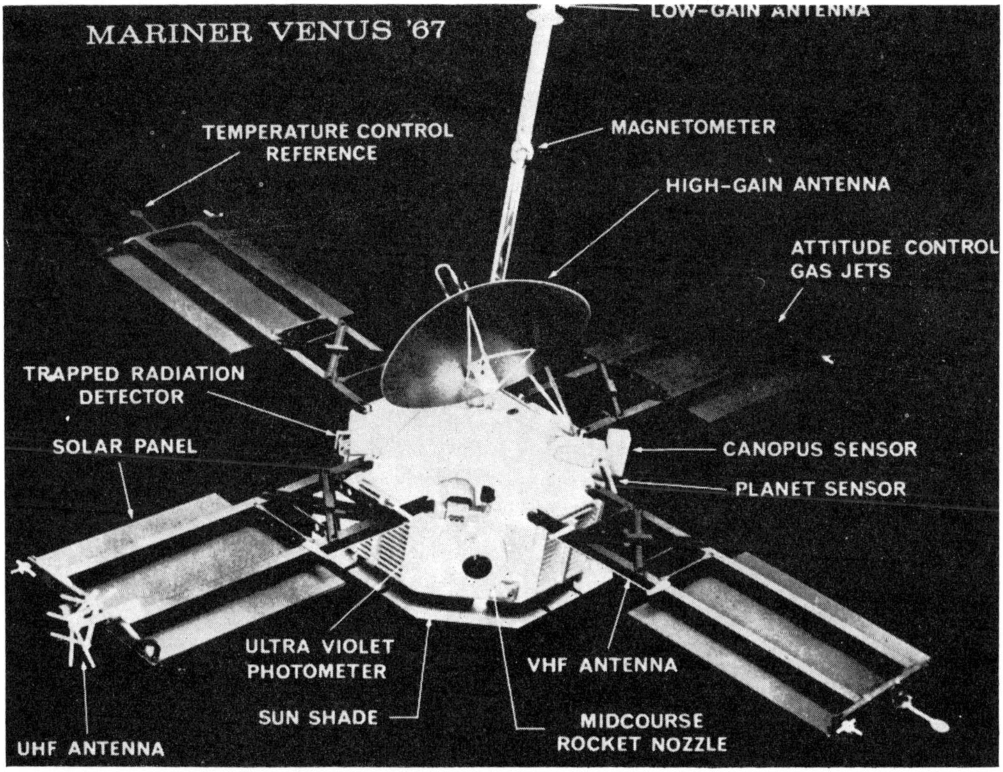

MARINER 5, which made a successful fly-by of Venus in November 1967. Primary Sun Sensors and a plasma probe are located under the bottom. Temperature control louvres can be seen below the beams supporting the solar panels.

later and, apart from the photographs already mentioned, sent back data indicating that there was no appreciable magnetic field or radiation belt. The trajectory was arranged so that the probe was occulted, i.e. passed out of view behind Mars, and the precise character of the disappearance and reappearance of its radio signals yielded valuable information on the atmospheric density. In October 1967 Russia soft-landed *Venus 4* near the equator of Venus. The probe transmitted during its descent through the atmosphere, and there was doubt whether its final data related to the actual surface. Later *Mariner* and *Venus* probes contributed significantly to our present knowledge of these planets (see **Mars** and **Venus**), and *Pioneer 10* was sent on a photographic reconnaissance of Jupiter.

Current U.S. plans include two nine-year

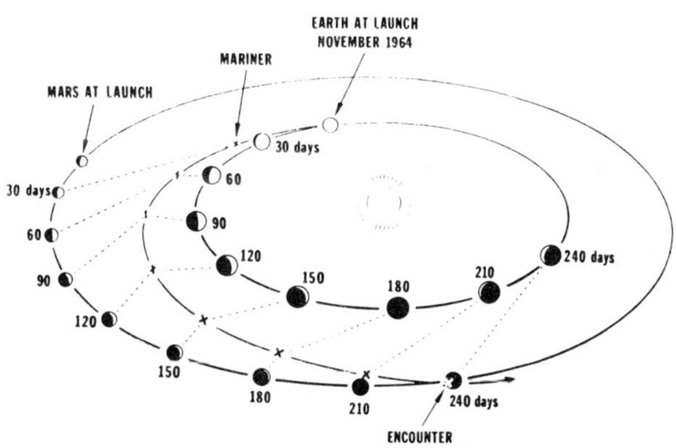

MARS PROBE TRAJECTORY as planned for *Mariner 4* and closely followed by it. It is now in solar orbit.

'Grand Tour' missions to distant planets. The first of these two begins in 1977 and will lead a craft closely past Jupiter and Saturn to Pluto. The second, scheduled for 1979, leads past Jupiter and Uranus to Neptune. Similar trips would ordinarily require about 40 years, but the positions of the planets in the late 1970s will be such that the gravitational influence of each accelerates the craft towards the next planet. Jupiter will be passed in January 1979 and March 1981, Saturn in August 1980, and Neptune in February 1988. (See also table under **Artificial Satellite** on p. 37.)

PLASMA. A fourth state of matter (neither solid, liquid, nor gaseous in the conventional sense). Complete **atoms** and molecules consist of **protons** carrying a positive electric charge, much smaller **electrons** with an equal but negative charge, and uncharged **neutrons.** When electrons are removed from an atom the remaining part has therefore a net positive charge and is called an ion. Ions can also arise through the separation of a molecule into constituents carrying an excess or deficiency of electrons and thus carrying opposite charges totalling zero. A gas which consists almost entirely of ionized atoms or molecules and the balancing electrons is called a *plasma.* It is a good conductor of electricity, and an electric current coupled with a magnetic field can accelerate the particles to very high velocities. This constitutes the basis of a method of rocket propulsion.

PLEIADES. The finest **galactic cluster** in the sky. The naked eye sees it as a group of stars in the constellation Taurus. The Pleiades are mentioned in the Bible and were named the *Seven Sisters* by the Ancients. Only six stars are now normally visible, and there are many old myths about the 'lost Pleiad'; presumably some change has taken place. On a really dark night the sharp eye may be able to detect as many as eleven or even fourteen. The cluster contains about 120 stars in all and is 126 parsecs away from us. It has a diameter of about four parsecs. A colour photograph appears between pages 120–121.

PLUTO is generally called 'the outermost planet', but this term is not always correct. It is true that Pluto's mean distance from the Sun is much greater than that of Neptune, but the eccentric orbit can bring it within the perihelion distance of Neptune. The next perihelion passage occurs in 1989, and from 1969 to 2009 Pluto will not mark the far boundary of the planetary system.

DISCOVERY. Even after Neptune had been found, there were still some unexplained perturbations in the movements of the outer giants. These perturbations were studied by Lowell and W. H. Pickering, and the position of a hypothetical planet was deduced. For a long time the new planet refused to reveal itself; it was eventually discovered in 1930, very close to the predicted position. The delay was due to the fact that Pluto is much smaller and fainter than Lowell had supposed.

ORBIT. Pluto's mean distance from the Sun is 3,666 million miles. The orbital eccentricity is very high, 0·248, and at perihelion the planet approaches the Sun to 2,766 million miles (three million miles closer than Neptune at its nearest), while the aphelion distance is 4,567 million miles. The inclination, 17°, is also much higher than that of any other planet. The sidereal period is 248 years, and the mean orbital velocity 3 miles per second.

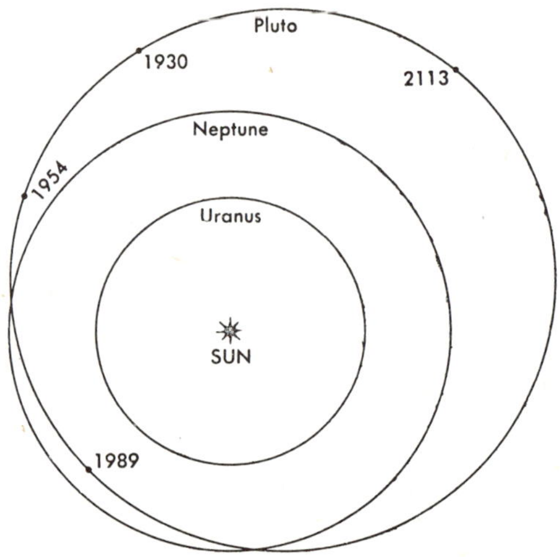

The eccentric orbit of Pluto. Owing to the inclination of the orbit, Pluto and Neptune can never collide.

DIMENSIONS AND MASS. Pluto is not a giant world, and in view of its remoteness its diameter is difficult to measure. According to recent measurements, Pluto is smaller than Mars, with a diameter of 4,000 miles. Assuming a normal density, this would make the mass less than 1/10 of that of the Earth.

This raises problems. A body of this insignificant nature could not perturb giants like Uranus or Neptune to any appreciable extent; yet it was from these very perturbations that Pluto was tracked down. Either we must suppose that its discovery was pure luck, or else we must suppose that Pluto is more massive than the diameter measures indicate.

It seems unlikely that both Lowell and Pickering could have independently reached the correct answer by sheer chance, and equally unlikely that Pluto has a high density. It has been suggested that the apparently small diameter is due to specular reflection, so that Pluto is in fact much larger than we think, and all we see is the light of the Sun reflected from a *part* of its surface.

ROTATION. Measurements of the brightness variations of the planet, made in 1955, indicate that the rotation period is about 6 days 9 hours, though this value is naturally uncertain. The variations are due probably to surface features of unequal albedo drifting across the disc because of the rotation, and there can be little doubt that Pluto, unlike the giants, presents a solid surface.

THE SUN AS SEEN FROM PLUTO, with its faint light reflected in frozen seas of methane. A speculative drawing by David Hardy.

TEMPERATURE AND ATMOSPHERE. Pluto has a mean temperature of −210° C., but the value is naturally higher near perihelion, while the aphelion temperature must be terribly low. A tenuous atmosphere may be retained, but none has so far been detected.

STATUS OF PLUTO. The small size, the strange orbit and the generally anomalous features of Pluto have led to doubts as to whether is should be ranked as a major planet. It is not much larger than **Triton,** and it has been suggested that Pluto is a former satellite of **Neptune** that has been parted from its primary.

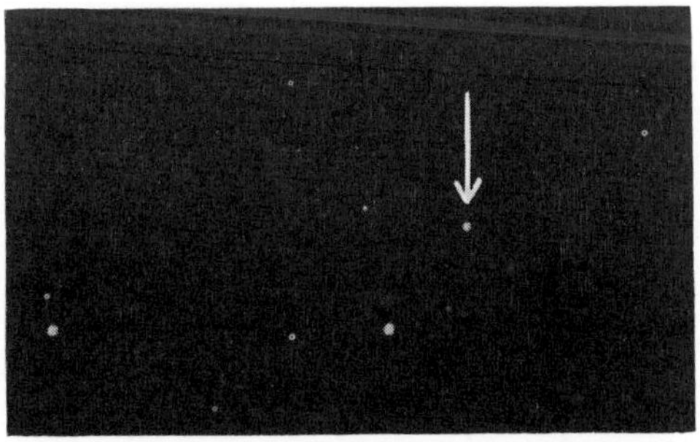

PLUTO'S MOTION IN 24 HOURS. (*Mount Wilson – Palomar*)

Such a theory is interesting, but is difficult to prove or disprove. Further research into the diameter and mass of Pluto may be significant, but is so difficult that it can be carried out only by the world's largest telescopes.

It must be admitted that Pluto is a peculiar world. It was discovered only about forty years ago, but even in that short time it has proved to be one of the most puzzling members of the solar system. (P.M.)

POLAR REGIONS. The areas near the poles of the axis of rotation of a planet. On the Earth, they are defined as the areas within the Arctic and Antarctic Circles, which are 23° 27′ from the North and South Poles respectively. The inclination of the Earth's axis to the 'upright' in its orbit is 23° 27′, so that at a **solstice** one polar region is bathed in sunlight 24 hours a day, while in the other the Sun does not rise.

The polar regions receive as much sunshine as any other place on the Earth, but most of it continuously near midsummer. The Sun's rays always strike the polar regions rather obliquely, so that the latter tend to be the coldest parts of the globe.

POLE STAR (also known as *Polaris*, or α *Ursae Minoris*). A fairly bright star which happens to be close to the north pole of the **celestial sphere.** Owing to this, it describes only a very small circle in the sky, and its bearing is always close to true North. It is a **binary star,** and one of its components is a Cepheid **variable star.** It is easily found, as it lies in almost a straight line with the 'Pointers' of the Plough or **Great Bear.**

POPULATION, STELLAR. See **Hertzsprung–Russell Diagram.**

PRECESSION. A slow change in the direction in space of the Earth's axis of rotation. It arises from the tendency of the gravitational attraction of the Sun and the Moon to pull the Earth's equatorial bulge into line with this attraction. The Earth behaves like a spinning top whose axis is tilted to the vertical at a fixed angle throughout its 'wobble'.

As a result of precession, the poles of the **celestial sphere** describe circles among the stars; a complete cycle requires nearly 26,000 years. It is pure chance that the pole is presently so close to as bright a star as Polaris; about 2900 B.C. α Draconis was the Pole Star; future candidates for the position are α Cephei and α Lyrae.

PRIMARY. The most massive body in any system of bodies which revolve about their common centre of gravity. The Earth is the Moon's primary; every planet is the primary of its satellites.

The term is used in double-star astronomy to denote the *brighter* star of a pair.

PRIME MERIDIAN. The arbitrary zero of longitude on the Earth's surface. It is defined as the **meridian** passing through the Airy transit circle at the Royal Greenwich Observatory.

PRISM. A glass block having flat surfaces inclined to one another. Light entering through one of the faces is bent by refraction; different colours are refracted by slightly different amounts (see **Refraction**) and the colours are therefore separated, and form a spectrum.

PROMINENCES. Masses of glowing gas projected above the solar chromosphere. (See **Sun.**)

PROPELLENT. A substance carried in a rocket to be expelled from the motor as a propulsive jet. A few rockets have only one propellent (*monopropellent*) but usually there are two. They may be either solid or liquid. If solid, they are mixed together and burnt *in situ*; if liquid, they are stored in separate tanks and fed into the motor, where they burn to form a gaseous exhaust jet. The chemical reaction which takes place is called oxidation, and the two propellents are the fuel and oxidant.

There is a theoretical maximum *exhaust velocity* attainable with any given pair of propellents; this could be reached only if *all* the chemical *energy* of the combustion were converted into kinetic energy of the molecules of the jet. The highest exhaust velocity achieved so far is about 10,000 ft. per second. If monatomic hydrogen could be stabilized in liquid form, the energy released when it forms molecules containing two atoms each would be sufficient to raise the theoretical maximum to 56,000 ft. per second. But such a fluid would have great risks of explosion attached to it, far more even than other monopropellents.

Nuclear reactors, on the other hand, can make use of almost any chemically inert working fluid as propellent. If ordinary molecular hydrogen is heated to 4,000° C. above absolute zero by such a reactor it can be expelled in a jet with a velocity of 35,000 ft. per second; this is sufficient to propel a single-stage rocket with a mass ratio of three to one into space. There is no reason why the engineering problems posed by this method should not ultimately be overcome. (See also **Rocketry.**)

PROPER MOTION. The continuous movement on the celestial sphere of a star, as seen from the Sun.

From the Earth, the motion of a star generally appears to be along a wavy rather than a straight line, as the observed shift is caused by a combination of proper motion and parallax.

Each star has its own proper motion, which is determined by comparison of photographs taken at long intervals, or more accurately by recording its meridian passages. The amount of the movement is very small: only 0·3 % of stars brighter than the 9th magnitude have proper motions greater than one minute of arc in 600 years. A few stars exceed this value considerably: the record is held by an insignificant star of the 10th magnitude with a proper motion of one minute of arc every six years.

PROTON. A fundamental particle existing in the nuclei of all **atoms.** It has a positive electrical charge and a mass almost equal to that of the hydrogen atom. The number of protons in the nucleus of an atom determines its atomic number and its identity as a chemical element.

PROTON-PROTON REACTION. One of the two principal nuclear processes which provide the energy of normal stars. (See **Stellar Energy.**)

PROXIMA CENTAURI. The star nearest to the Earth (excepting, of course, the Sun). Its distance is 4¼ light years or 25,000,000,000,000 miles. A red dwarf of apparent magnitude 11·3, it is an insignificant member of the multiple system α **Centauri,** whose two bright components are relatively close together. Proxima is at an immense distance from these two (it is seen 2° away in the sky) and although it is slowly describing its orbit around the centre of the system it will be a long time before it becomes remoter than the bright pair. (See colour plate facing page 64.)

PTOLEMAIC SYSTEM. According to most of the old astronomers, the Earth lay at rest in the centre of the Universe while the heavenly bodies revolved round it at various distances. This system was summed up by Claudius Ptolemaeus (Ptolemy) in his famous 'Almagest', and is thus known as the Ptolemaic System, though as a matter of fact Ptolemy himself was not its originator. (See **Cosmology.**)

PULSARS. Celestial radio sources which emit pulses of emission with a strictly regular periodicity. The first pulsar, known as CP 1919 (Cambridge Pulsar at 19 hr. 19 min. right ascension), was detected in 1967 by radio telescopes at the Mullard Radio Astronomy Observatory near Cambridge. Since then more than thirty of these pulsating radio sources have been discovered. Their most peculiar characteristic is the astonishing regularity with which the pulses repeat, once allowance has been made for **Doppler effects** due to the motion of the Earth about the Sun. Thus CP 1919 emits a pulse of radio emission every 1·33730113 seconds, or at least that is the limit of the accuracy with which it can be measured with atomic clocks! The periods of this fluctuating emission vary from source to source, between about 1/30 of a second to 3 seconds.

Another remarkable feature of the sources is that, while they flash with such strict constancy, the actual strength of the pulses of radio waves being emitted varies in a completely irregular way. The amplitude of the pulse received by the radio telescopes sometimes dies completely but always returns exactly on time when signals are again detected. In addition to the general periodic property of the pulses, a fine detailed structure of the pulses themselves can be observed.

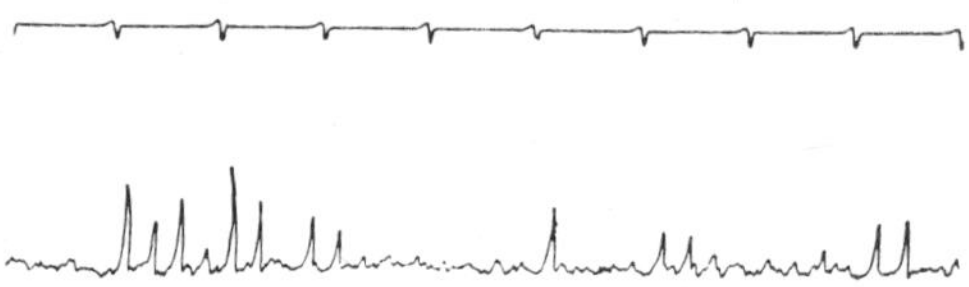

PULSAR PULSES. Sequence of pulses received from pulsar CP 0950 with, above, a laboratory time signal at second intervals for comparison.
(*A. Hewish/Mullard R.A.O., Cambridge*)

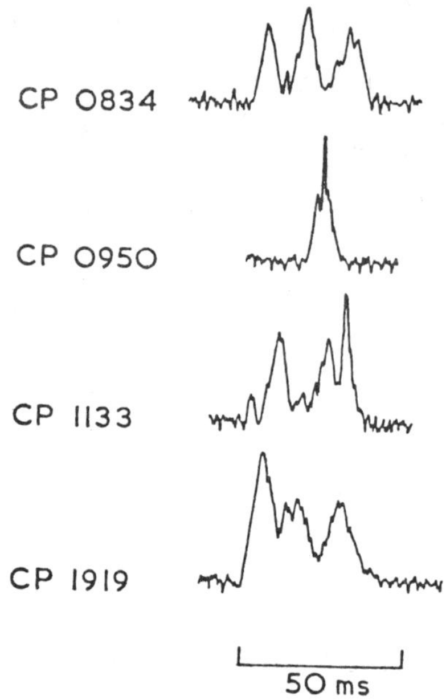

Resolution of individual pulses, showing fine structure, for four pulsars.
(*A. Hewish/Mullard R.A.O., Cambridge*)

The distances of pulsars have been the subject of some controversy amongst astronomers, but all indications now suggest that they are members of our Galaxy. By using radio interferometers (see **Radio Astronomy**) the positions of several pulsars have been determined with sufficient accuracy to enable their optical images to be found. Thus **CP 1919** has been tentatively identified with a faint yellow star, and two other pulsars, **NP 0532** and **NP 0527**, are situated near the **Crab Nebula.** One of these latter pulsars has recently been identified with an optical object whose light fluctuates with a period identical to that of radio and X-ray pulses. Other recent observations seem to suggest that the period of the pulses is slowing down, but at such an extremely slow rate that it can be detected only after a year's observations. In this time a typical pulsar has pulsated about ten million times but the period has increased by only 1 part in ten thousand.

Several theories have been proposed to explain the phenomenon, but so far there is no agreement. It is generally thought that the sources are **neutron stars,** since their size must be extremely small to produce such short pulses. Attempts at accounting for the clock mechanism which controls the periodicity range from orbiting binary systems to rotation or vibration of single stars. Whatever the mechanism, there are reasons for thinking that pulsars are produced in **supernova** explosions, and are now, in some cases, the only remnant. (G.T.B.)

Q

QUADRANTIDS. A **meteor** shower with quite a small orbit; its displays occur on January 3.

QUADRATURE. See **Conjunction.**

QUANTUM. Electromagnetic radiation is not emitted continuously but is always divided into separate small 'packets' of energy named *quanta*. The wavelength of the radiation determines the energy of the quantum, i.e. the size of the packet. One quantum is the smallest amount of energy which can be transmitted at any given wavelength. (See also **Electromagnetic Radiation** and **Photon.**)

QUARK. In the same way as a nucleus of an atom is made up of protons and neutrons, it is now possible to consider the neutrons and protons themselves to be made up of other particles called quarks. There are in fact three quarks, and the name is derived from a sentence in James Joyce's *Finnigan's Wake*, 'Three quarks for Muster Mark'. Apart from protons, neutrons and electrons, physicists have discovered a bewildering number of other sub-atomic particles, the so-called *elementary particles.* It now appears that quarks may be so basic that they may be the true 'elementary particles' out of which all other particles can be constructed.

These particles are expected to have charges of one and two-thirds that of the electron; *all* other known particles have charges equal to integral multiples of the electronic charge. Further, at least one of the quarks is stable and there is no reason to suppose that it cannot have a mass very much larger than any known particle.

It should be stressed that, so far, quarks have not been found experimentally, but the theory which predicts them also predicts other results which have been verified experimentally. This theory has brought a definite order to the astonishing number of different 'elementary particles' found in the 1950s, in that it shows that all these particles can be divided into a small number of groups.

It can therefore be seen that although, at this time, no experimental evidence for its existence has been found, the quark can still be regarded as a very useful mathematical concept.

QUARTZ CLOCK. A type of clock used by observatories for time determinations of the highest accuracy. It relies upon a small disc of quartz, which when set in vibration oscillates at a particular 'natural' rate with great constancy. The quartz is sandwiched between two metal plates; these are connected to an electrical system which both maintains and counts the vibrations of the quartz disc. The counting mechanism constitutes the 'clock'. Quartz clocks are rather temperamental, and several are generally run together to check one another. Although very accurate over short periods – their daily error does not exceed 0·00001 of a second – pendulum clocks are still better time-keepers over periods of a year or more.

QUASAR. The identification of the quasars has, without doubt, been one of the most spectacular observational discoveries of modern astronomy. These objects appear as faint starlike images when photographed by large optical telescopes, yet they are emitting enormous quantities of radio and optical radiation. These strange objects were first called 'quasi-stellar radio sources', 'quasi-stellar objects' or more recently 'quasars'.

In the 1950s, mainly through the work carried out at Jodrell Bank, the structures of many radio sources, especially their diameters, were observed. This programme revealed a source, 3C 48 (see **Nomenclature**), which at the time could not be resolved, but later was shown to have a diameter of less than four seconds of arc. Later, when the first accurate position of this source became available, it was identified with a sixteenth-**magnitude** starlike object, and 3C 48 became the first quasar. Later the position of the radio source 3C 273 was measured with unsurpassed accuracy, using the method of lunar **occultation.** This source was found to have two components and one of them coincided with the image of a thirteenth-magnitude starlike object.

The next major advance was the optical spectrum of 3C 273, which showed strong emission lines (see **Spectroscopy**); these lines, at first sight, appeared impossibly difficult to identify until it was realized that they belonged to familiar elements like hydrogen, except that all the lines show the *same* **red shift.** Even though large numbers of these objects have now been identified, some with red shifts greater than two, the nature of this red shift is still under discussion and no one is prepared to say with certainty whether the red shift is due to the conventional *Doppler* shift of a distant receding galaxy or whether the shift is gravitational (see **Gravitational Collapse**). The properties which characterized 3C 48 and 3C 273 were a small radio diameter, starlike appearance, a red-shifted emission line spectrum and an **ultraviolet excess** in the optical continuum (see **Spectroscopy**); of these the ultraviolet excess has proved particularly useful in identifying new quasars, once accurate radio positions are found.

Another characteristic feature of quasars is that their optical radiation is *variable*; large changes can occur in times as short as one day in the case of 3C 446. The variability of these objects is equivalent to switching on and off the light of some one hundred million suns within short times. This variability sets a limit to the size of the region which is emitting this energy. Essentially the size of the emitting region can be no greater than the distance light travels in the time over which the object varies. Cyclic variations at radio frequencies also take place but, so far, radio variations have not been found to occur over such short periods found in the case of optical variations.

Objects have now been found which are similar to quasars except that they are not emitting radio energy. Hence the optical properties which characterize quasars can be summarized as follows:

(1) Starlike objects sometimes identified with radio sources.

(2) Large ultraviolet excess.

(3) Broad emission lines, with some absorption lines present.

(4) Large red shifts of the spectral lines.

(5) Variable light.

If the red shift of these objects is interpreted as a Doppler shift due to the expansion of the Universe (see **Cosmology**), the distances are so great that they are emitting an amount of radio energy comparable to that of the brightest radio galaxies; further, their optical power is about one hundred times that of a giant galaxy!

The problem of this enormous energy release has, so far, not been resolved, although there have been qualitative attempts at explaining it. Since the origin of these objects must involve an enormous amount of energy, and the variability indicates that these energies are released in very short times, various types of 'violent events' have been suggested. Recently quasars have been found to resemble the nuclei of **Seyfert galaxies** in many respects and certainly some type of explosion is taking place in these galaxies; in fact, it has been suggested that if one **supernova** explosion sets off a chain reaction, a vast quantity of energy could be released in a very short time. Another possible way of producing these large quantities of energy is gravitational collapse.

It is perhaps true to say that, although it is now several years since the first quasar was identified, as more and more observations have become available, the theoretical explanation of these objects has become more difficult. Perhaps the only thing that can be said with reasonable certainty on the theoretical side is that, like radio galaxies, the radio emission is consistent with the radiation being **synchrotron** emission. (T.M.H.P.)

R

RADAR (short for *Ra*dio *D*etection *a*nd *R*angefinding) is essentially an electronic method of seeing.

With the exception of self-luminous objects like the Sun or an electric lamp, the human eye can see objects only if (*a*) they are illuminated by a source of light, such as daylight or a searchlight; (*b*) some of the illuminating light is reflected or scattered by the objects; (*c*) the eye receives enough of the scattered light to evoke a distinct sensation. Radar substitutes a beam of extremely short radio waves (or sometimes a **laser** beam) for the searchlight, and a strongly directional receiving antenna and amplifier for the eye's lens and retina.

The radar beam takes the form of short pulses, and the time which elapses between the emission of a pulse and the detection of an echo from an object is a measure of the object's distance. If the echoed frequency differs from that of the beam by **Doppler shift,** the motion of the object towards or away from the observer can be calculated. Thus radar can detect the position and relative motion of any target which is sufficiently strongly illuminated by the beam and which reflects enough radiation at the frequency which the beam employs.

Many systems use a beam which sweeps round the sky or the horizon, and the received reflections are plotted on a screen by a rotating streak which builds up a map-like picture of the scanned area. The luminous spots of which this picture is composed fade between sweeps, so that a moving object will give rise to a fresh spot in a new position at each sweep. Metallic objects show up best on radar screens, but rainclouds, land formations, and most other solid bodies also give radar echoes. As is the case with light, radar waves travel essentially in straight lines, and their intensity falls off in proportion to the square of the distance from the source.

Basic *primary radar* is therefore inadequate for detecting very small, very distant targets such as a planetary probe, and *secondary radar* is used to increase sensitivity. Here the beam is used not to provide scattered echoes from the target but to trigger off a transmitter in the target which, using its own power source, can return a much stronger signal and even one beamed in the right direction to the first location. *Beacon tracking* dispenses with the illuminating beam altogether and relies solely on receiving continuous or intermittent signals from a transmitter in the target. It is obvious that secondary radar and beacon tracking cannot be applied to hostile military targets.

Simple pulse radar, whilst forming the vast bulk of radar installations in use at present, is being superseded for the more advanced applications by systems with much more complex types of signal, complemented by methods of signal processing which owe much to the relatively recent applications of information theory to radar system analysis. These modern systems may be designed either to obtain extreme concealment of the transmission for a military application, or to use advanced modulation and detection methods to obtain maximum sensitivity and minimum false alarm rate. CW and Doppler methods are being used to achieve remarkable precision in range rate measurements which are needed for space applications, and computer-controlled systems of quite frightening complexity are required for anti-ballistic-missile systems. It is probably fair to say that as much advance in radar techniques has taken place in the last thirty years as has taken place in powered flight over the last sixty, and an article of this length cannot even scratch the surface of what is now an immense field of technology. The following article describes the applications to Astronomy.

RADAR ASTRONOMY. The application of **radar** to bodies outside the Earth. Since celestial bodies are very distant and appear very small, extremely powerful transmitters and receivers are needed if the returning radio waves are to be detectable. Echoes from the Moon were first received shortly after the second world war but it was not until 1961 that there was successful radar contact with a planet. Venus was first, and was followed rapidly by Mercury and Mars. If the same equipment were used with the Moon and Venus, the echo from the Moon would be ten million times stronger than that from Venus. With the largest equipment now available, such as the 1,000-foot diameter radar telescope at Arecibo in Puerto Rico, Jupiter is within range, but to reach all the planets out to Neptune will require an increase in sensitivity of another ten million times.

The time taken for a short pulse of radio waves to travel to Venus and back can be measured very accurately. The waves travel at the speed of light, which is an accurately known physical constant, so the distance of Venus can be calculated. From optical observations of planets over many years astronomers know the *relative* dimensions of the solar system very accurately. Using the radar measurement of the distance of Venus in kilometres, these relative dimensions can be converted to kilometres. In particular we obtain the value of the **astronomical unit** (the mean distance of the Earth from the Sun) much more accurately than by other methods.

The various parts of a planetary surface are at different distances, so the radio waves that they reflect arrive back on Earth at different times. The motion of the planet causes a change in the wavelength of the radio waves by the **Doppler effect.** This change varies over the planetary surface because of rotation, and if the transmitted signal is of just one frequency the echo has a spread of frequencies. The effects of rotation and differing distances vary over the planetary surface in different ways and so from the time of arrival of an echo and its frequency the place on the planet which reflected the echo can be determined. In this way spots have been found on planets which reflect radio waves more strongly than surrounding regions. The strengths of the reflections are related to the properties of the surface layers of the reflecting body and much valuable information about the Moon and nearby planets has been obtained.

From the way in which the frequency spread of the echo varies as a planet moves around the Sun the *rotation period* (or day) of a planet can be found. Venus is covered in clouds which absorb light but allow radio waves to pass freely, so radar is the only way of measuring its period. The results show that this is about 250 days and, what is most surprising, that Venus rotates in the opposite direction to nearly all the other bodies in the solar system. Mercury was thought until recently to rotate in 88 days, which is the same as its orbital period, but radar has shown that the rotation is about 59 days and is probably two-thirds of the orbital period.

The passage of **meteors** through the atmosphere of the Earth causes **ionization** along the meteor's track. These ionization tracks reflect radio waves and radar studies have given us much knowledge about the orbits of meteors in the solar system. (P.R.O.)

RADIAL VELOCITY. The speed of approach or recession of a body from the point of observation.

Radial velocity with respect to the Earth can be determined by measuring the **Doppler shift** between lines of the same elements in the spectra of the star and of a laboratory source on the Earth. It is reckoned to be positive if the star is receding, negative if it is approaching.

RADIANT, METEOR. The point in the sky from which the meteors in a particular shower seem to emanate. (See **Meteor.**)

RADIATION, ELECTROMAGNETIC. See **Electromagnetic Radiation.**

RADIATION BELT. See **Van Allen Belts.**

RADIATION PRESSURE. The force exerted by light or other **electromagnetic radiation** in a direction away from its source.

The Sun's radiation pressure is of special interest, as it has observable effects. The pressure is small for any but tremendously intense radiation, and for large bodies its force becomes entirely insignificant compared with the gravitational force acting in the opposite direction. However, radiation pressure is proportional to the illuminated area of a body, while gravitation depends on the mass, and the latter decreases more rapidly than surface area with decreasing size. Consequently, for bodies below a certain critical diameter the radiation pressure has the upper hand, and such particles are repelled from the Sun. This phenomenon is responsible for the tails of comets. The particles in those tails must be below the limiting size of about 0·0015 millimetres but more than 0·0007 mm. across, at which value radiation pressure again becomes ineffective as the waves constituting the radiation do not 'break' on so small an obstruction and so exert no pressure on it.

RADIO ASTRONOMY. Until recently, astronomical research has been carried out with telescopes and other instruments receiving light waves emitted by the stars in the visual part of the **electromagnetic spectrum.** Auxiliary instruments, such as photo-electric cells and photographic plates, have extended these studies somewhat beyond the visual limits into the infra-red and ultraviolet regions, but great extension is impossible because of the absorption caused by water vapour and fine dust in the Earth's atmosphere. There was, until lately, little astronomical interest in a second, more extensive gap or window in the atmosphere, and yet, radio astronomy has, without doubt, led to the most spectacular discoveries in recent astronomy: it is through the development of radio telescopes that **quasars, radio galaxies** and, most recently, **pulsars** have been discovered.

This other gap exists in the radio wave region. At its short-wave end it also is limited by atmospheric absorption at a wavelength of a few centimetres, and at the long-wave end by reflection from the Heaviside layer at a wavelength of about twenty metres. Any radio wave generated in space within this waveband can be received on the Earth.

THE NATURE OF RADIO WAVES FROM SPACE. Such radiation was first detected in 1931. In 1948 it was found that some of the radio waves were coming from discrete patches in the sky. During the next few years numerous other discrete sources of radio waves were discovered, but for some time it was not at all clear whether these sources were members of our own Galaxy or whether they were extragalactic in origin. It is now known that most of these sources are radio galaxies, a good proportion being quasars.

RADIO ASTRONOMY

Such has been the development and interest in radio astronomy during the last twenty years, and the large amount of observational data which this interest has brought with it, that **radio galaxies, quasars** and **pulsars** demand separate entries. These subjects are dealt with under their respective headings. **Supernovae** are also found to emit radio waves (e.g. the **Crab Nebula**), although this is not at all surprising as radio radiation is thought to be mainly **synchrotron radiation** (see also **Radio Galaxies**). As this process requires charged particles moving near the speed of light, it is probable that these particles are generated in a supernova explosion.

As yet, there is no way of obtaining the distance of a radio source purely from the radio observations and we are almost entirely dependent on the **red shift** of the corresponding optical component (if one exists!). Hence the optical identification of these radio sources has required a close co-operation between the radio and optical astronomers.

RADIO TELESCOPE AERIAL at Pulkovo Observatory.

RADIO TELESCOPES. Once the existence of discrete radio sources was established, the next step was to examine their structure, and to do this it was obvious that radio telescopes with greater resolution (i.e. ability to see fine detail) had to be developed. This development has taken place along with new techniques which have become necessary.

Basically a radio telescope antenna (this corresponds to the objective in an optical telescope) can be one of two types:

AERIAL ARRAY at Kharkov Radio Observatory. 2,000 aerials are spread over 40 acres in the form of a vast T.

(1) a huge concave paraboloidal dish, or

(2) an array of individual dipoles, which is similar to an outdoor television aerial.

The simplest and most direct way of investigating the structure of a radio source is to scan the source with a radio antenna which 'looks' at only a very small area of the sky. This is the so-called *pencil beam*. Only a few sources can be resolved using this method, however (i.e. the pencil beam cannot be made narrow enough), and most information has come from using, instead of just one antenna, two or more reflectors. These are called *radio interferometers*. Basically the instruments work on the same principle as the stellar **interferometer.** So far, the most detailed information has come from the one-mile interferometer at Cambridge. This telescope consists of three reflectors, two of which are fixed while the other is a movable dish.

THE SPECTRAL LINE EMISSION FROM NEUTRAL HYDROGEN GAS. A great deal of attention has lately been given to another type of radio emission from space, generated in the neutral hydrogen gas in the Milky Way on a single wavelength of 21 centimetres. This is a spectral line emitted when the spin of the nucleus in the ground state of a neutral hydrogen atom reverses (see **Spectroscopy**). The hydrogen clouds are tens of thousands of light years distant and contain only about one hydrogen atom per cubic centimetre. In an undisturbed atom a change of spin is only likely to happen once in about eleven million years. It seems, however, that the atoms in their random motions collide every fifty years, and in a collision there is a one-in-eight chance that this transition will occur.

The hydrogen clouds are in motion relative to the solar system and this 21-centimetre spectral line consequently shows a **Doppler shift** in its frequency. This shift has made possible the study of the motions of the hydrogen clouds in the Milky Way, which are obscured from the view of the optical telescopes by dust. Whereas a few years ago there was a good deal of speculation as to the exact structure of the Milky Way system, these uncertainties have been very largely removed. In a few years we have acquired a remarkably complete description of the spiral formation of our Galaxy. The extension of this work to the extragalactic nebulae by the use of larger radio telescopes is eagerly awaited.

THE DETECTION OF METEORS BY RADIO. Many millions of meteors enter the Earth's atmosphere every day; but their trails last so short a time that the accurate measurement of the velocity and direction by visual means is difficult. Photographic techniques are accurate, but unfortunately the interesting events are often obscured by cloud or bright moonlight. However, a part of the energy of the meteor is used in ionizing the air through which it passes, and the resultant column of electrons reflects radio waves. In the last few years, radio echo techniques for the study of meteors have been perfected and the individual orbits of several hundred meteors a day can be measured with a single apparatus.

A great advantage of the radio echo technique is that the study of meteors can be carried out without hindrance by cloud or daylight. The discovery that great streams of meteors are active in the summer daytime is one of the most dramatic results of this new

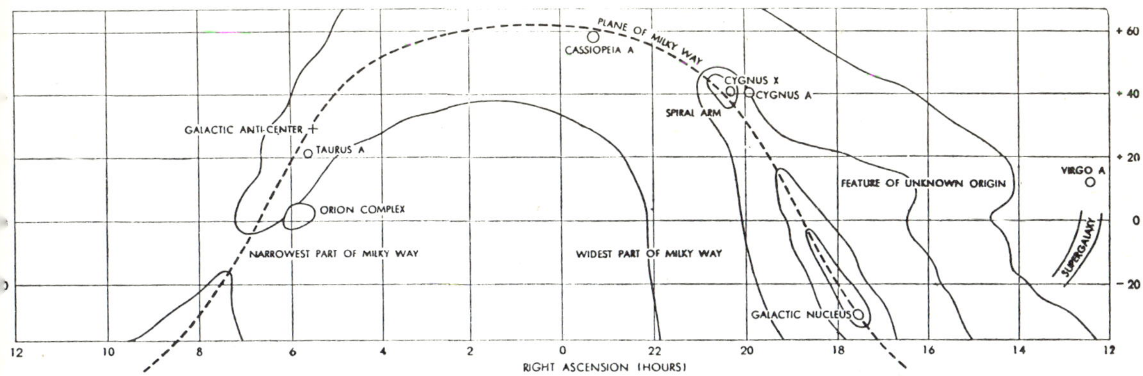

THE RADIO SKY as it would appear to our eyes if they were sensitive to radio waves instead of light. The bright patch on the right lies in the direction of the centre of our galaxy and is unobscured by the dark nebulae that absorb so much of the visible part of the spectrum. Few of the individual radio stars correspond to visual objects.

Left:
THE HORN ANTENNA of the Bell Telephone Laboratories at Holmdel, N.J., looks more like a bizarre, patent windmill than a very sensitive receiver of radio waves from space. It was originally designed to collect signals from Echo satellites, and later communicated with Telstar. It was subsequently used in a successful search by Arno Penzias and Robert Wilson for the background radiation at 3·2 cm. which is believed to have originated at the time of the primordial 'big bang'. This discovery is in good agreement with certain cosmological models (but not with the 'steady state' theory) and any alternative explanation of this radiation must account for the fact that it arrives uniformly from all directions in space.

technique. These daytime meteor streams move around the Sun in orbits of only a few years' period. One stream is almost certainly moving in the orbit of Encke's Comet, but for the remainder no relations with other bodies in the solar system have been established.

Radio investigation of meteors can be used to measure the physical and meteorological conditions in the atmosphere 50 to 120 kilometres above the Earth's surface. Pressures, temperatures, winds and other topics come within the province of these investigations. (See also **Meteor.**)

STUDIES OF THE SUN. The **Sun** itself emits radio waves. These appear to originate in the corona, but when a spot appears on the surface much more intense waves are emitted, and these are known to originate in the region of the spot itself. Occasionally, when the spot activity is considerable, a solar flare forms in the region of a group of spots. Flares have much influence on the Earth's ionosphere; one result is the appearance of aurorae. Radio investigations of the Sun and the ionosphere are being actively pursued, and the surprisingly high temperature of the Sun's corona derived by optical methods has been confirmed.

RADIO TRACKING. Ever since the first **artificial satellites** were launched, radio telescopes have played an important part in tracking them, in receiving signals from them and in sending signals to the satellites to activate part of the instrumentation. The range over which signals from a space vehicle can be received and disentangled from the background of random radio noise depends on the strength of the transmitter and the diameter of the paraboloid bowl or 'dish' of the radio telescope.

RADIO GALAXIES. In the late 1940s large numbers of discrete or localized radio sources were discovered; these were subsequently called *radio stars.* With the discovery in 1954, however, of the optical counterpart of Cygnus A came the realization that these 'radio stars' were in fact *galaxies* emitting radio waves ranging in intensity from ten to a million times more than that of our own Galaxy. The work during the last twenty years has been to examine the structure of these sources, the relation between the radio source and the visible galaxy and to enquire into the physical conditions within the source, its origin and evolution.

The structure of radio galaxies has been investigated using three techniques: that of interferometry, pencil beam antennas (see **Radio Astronomy**) and the method of lunar **occultation.** Most of the information of radio source structure has come from interferometric surveys carried out at Cambridge, Jodrell Bank, Nancay and Owens Valley observatories. The most concrete and perhaps significant discovery that has emerged from these surveys is that about 70 % of the sources investigated have two or more major components; most of these consisting of two well-separated components, one on either side of the parent galaxy, similar to Cygnus A. A few sources show a core-and-halo type of structure, where a bright nucleus lies within an extended, low-intensity halo.

The most common type of galaxy identified with radio sources is the D system (a giant elliptical galaxy with an extended envelope). Other types of galaxies associated with radio sources include dumb-bell and N type systems, i.e. those with brilliant, star-like nuclei and small envelopes. *No* strong radio source has so far been identified with a spiral or an irregular galaxy. In almost every case, the optical component of a radio galaxy shows evidence of some kind of 'violent event' taking place in the nucleus. This, as will be seen in the next paragraph, is probably connected with the mechanism by which the galaxy produces such enormous quantities of radio energy.

It now seems certain that the mechanism by which the radio emission is produced is the **synchrotron** process. This process is well understood and predicts that, provided the magnetic field is not completely 'tangled' like a skein of wool, the radio emission should be polarized. This polarization was indeed found, thereby providing good evidence as to the correctness of the hypothesis. The synchrotron process requires particles moving near the speed of light and it seems probable that their existence is connected to the violent events described earlier.

At present there is no comprehensive theory of the origin and evolution of radio galaxies which explains *all* the various observations. However, since we observe radio sources with similar shapes, but with a wide range of sizes, from double sources contained within the optical galaxy to components having separations of a hundred kiloparsecs or more on either side of the galaxy, it is tempting to suggest that the two components were ejected from the parent galaxy by some type of violent event; in this case we are merely observing sources at different stages of their evolution.

The origin of a radio source involves an enormous amount of energy, sometimes in excess of the maximum nuclear energy release of one hundred million suns. This type of large energy release is also found in **quasars,** and it seems likely that the origin of radio sources is connected, in some way, to these objects. (T.M.H.P.)

RED SHIFT. The displacement of spectral lines towards the red end of the spectrum, measured by *z*, the ratio of the change of

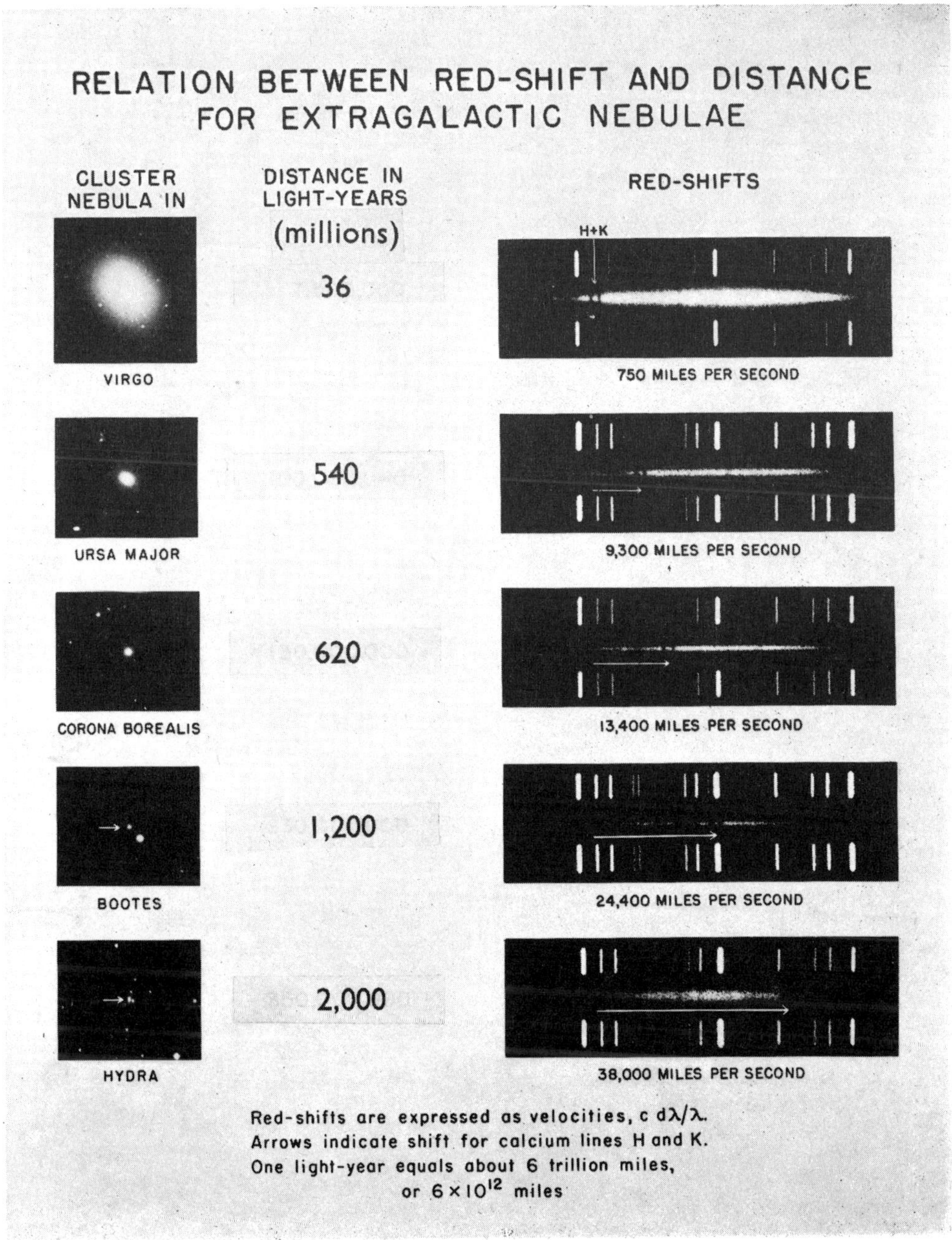

The above plate illustrates the shift of spectral lines towards the red end of the spectrum, which is greatest for the most distant sources. If it is a true Doppler shift, it must be due to the speeds with which the galaxies recede from us and from each other – the expansion of the Universe.

(*Mount Wilson – Palomar*)

wavelength of a spectral line to the undisplaced wavelength. There are two known causes, which both produce the same z for all lines, the **Doppler effect** due to motion away from the observer, and the *gravitational red shift* of the light from massive, very dense objects. All galaxies whose spectra have identifiable lines exhibit red shifts with the exception only of a few local objects. Since galaxies are known to have very low densities their gravitational red shifts are negligible and the observations are interpreted as being due to recession of the galaxies from each other. In the 1920s Hubble measured the distances of galaxies of moderate red shifts and, assuming the red shifts to be due to recession, found the velocity to be proportional to the distance. The best value of the recession velocity per unit distance is 100 km. per sec. per million parsecs (i.e. 20 miles per sec. per million light years); this is known as *Hubble's constant.* Current theories of **cosmology** predict, for red shifts of about one and larger, departures from this law, which vary from one model to another. Because of this the red shift itself is frequently used as a measure of distance. (P.R.O.)

RED SPOT. The only spot on **Jupiter** which seems to be semi-permanent. It was seen in 1831, perhaps much earlier; it became very prominent in 1878, when it was 30,000 miles long and 7,000 miles wide. Since then it has been under fairly continuous observation. Occasionally it disappears temporarily, but it always returns; and it was still red and prominent in 1969.

The nature of the Spot is uncertain. It may be a 'raft', i.e. a solid or semi-solid body floating in Jupiter's outer gas; this theory was supported by the late B. M. Peek, who held that the occasional disappearances were due to the temporary sinking of the Spot into denser gas. Alternatively, the Spot may be a 'Taylor column' of stagnant gas associated with a subsurface feature. It is notable that the Spot is not fixed in position, and drifts about in longitude. It is associated with the Red Spot Hollow in the south equatorial belt, which also vanishes sometimes (as in 1969). During the visibility of the South Tropical Disturbance (1901–40), there were marked interactions when the Disturbance overtook and passed the Red Spot.

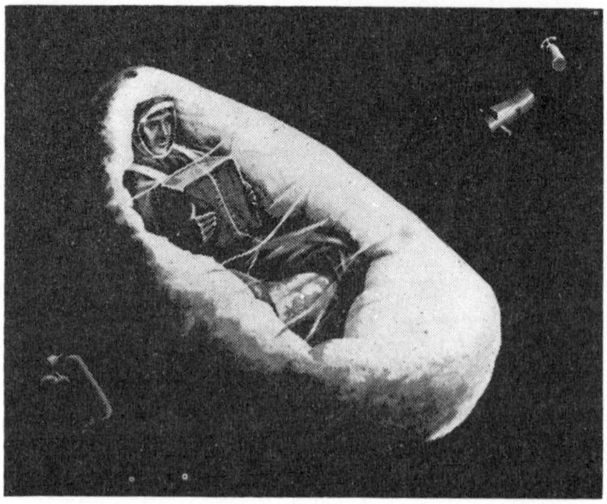

EMERGENCY RE-ENTRY. The Re-entry Systems Department of the General Electric Company has studied the method illustrated above. The astronaut is encapsulated in plastic foam which separates his body from the heat shield. The front of his body is protected by an aluminized plastic sheet through which his attitude/de-orbit package protrudes to slow him down for re-entry. The final descent is made by parachute.

RE-ENTRY. The return into the Earth's atmosphere of a space vehicle or missile. A body re-entering at high speed is liable to be partly or wholly vaporized by the heat generated by atmospheric friction, as if it were a meteorite. This may be prevented by slowing it down with retarding rocket pulses, by providing it with heat shielding, or by the technique of **braking ellipses.**

The favoured method is to have a shield of high thermal capacity sandwiched between an insulating layer on the cabin side and an *ablation layer* on the outside. The latter is made of a composite material (which may include asbestos, fibreglass, ceramic substances or nylon) which vaporizes under extreme heat. Heat of vaporization is therefore continually taken from the layer during re-entry, and swept away with the resulting gases. As the gases are constantly replenished during the ablation, they form a further protective layer round the shield.

During re-entry the sheath of ionized gas which is formed around the vehicle can black out radio communications for several minutes.

REFRACTION. The change in the direction of a ray of light as it passes from one transparent substance into another.

If the boundary between the substances is sharp, as in the case of a glass lens in air, the path of the light is bent through a definite angle at the surface. Where the transition is gradual, e.g. from a dense layer of air into a more rarefied one, the light path is bent into a curve.

The twinkling of stars is due to irregular refraction by moving streams of air of different densities, and is a serious handicap to astronomers. This is discussed under **Seeing.** Atmospheric refraction also affects the apparent **altitude** of objects above the horizon, and in extreme cases gives rise to mirages.

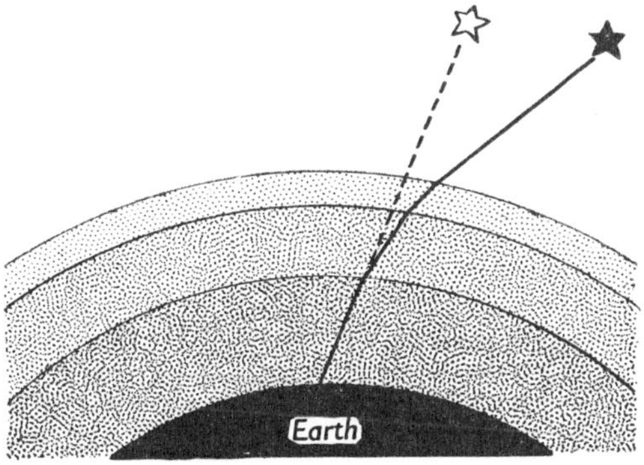

Light from a star is bent as it passes from one layer of air into a lower, denser one. We 'see' a celestial object in the direction from which its light appears to come to the point of observation. The layers of air are not, of course, distinct as in the diagram.

Most light-sources emit light which consists of a mixture of wavelengths (i.e. colours), and shorter wavelengths are refracted more than longer ones. This may be used to separate the constituent colours of a beam of light to form a spectrum, of which the rainbow is a natural example. (See also **Prism** and **Lens.**)

RELATIVITY, THEORY OF. A theory, now universally accepted, proposed by Albert Einstein to account for and predict the results of certain physical observations which could not be satisfactorily explained by earlier theories.

The main stumbling block to the old system of physics was an experiment made in 1887 by Michelson and Morley to demonstrate once more that the Earth moved. It was argued that light was a wave motion, and that there must therefore be some medium in which the waves could travel, in the same way as sound waves need some medium, such as air, in which to propagate. As light seemed to travel everywhere in space, the physicists of the 19th century postulated an intangible aether, pervading the whole of space. There was every reason to suppose that light travelled at the same velocity in the aether whatever the motion of the source; and that the Earth's velocity relative to the aether could be found by measuring the apparent velocity of light in different directions on the Earth. Consternation reigned when the experiment was performed, for it showed beyond doubt that *the Earth was always at rest in the aether!* This result struck at the very foundations of the physics which had stood for so long that its validity was accepted without question. Many scientists attempted to reconcile the experimental result with theory, but in vain.

Einstein, with characteristic irreverence for 'established' physics, and realizing the fundamental importance of the Michelson–Morley experiment, used its result, stated in the form that *the velocity of light appears the same to all observers*, as the fundamental postulate of a new theory; he used also the axiom that *it is impossible to detect absolute rest in space*, i.e. that all velocities must be measured with respect to some other body. On these two foundations Einstein erected his Theory of Relativity. The first instalment, the Special Theory, published in 1905, showed that distance and time were interrelated, and that extreme velocities comparable with that of light could not be added according to the simple rules governing the addition of everyday velocities. Other results were that no velocity greater than that of light could be observed; that the mass of a body increases with its velocity, and that mass and energy are equivalent. The latter relation is written

$$E = mc^2$$

(energy = mass × square of the velocity of light).

The theory aroused little interest at first, as there seemed little prospect of obtaining velocities high enough to test its validity: at low velocities relativistic theory gives results which are virtually indistinguishable from those of 'classical' mechanics. Also, no one had ever witnessed the interconversion of mass and energy; in all operations mass and energy had separately been conserved. However, the second instalment of the theory, the General Theory, published in 1916, had more immediately tangible results. The General Theory incorporated gravitation into relativity theory, proved that it was identical with acceleration and showed that orbits should undergo a continuous precessional motion, such as that of the perihelion of the orbit of Mercury; it gave the exact amount of this precession, removing an anomaly which had long puzzled astronomers. It showed also that light rays should be deflected in passing through a gravitational field. Here was a prediction which could be verified: at a total eclipse of the Sun stars can be seen shining close to the Sun's limb, and the theory predicts that, owing to deflection of the starlight by the Sun's gravitational field, the stars should appear to be shifted outwards slightly. A successful expedition to Brazil to observe the eclipse of May 29, 1919, confirmed this prediction completely.

Many further confirmations of the theory of relativity have since been forthcoming. Light which is emitted from a source in a very strong gravitational field, such as that of a white dwarf star, is found to be slightly reddened as expected by the theory. The increase of mass with velocity is a very important factor which has to be taken into account in designing machines for use in nuclear physics research. The relation between mass and energy has been convincingly demonstrated in the atomic bomb, which obtains its great energy by the annihilation of part of the matter of which it is made.

Except in a few instances and where very great velocities are concerned, the Newtonian law of gravitation and 'classical' mechanics remain excellent approximations to the truth; the theory of relativity is necessary to explain the shortcomings in the older theories only when they are applied to velocities and conditions outside the ordinary range of human experience. (R.G.)

RESOLVING POWER. The ability of an optical system to distinguish separately two closely spaced sources of light. (See **Telescope.**)

RETROGRADE MOTION. Orbital motion within the solar system in a direction opposite to that of the planets, i.e. clockwise as seen from north of the ecliptic. Halley's Comet, some of the satellites of the giant planets, and certain meteor streams have retrograde motion, but the vast majority of bodies in the system revolve in *direct* orbits.

RIGHT ASCENSION. The sidereal time that elapses between the meridian passage of the First Point of Aries and of a celestial body is that body's Right Ascension. It corresponds to longitude on the Earth. (See **Celestial Sphere.**)

ROCHE LIMIT. If a satellite approaches its primary within a certain critical distance, tidal forces shatter it into fragments. This critical distance is the Roche Limit, and is 2·44 radii of the primary from its centre. It only holds for satellites which are held together chiefly by their own gravitation fields and not by cohesion of their materials and therefore does not apply to artificial satellites, or to such small bodies as Phobos, which is only just outside the Roche Limit.

Saturn's ring system is entirely within the Roche Limit, measuring 2·30 radii of Saturn: very possibly it has resulted from the tidal disruption of one or more satellites. The closest of Saturn's satellites, Janus, is 3 radii from the centre of the planet.

ROCKETRY. A rocket is essentially a tube from which mass is ejected at high velocity at one end, so that the tube is driven in the opposite direction by the recoil. Three broad classes exist, deriving their energy from chemical, nuclear or (indirectly) electrical sources. This article describes the basic principles common to all types, and discusses chemical rockets. Separate entries will be found for **Electric Propulsion** and **Nuclear Propulsion.**

In a chemical rocket, propellents burn or react with each other to form gases at high temperatures and at pressures up to forty atmospheres. A nozzle allows the gases to expand and thereby to change the heat energy into energy of motion, so that the gases are ejected at high velocity. This exhaust velocity depends on the temperature of the gases before they enter the exhaust nozzle. Unlike the air-breathing jet engine, a rocket motor does not use atmospheric oxygen for the combustion of its propellents and can therefore function in the vacuum of space.

Rocket motors can employ solid or liquid propellents.

SOLID PROPELLENT ROCKETS are much the less complicated. The propellent tank and the combustion chamber are one and the same compartment: for instance, a gunpowder rocket has a solid mass of propellent hollowed out into a cone-shaped combustion chamber. When ignited, the gunpowder burns only on the inner surface, and combustion proceeds until all the propellent has been converted into hot gases, which escape through an expansion nozzle and in doing so are accelerated to supersonic velocity.

Modern solid propellent rockets use smokeless powders such as cordite. For 'unrestricted burning' rockets the cordite is extruded under pressure through dies which shape the surface of the charge so that burning can take place over a large area and not just at one end or in a hollow cone as in the early gunpowder rockets.

The 'restricted burning' rocket uses a charge which completely fills the inside of the tube. Between the charge and the walls of the tube an inhibitor is used to prevent the passage of the flame. On ignition, the charge burns on the face only and continues to do so until it has all been consumed, like a cigarette. Plastic propellents are often used because they do not deteriorate with storage.

Solid propellent rockets find extensive use as boosters for missiles and research rockets, as the main power plants of small guided missiles, and for the assisted take-off of aircraft.

GUNPOWDER ROCKET

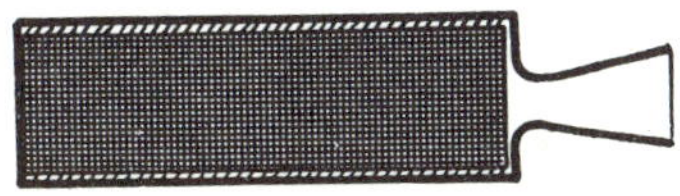

RESTRICTED BURNING

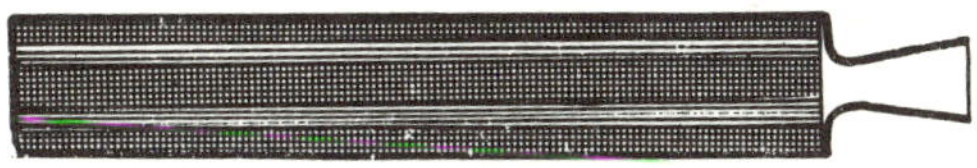

UNRESTRICTED BURNING

LIQUID PROPELLENT ROCKETS. The simplest type makes use of a 'mono-propellent', a single liquid which can be made to break up chemically with a considerable release of energy in suitable circumstances; but mono-propellents are often unstable, and sometimes explode accidentally. An example of a mono-propellent is hydrogen peroxide. It is decomposed in the 'combustion' chamber to give superheated steam and oxygen.

Most rocket units employ two propellents: a fuel and an oxidant. The oxidant is often liquid oxygen, nitric acid or hydrogen peroxide; examples of fuel are petrol, alcohol, aniline and hydrazine. Some pairs of propellents ignite spontaneously when they come into contact in the combustion chamber; some do not. Self-igniting propellents have one disadvantage in that spillage or leaks are very dangerous.

There are several possible methods of feeding the fuel and oxidant from the storage tanks into the combustion chamber. One is to pressurize the tanks with some inert gas, so that when the valves are opened the liquids are forced out; but in the case of large rockets and aircraft power plant, where the duration of thrust may be considerable, pressurized tanks are not acceptable. Pump feeds are employed, and the extra weight of the pumps is offset by the saving in the weight of the propellent tanks themselves because, not having to withstand high internal pressures, these need not be so strong. The pumps are normally operated by a turbine; this is driven by gases from the decomposition of concentrated hydrogen peroxide or from combustion of the propellents themselves. In the latter case it is necessary to cool the evolved gases before they are applied to the turbine blades.

When the liquids enter the combustion chamber they have to be burned with maximum efficiency. Special injectors ensure that they are finely atomized and intimately mixed in the correct proportions. Some injectors rely upon simple impinging jets, others on multiple sprays or on impinging cones of liquids.

of metal capable of absorbing most of the heat transmitted to the motor walls during the firing without becoming too hot. Obviously this method is only practical for motors which have short firing times. The second method was to use a combustion chamber lined with material of poor conductivity, so that the heat transmitted to the structure would be minimized. Difficulties arose owing to the different expansions of the liner and the metal motor wall, and because shocks in the motor could break the brittle material and cause it to flake from the metal.

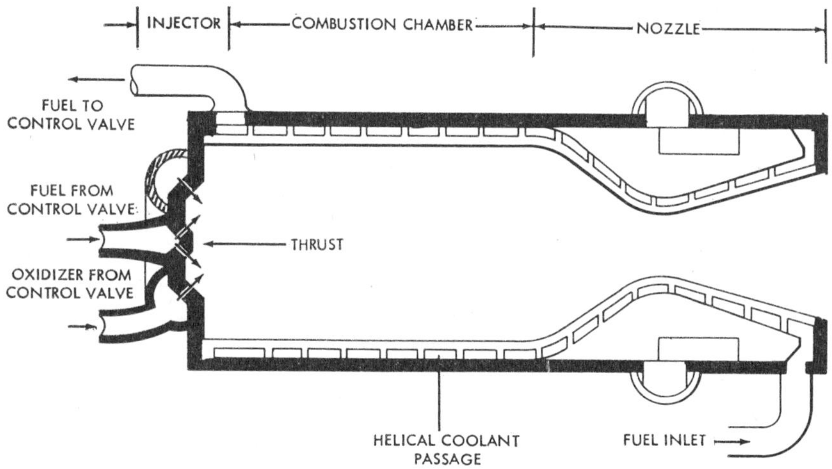

SECTION THROUGH LIQUID PROPELLENT ROCKET MOTOR with regenerative cooling.

In addition to feeding and mixing the propellents, the injectors are often designed to give a controlled sequence of injection. Build-up of one of the propellents in the chamber before the other arrives must be avoided or a damaging explosion may take place. Again, it is inadvisable to start the motor with the propellents being injected at the full operating pressure, and a two-stage build-up of pressure is used in several designs.

When the propellents burn inside the combustion chamber, extremely high temperatures are developed and some method of cooling the motor is essential. Early motors made use of one of two simple methods. The first was christened the 'heat sponge'. In effect the motor consisted of a bulky mass

This method has been considerably improved, and is used in a number of short-life liquid propellent units.

It was, however, *regenerative cooling* which made really large rockets practicable. In this system one of the propellents acts as a coolant and flows round the combustion chamber and nozzle in the space between the double walls before entering the chamber. Heat which would otherwise be lost is carried back into the combustion chamber, and at the same time the coolant keeps the temperature of the chamber walls at an acceptable value. In some motors, supplementary injection of coolant takes place from a series of orifices at strategic positions in the combustion chamber and nozzle to form a cool layer of gases next to the metal wall; this is known as *film cooling*. A further development is to make the walls of the combustion chamber of porous material, so that as the coolant flows round them some seeps through to the inside. Much of the heat which would otherwise reach the wall is used to boil the coolant as it seeps into the chamber. This is called *transpiration cooling*.

Liquid propellents are classed as *storable* or *cryogenic*. Though storable propellents (both fuels and oxidants) may be very unpleasant or corrosive, in principle they may be stored indefinitely in the vehicle tanks, so a vehicle using them may in theory have the same readiness advantages as one using solid propellents. In practice, however, trouble may arise with seals, and as the vehicles are generally more complex than solid-propellent vehicles they will probably require more servicing, so despite some performance advantage over solids, storable propellents have snags. Cryogenic (liquefied gas) propellents suffer from boil-off, and so after the vehicle is fuelled it must be fired before the boil-off reaches the point where insufficient is left for the task in hand. Despite this, cryogenic propellents are generally used for space missions because they offer higher performance and lower cost than most storable combinations. Liquid oxygen is widely used as an oxidant with the very stable hydrocarbon fuels, e.g. in the Saturn 5 Apollo launch vehicle. All-cryogenic combinations such as liquid hydrogen/liquid oxygen are about 40 % more effective and come close to the ultimate in chemical

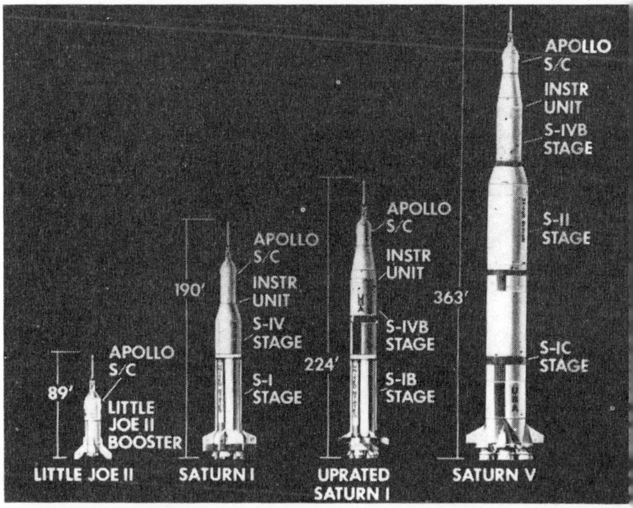

Top: A family of Atlas launch vehicles on the production line at Convair, General Dynamics.

Middle: Various configurations for the Apollo programme. Little Joe II was used in sub-orbital tests of the escape tower system at the top of the service and command modules (S/C). The uprated Saturn I lifted the first American manned flights, and Saturn V launched the manned Moon flights.

Bottom: Final stage in the manufacture of the Saturn IB booster – 80 feet long, 21 feet in diameter, and weighing 93,000 pounds.

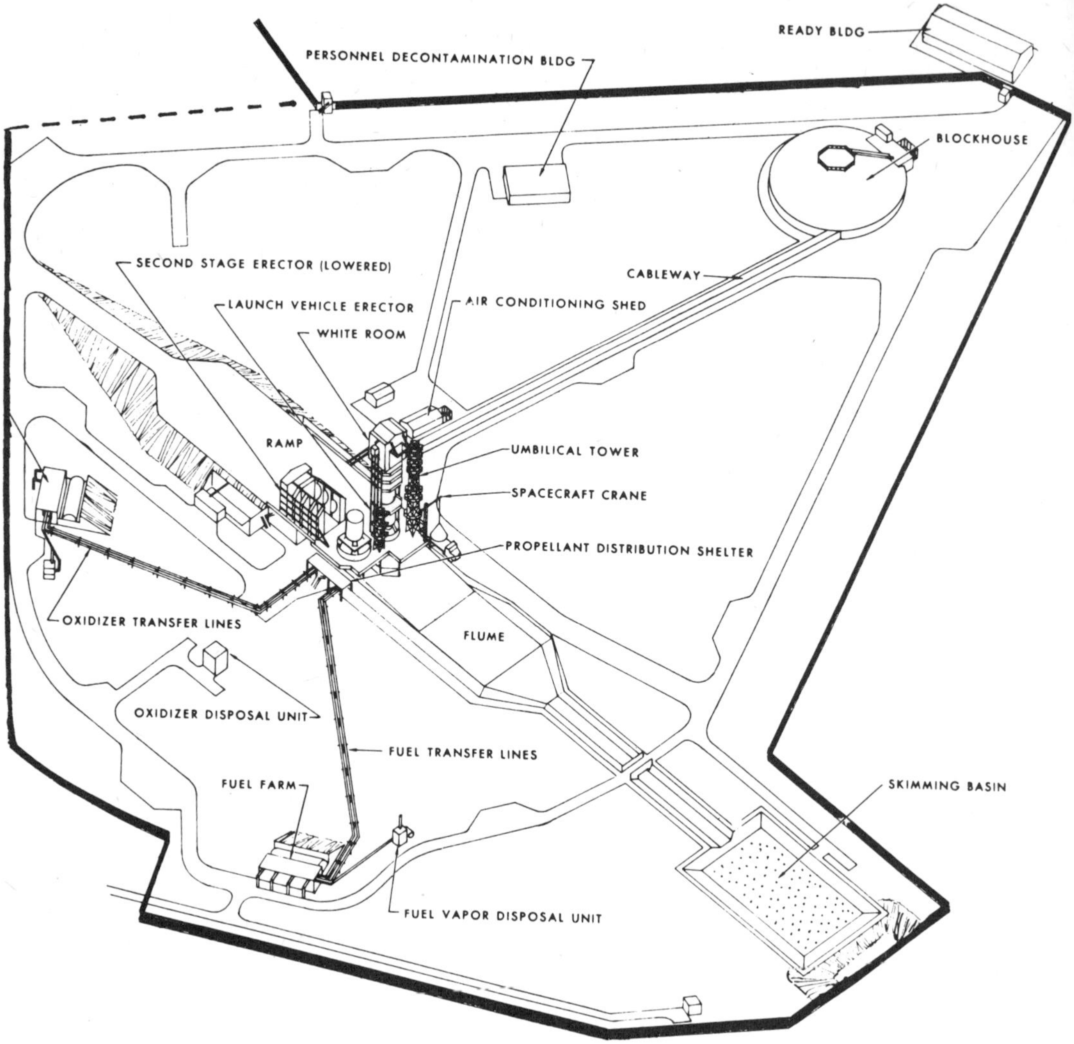
READY BLDG
PERSONNEL DECONTAMINATION BLDG
BLOCKHOUSE
SECOND STAGE ERECTOR (LOWERED)
CABLEWAY
LAUNCH VEHICLE ERECTOR
AIR CONDITIONING SHED
WHITE ROOM
RAMP
UMBILICAL TOWER
SPACECRAFT CRANE
PROPELLANT DISTRIBUTION SHELTER
OXIDIZER TRANSFER LINES
FLUME
OXIDIZER DISPOSAL UNIT
FUEL TRANSFER LINES
FUEL FARM
SKIMMING BASIN
FUEL VAPOR DISPOSAL UNIT

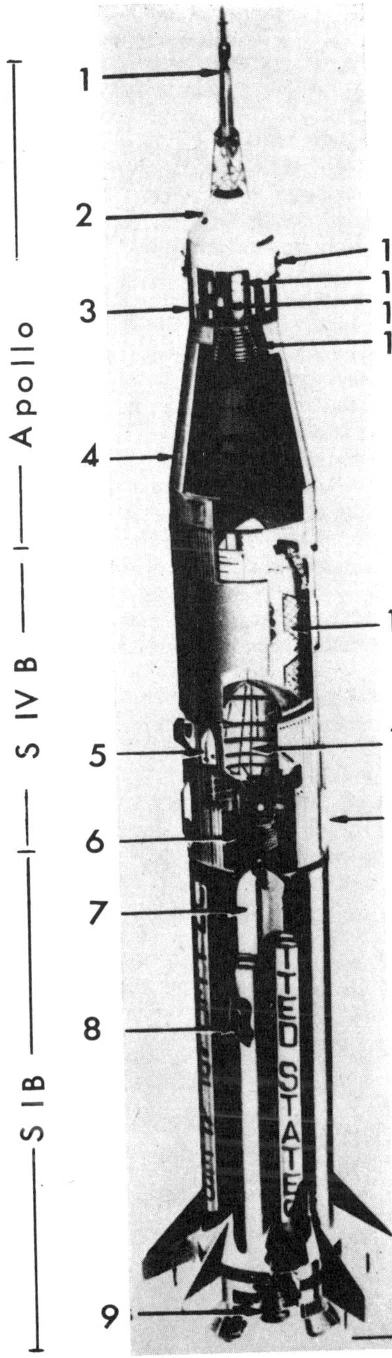

Above: SATURN IB-IVB-APOLLO CONFIGURATION. (1) Launch Escape System; (2) Command Module; (3) Service Module; (4) Lunar Module Adapter; (5) APS Module; (6) J-2 Engine nozzle; (7) Liquid oxygen tank; (8) Fuel tank; (9) H-1 Engine nozzles; (10) Reaction motors; (11) Propellent tank; (12) Helium tank; (13) S.M. propulsion engine nozzle; (14) Hydrogen tank; (15) Liquid oxygen tank; (16) Retro rocket.

Opposite: SPACEPORT AT CAPE KENNEDY. An aerial view of three vehicle assembly buildings, and a plan of a typical launch complex (No. 19).

propellents, liquid hydrogen/liquid fluorine, with its very low molecular weight for the exhaust gas.

The velocity which a rocket can achieve with a given payload and given propellents can be increased by *staging*. This is done by stacking rockets one on top of another (tandem staging) or strapping them side by side (parallel staging) or by a combination of both methods, as on the Soviet Vostok launch vehicle. The first stage of a tandem-staged vehicle is the rearmost part and is fired at take-off. When its propellents are depleted the motor and tank structure is jettisoned and the next stage takes over, and so on until the last stage has fired. Because the empty tankage is discarded step by step, this has the effect of increasing the effective mass ratio of the vehicle. Theoretically the staging principle may be carried as far as one likes, but even very small final stages require very massive earlier stages, and at each step cost and complexity increase while reliability decreases.

The problems of bringing rockets back into an atmosphere from space are discussed under **re-entry.**

GENERAL PRINCIPLES. The reaction force or thrust experienced by a rocket with exhaust velocity V_e is given by

$$F = \dot{m}V_e,$$

where F is the thrust and $\dot{m}$ is the mass expelled in unit time. For motion in a straight line,

$$\log_e \frac{m_o}{m_t} = \frac{V_t - V_o}{V_e}$$

where m_o = initial mass,

m_t = mass at time t,

V_o = initial velocity,

V_t = velocity at time t,

and V_e = exhaust velocity.

If t is the time when all the propellent is consumed, $V_t - V_o$ will be the maximum velocity increment that the rocket can reach, and $\frac{m_o}{m_t}$ is defined as the *mass ratio* of the rocket, the ratio of the initial to the final mass. It will be seen that if the mass ratio is high enough the final velocity can be greater than the exhaust velocity. For example, if the mass ratio is 3, and the exhaust velocity is 8,000 ft./sec. then application of the formula indicates that the final velocity will be 8,800 ft./sec. If the mass ratio is doubled, and the exhaust velocity held constant, the final velocity would be 14,400 ft./sec. Conversely, if the mass ratio is held at 3, it would be necessary to use an exhaust velocity of 13,100 to achieve the same vehicle velocity.

Clearly, for high-performance vehicles, it is desirable that the exhaust velocity and the mass ratio should be large. A useful yard-stick for assessing the performance of a rocket system is its *specific impulse*, usually denoted by I_{sp}, which is defined as the thrust produced per unit mass flow, and has the dimensions of *time*. It works out to be V_e/g seconds, where g is the acceleration due to gravity.

ROCKOON. A device consisting of a rocket suspended from a balloon which is carried to high altitude before the rocket is launched.

ROTATION. In astronomy, the spin of a body about an axis through its centre of gravity. It must not be confused with *revolution*, which is a movement in an orbit about an external point or object.

RR LYRAE. The star which gives its name to a class of variable stars. All RR Lyrae variables are found to be of the same absolute magnitude, just about zero. (See **Variable Stars** and **Stars, Distances and Motions.**)

S

SAROS. A cycle first used by Chaldean astronomers to predict eclipses. An eclipse of, say, the Sun can only occur if the Moon is New and near a **node**; otherwise it passes 'above' or 'below' the Sun. After 18 years and $10\frac{1}{3}$ days (or $11\frac{1}{3}$, depending upon leap years) the Moon is again New and the Sun is again in virtually the same position with respect to the node. The conditions for eclipse hence recur after this period or Saros. By chance, the distance of the Moon also happens to be almost the same at the beginning and end of a Saros, and so eclipses of successive cycles are similar in type – annular or total.

Owing to the odd third of a day in the Saros, succeeding corresponding eclipses are seen in different longitudes, shifting one-third of the way round the Earth at each return; hence an eclipse recurs at the same place only at intervals of three Saros cycles. (See also **Eclipse.**)

SATELLITE. Most of the major planets are accompanied by one or more satellites. So far as is known at present, the Earth has one satellite (the Moon); Mars two, Jupiter twelve, Saturn ten, Uranus five and Neptune two. A satellite of Venus was reported periodically during the 15th and 16th centuries, but is definitely non-existent, and this applies also to an extra satellite of Saturn, Themis, reported in 1904.

Of these 32 satellites, the only one known to possess an atmosphere is Titan, the seventh member of Saturn's family, which is surrounded by a tenuous mantle composed chiefly of methane. Titan has a diameter of 3,500 miles, and is the largest satellite in the solar system. It has, however, only 1/20 the diameter and 1/4700 of the mass of its primary, Saturn, whereas the Moon has $\frac{1}{4}$ the diameter and 1/81 the mass of the Earth.

The origin of the satellite systems is uncertain, but it has been suggested that the smaller bodies – both satellites of Mars, Jupiter VI to XII, Phoebe in Saturn's system and Nereid in Neptune's – may be captured **asteroids.**

Future research may result in the detection of additional satellites, particularly with regard to the four giant planets. (See tables under **Solar System.**)

SATELLITE, ARTIFICIAL. See **Artificial Satellite.**

SATURN is the second of the giant planets. In size and mass it is inferior only to Jupiter, and even at its tremendous distance from the Earth – never much less than 740 million miles – it appears a conspicuous object. When at its brightest, the magnitude is −0·2, so that Saturn is outshone only by four planets (Mercury, Venus, Mars and Jupiter) and two stars (Sirius and Canopus).

Saturn is without doubt the most beautiful object in the heavens. A moderate telescope will show the ring system well, and the planet presents a spectacle that is not only unrivalled, but unique.

ORBIT. The perihelion and aphelion distances from the Sun are 938 and 835 million miles respectively, giving a mean value of 886 million miles. The orbital eccentricity is 0·056, slightly greater than that of Jupiter, and the inclination is 2°.5, while the mean velocity is 6 miles per second. Saturn has a synodic period of 378 days, and thus comes to opposition almost every year. The period is 29·46 years.

Owing to its remoteness, Saturn appears to move very slowly among the stars. The Ancients considered that its dull yellowish glare made it look heavy and baleful.

SATURN taken by the 200-inch in blue light. A colour photograph faces page 240.

DIMENSIONS AND MASS. The polar compression of Saturn is greater than that of any other planet. The equatorial diameter is 75,100 miles, the polar only 67,200. The volume is 763 times that of the Earth, but the mass is only 95 times that of our own globe, indicating a very small density. Saturn has in fact only 0·7 of the density of water, and the surface gravity is only 1·16 times that

These beautiful photographs of Saturn were obtained by Barnard as long ago as 1911. Cassini's division in the ring system is well shown.

on Earth. The escape velocity is 22 miles per second.

TELESCOPIC APPEARANCE. The rings of Saturn were first seen by Galileo, but not clearly enough for him to tell what they were; he believed Saturn to be a triple planet, and it was left to Huyghens, later in the 17th century, to solve the mystery. Little detail can be seen on the globe, even with large instruments; belts are visible, and occasional spots, but the rings are bound to occupy the main attention. A 3-inch telescope will show them adequately, and in a larger instrument they look superb. They lie in the plane of the equator, and twice in the 29½-year sidereal period they are favourably presented; 7½ years later they are seen edge-on, and owing to their thinness they then vanish except with a powerful telescope. The system presented itself fully in 1958, and the last edge-on view was that of 1966.

CONSTITUTION OF THE GLOBE. Saturn is essentially similar to Jupiter, though its size and mass are much inferior (see **Jupiter**). On Wildt's model, the rocky core will be 28,000 miles in diameter, with an ice layer 8,000 miles deep; on Ramsey's, the core of metallic hydrogen will account for 70 % of the planet's radius. The outer layer is not unlike Jupiter's, but with more methane at the expense of ammonia. The greater apparent abundance of methane is due probably to the lower temperature, $-153°$ C. at maximum.

SURFACE FEATURES. The curved belts are not difficult objects, but lack detail. Spots are very rare. The last major outbreak occurred in 1933, when a prominent white spot was detected near the equator. It was not long-lived, but remained conspicuous for a few weeks. Spots of lesser importance have also been seen.

The equatorial zone is nearly always the brightest part of the planet, the poles being relatively dusky. Green and brown hues have been reported from time to time, but are faint and elusive.

ROTATION. The lack of well-marked features on Saturn suitable for determining the rotation period means that the values given are less accurate than in the case of Jupiter. The equatorial period is about 10 hours 14 minutes; near the poles this is longer by perhaps 20 minutes.

THE RINGS. The ring system is of vast extent. From tip to tip it measures 169,300 miles, but the thickness is remarkably small; certainly less than 50 miles, probably no more than 10.

There are three main rings, A, B and C. Ring A, the outer, is separated from B by a well-marked gap known as Cassini's Division, discovered by G. D. Cassini in 1675. Ring B is appreciably brighter than A, while the inner ring, known as C or the Crêpe Ring, is much duskier. A dusky ring outside Ring A has been reported at various times, but has yet to be confirmed.

As early as 1857, Clerk Maxwell proved mathematically that a solid or fluid ring would be disrupted by the powerful pull of Saturn, so that the only system of rings

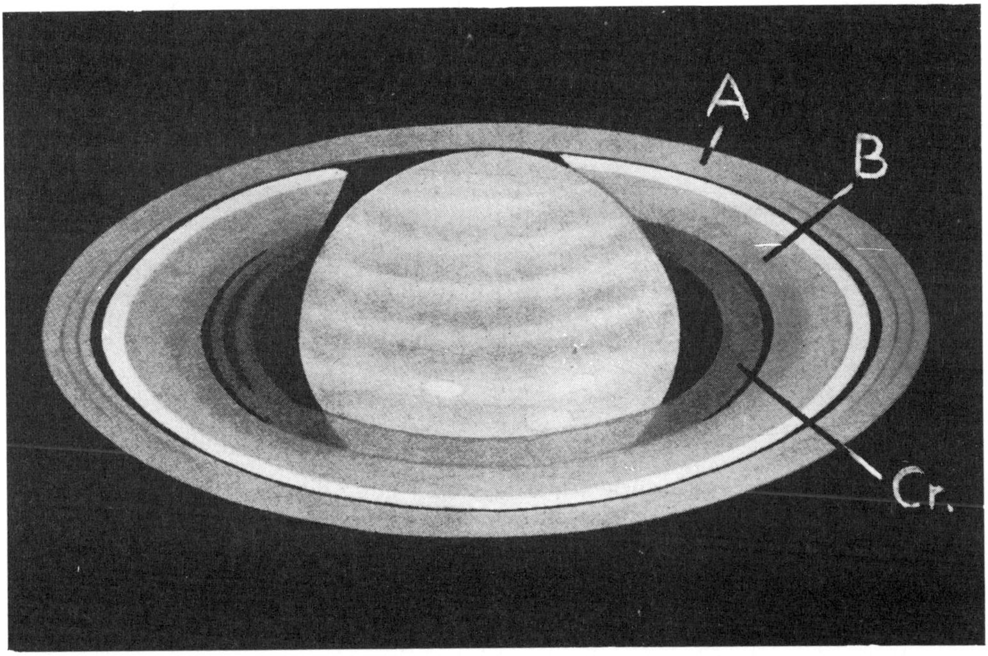

A drawing of Saturn, showing Rings A and B, with the Cassini Division between them, and the more transparent Crêpe Ring (Cr). The narrower divisions are difficult to observe and appear only in parts of the drawing. The tilt of the ring system remains constant as Saturn moves around the Sun, but the angle at which we see it from the Earth changes. Every fourteen years the rings appear fully opened; the next time this happens will be in 1972.

which can exist is one composed of an indefinite number of particles, revolving round the planet with different velocities according to their respective distances. This theory was confirmed spectroscopically by Keeler in 1895. The composition of the particles remains uncertain. It is significant that not even the two bright rings are completely opaque; in 1917 it was noticed that an occulted star remained faintly visible even when behind Ring B.

The Cassini Division is prominent enough to be seen in a small telescope when the ring system is fully presented. There is also a division (Encke's) in Ring A. Other divisions have been reported, but according to Kuiper these are mere 'ripples' and can hardly be regarded as true divisions. All these phenomena are due to the gravitational influence of Saturn's satellites, and the effect is analogous to the **Kirkwood Gaps** in the asteroid zone (see **Asteroids**).

It is curious that the Crêpe Ring, which is by no means a difficult object, should have remained undetected until the middle of the 19th century. It has been suggested that it has increased in brilliance, but such an increase would be very difficult to explain.

SATELLITES. Saturn is attended by ten satellites. Details are given in the table under **Solar System.**

Titan has an escape velocity of over 2 miles per second. It is the largest satellite in the solar system, and is the only one known to have an atmosphere; a tenuous mantle, consisting chiefly of methane, was discovered by Kuiper in 1944. *Iapetus* is variable in brilliance, being at its brightest when west of Saturn; clearly it has a surface which is not uniform in reflective power, and the regularity of its variations is an extra proof that it, like its companions, keeps the same hemisphere turned towards its primary (see **Tidal**

Friction). *Phoebe* has retrograde motion, and alone among the satellites has a highly inclined orbit. Like the minor members of the Jovian family, it may well be a captured asteroid.

EXPEDITIONS TO SATURN. The great distance of Saturn means that no journeys there will be possible until we have progressed far beyond our present technical level. It has been suggested that Saturn's Titan will be visited before any of the satellites of Jupiter, owing to the presence of a methane atmosphere, but at the moment it is pointless to speculate. (P.M.)

SCATTERING. Light which falls on a small particle is deflected and leaves in all directions; this phenomenon is called scattering. Blue light is scattered more readily than red. The particle must be of a diameter small compared with the wavelength of light – of the order of 1/100,000 of an inch.

SCHMIDT CAMERA. A reflecting **telescope** of short focal length and having a lens at the top of its tube, used to photograph the sky.

An ordinary reflector cannot photograph more than about one square degree of sky at a time; it has a parabolic mirror which gives sharp images on its own axis, but definition falls off very seriously away from the axis.

In 1930 Bernhard Schmidt invented the optical system which bears his name. A spherical mirror is used instead of a parabolic one, and a glass correcting plate or lens of complex shape is placed at the centre of the sphere of which the mirror forms a part. The spherical mirror cannot by itself form sharp images of stars, but the correcting plate alters the paths of the light rays in just such a way as to make the mirror act like a paraboloid. A paraboloid has one, and only one, axis about which it is symmetrical, but a sphere is symmetrical about any diameter and therefore gives images of stars which are equally sharp whether they are on the axis of the telescope or not. The correcting plate permits the telescope to take photographs of areas of the sky five to ten degrees in diameter, which are wonderfully sharp right to the edge.

The focus of a Schmidt camera is not flat but curved, and the photographic plate must be similarly curved; sometimes specially

MEMBERS OF THE SOLAR SYSTEM

Upper left:

COMET HUMASON 1961 *e*. The telescope followed the comet as it moved against the background of stars during exposure, and thereby elongated the star images into short streaks.

Upper right:

MARS, with the southern ice cap at about 10 o'clock.

Middle left:

JUPITER, with the Red Spot in the southern hemisphere (*upper half in the photograph*) and the belts clearly visible.

Middle right:

SATURN presenting a fairly open view of the ring system. The Cassini Division and the shadow of the planet on the rings are apparent.

(*Mount Wilson–Palomar*)

Bottom:

A YELLOW DWARF STAR of spectral class G 2 seen from one of its planets at a distance of 93 million miles. The star is about to set and appears reddened owing to light scattering by the planet's considerable atmosphere, which consists largely of nitrogen. Clouds of H_2O vapour drift across the sky. Life forms are silhouetted against the horizon, and in fact almost the entire land surface of this planet is at present covered by living structures and their remains. This view is taken from a position in 51°06′ N., 00°19′ W., well clear of the planet's polar ice cap.

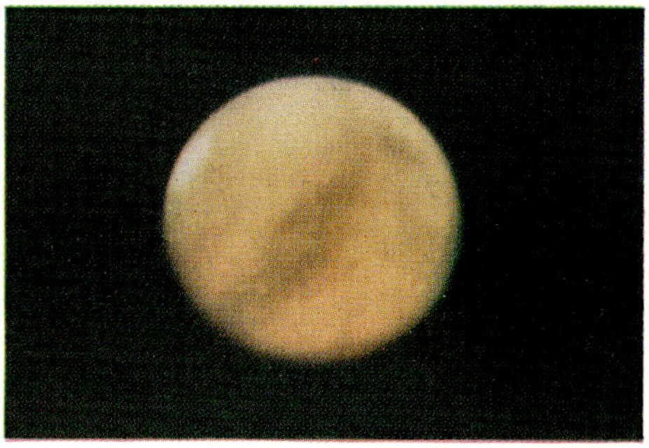

Right: THE NORTHERN HEMISPHERE DURING THE SEASONS. In summer, most of the northern half of the Earth is illuminated by the Sun; in winter, most of it is in darkness, and therefore receives less heat from the Sun. Moreover, the Sun's rays strike it more obliquely during winter. Thus shorter and less effective illumination causes the winter climate to be colder. Conditions are more even in the tropics throughout the year.

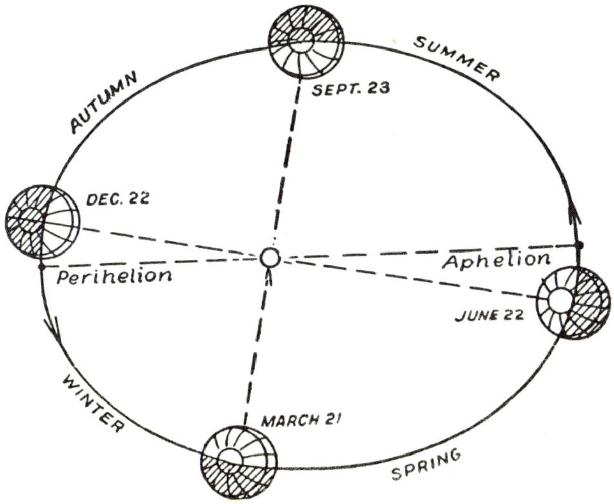

moulded plates are used, and sometimes flat plates are bent in a special holder during the exposure. The curvature is quite small and the glass 'springs' to the required extent.

SCINTILLATION COUNTER. An instrument which detects energetic sub-atomic particles and rays. It has a screen coated with phosphorescent zinc sulphide; a tiny flash of light signals the arrival of a particle.

Opposite:

Top: NGC 3034 (Messier 82), an irregular galaxy in Ursa Major. A black-and-white photograph will be found under **Galaxies,** where it may be compared with types of more definite structure.

Bottom: NGC 6853 (Messier 27), the Dumb-bell Nebula in Vulpecula. This planetary nebula is a sphere of gas thrown off by the hot star in its exact centre. Ultraviolet radiation from the star makes the gas fluoresce. The outer, cooler regions of the gas glow red, and show up best where they are seen edge-on. Both photographs were taken with the 200-inch at Mount Palomar.

SEASONS. The seasons of the year are determined by the direction of the Sun relative to the tilt of the Earth's axis, and therefore by the position of the Earth in its orbit. When the North Pole is tilted towards the Sun, the northern hemisphere is enjoying its summer (despite the fact that the Earth is then farther from the Sun than at other times of the year). Six months later the North Polar Region faces away from the Sun and is not illuminated at all, and the rest of the Northern Hemisphere receives its sunlight more obliquely and for a smaller portion of the day: it is winter in the North and summer in the South.

SEEING. The turbulence in the Earth's atmosphere, which causes stars to 'twinkle' and makes their telescopic images unsteady.

The atmosphere is never at rest, and the layers are constantly shifted by winds which vary with altitude and locally affect the density of the air. This entails irregular **refraction,** the rays of light from a star being bent this way and that. The image in a telescope often appears to jump about and 'boil'. The upper air currents may be seen by pointing the telescope at a planet with a fair-sized disc and withdrawing the eyepiece slowly: this focuses the telescope on high layers of the atmosphere and the winds at different levels can be seen streaming across the out-of-focus planetary image.

Much of the seeing, however, arises near the telescope itself through convection currents, as the instrument retains the heat of the day longer than the surrounding air. Perfect seeing is encountered only on very rare occasions, even at observatories placed high up on mountains.

SEYFERT GALAXIES were first discovered by Seyfert in 1943. They have small, very bright nuclei, but the distinguishing mark which puts these galaxies in a class by themselves is in the spectra of their nuclei, which indicates that some kind of violent event is taking place.

SHELL STAR. A star surrounded by a very voluminous, tenuous shell of gas. These shells have in many cases been thrown off by the rapid rotation of the central star. (See also **Wolf-Rayet Star.**)

SHOCK WAVE. A sudden change in the pressure and density of a gas, caused by the passage of some missile at a speed greater than that of sound in the gas. (See **Air Resistance.**)

SIDEREAL PERIOD. The time between two successive passages of a planet or satellite through the same point on its orbit. It is also called the *period of revolution* or simply the *period.*

SIDEREAL TIME is measured by the stars and not (like civil time) by the Sun. In a sidereal day the stars appear to have made one circuit of the sky. The Sun, however, has not quite completed a circuit as the motion of the Earth in its orbit has displaced the Sun's apparent position slightly eastwards, and the Earth takes about four minutes more to 'catch up' this movement.

The sidereal day is hence slightly shorter than the solar day; it is 23 hours 56 minutes 4·09 seconds of solar time. The two are in step at the autumnal equinox and sidereal time gradually gains on solar time to the extent of a whole day by the next autumnal equinox.

The *sidereal year* is the time between two successive occasions on which the Earth is in the same direction from the Sun. (See **Year.**)

SILICON. A chemical element whose atomic nuclei each contain 14 protons; it resembles **carbon.** It is very common in the Earth's crust and probably in the interior as well, is not well represented in the Sun and stars, but occurs in interstellar dust.

SKY, COLOUR OF. The sky is blue. Sunlight entering our atmosphere is **scattered** by the air molecules. Blue light is scattered much more than red, so that the sunlight is reddened (instead of almost white it appears yellow) and the blue light taken from it is thrown back from molecules in all directions in the sky. As one goes higher in the atmosphere the sky becomes darker as there are fewer air molecules to scatter light.

SKYLAB. A U.S. space laboratory in Earth orbit manned by astronauts who transfer to and from it in docking **Apollo** command modules. (See also p. 258 and under *Skylab* in **Appendix.**)

SOLAR APEX. The Sun, and the entire solar system with it, is moving relative to the neighbouring stars with a speed of about 12 miles per second towards a point in the constellation Hercules. This point is the *solar apex*, and the point diametrically opposite to it from which the Sun is receding is the *solar antapex*.

This motion is quite distinct from that which the Sun, together with the neighbouring stars, performs about the centre of our Galaxy at a far greater speed (about 200 miles per second).

SOLAR SYSTEM. The solar system in which we live consists of one star (the **Sun**), nine planets possessing a total of 32 discovered satellites, and a great many smaller bodies – **asteroids, comets, meteors** and those causing the **zodiacal light.**

The nine planets are described in separate articles, together with their satellites; the **Moon,** however, is discussed under its own heading. The striking characteristics of the system as a whole are given in **Solar System, Origin of.**

SOLAR SYSTEM, ORIGIN OF. The solar system is far from being a haphazard collection of bodies revolving in random orbits about the Sun. A satisfactory theory of

TABLE OF PLANETS

PLANET:	Mercury	Venus	Earth	Mars	Jupiter	Saturn	Uranus	Neptune	Pluto
Diameter (*miles*)	3,100	7,700	7,900	4,200	88,000	75,100	29,300	31,000	4,000?
Mass (*Earth* = 1)	0·05	0·81	1·00	0·11	318	95	15	17	?
Density (*Water* = 1)	5·4	5·0	5·5	3·9	1·3	0·7	1·3	1·6	?
Surface gravity (*Earth* = 1)	0·37	0·87	1·00	0·38	2·64	1·16	1·1	1·4	?
Escape velocity (*miles/sec.*)	2·6	6·3	7·0	3·1	37	22	13	14	7?
Rotation period	58½ days	243 days	23h 56m	1d 0h 37m	9h 55m	10h 38m	10h 40m	14h	6d 9h
Solar distance, mean (*million miles*)	36	67	93	142	483	886	1,783	2,793	3,666
Sidereal period	88 days	225 days	365 days	687 days	11·9 years	29·5 years	84·0 years	164·8 years	248·4 years
Eccentricity	0·206	0·007	0·017	0·093	0·048	0·056	0·047	0·009	0·248
Inclination to Ecliptic	7·0°	3·4°	0°	1·9°	1·3°	2·5°	0·8°	1·8°	17·3°
Synodic period (*days*)	116	584	—	780	399	378	370	367	367
Incl. of equator to orbit	?	?	23·5°	25·2°	3·1°	26·7°	98·0°	29°	7°?
No. of Satellites	0	0	1	2	12	10	5	2	?

origin should at least explain the following conspicuous regularities in the system:

1. The major **planets** all have nearly circular orbits lying in approximately the same plane, and all revolve round the Sun in the same direction, which is also the direction in which the Sun rotates on its own axis.

2. All the large **satellites,** with the exception of the Moon and **Triton,** revolve in nearly circular orbits in the equatorial planes of their respective **primaries.**

3. The distances of all except the two outermost major planets from the Sun are represented by **Bode's Law.**

4. The major planets fall into two groups. Mercury, Venus, the Earth and Mars are relatively small and dense, and have few satellites and long rotation periods; Jupiter, Saturn, Uranus and Neptune are relatively large and have low densities, many satellites and short rotation periods. About Pluto little is yet known.

Two basic types of theory of the origin of the solar system have been proposed: one seeks to show that the system could have arisen from a nebula, or a cloud of dust and gas, while the other invokes some catastrophic event, such as the close passage of another star causing the Sun to eject matter, or the spontaneous disruption of the Sun itself. All theories of the latter class are open to the fundamental objection that the material ejected from a star would be at enormous temperature and pressure, and on being released into space it would certainly explode and could not possibly condense to form planets.

TABLE OF SATELLITES

Planet	*Satellite*	*Mean distance from primary (million miles)*	*Period (days)*	*Diameter (miles)*	*Stellar magnitude (maximum)*
Earth	Moon	0·24	27·32	2,160	−12·5
Mars	Phobos	0·0057	0·32	10	11
	Deimos	0·0146	1·26	5	12
Jupiter	V	0·11	0·50	100?	13
	I Io	0·26	1·77	2,310	5·3
	II Europa	0·42	3·55	1,950	5·7
	III Ganymede	0·67	7·15	3,200	4·9
	IV Callisto	1·17	16·69	3,220	6·1
	VI	7·1	251	70?	15
	X	7·2	254	10?	19
	VII	7·3	260	30?	17
	XII	13·0	620	15?	19
	XI	14·0	692	15?	18
	VIII	14·6	739	15?	17
	IX	14·7	745	15?	18
Saturn	Janus	0·09	0·74	250?	14
	Mimas	0·12	0·94	300?	12·1
	Enceladus	0·15	1·37	300?	11·7
	Tethys	0·18	1·89	700?	10·6
	Dione	0·23	2·74	700?	10·7
	Rhea	0·33	4·52	1,000	10·0
	Titan	0·76	15·95	3,500	8·3
	Hyperion	0·92	21·28	200?	15
	Iapetus	2·21	79·33	800?	10·8
	Phoebe	8·05	550·45	200?	14
Uranus	Miranda	0·077	1·41	80?	17
	Ariel	0·119	2·52	300?	16
	Umbriel	0·166	4·14	250?	16
	Titania	0·282	8·71	600?	14
	Oberon	0·364	13·46	600?	14
Neptune	Triton	0·220	5·87	3,000?	13
	Nereid	3·46	359·4	130?	20

The theories involving a nebula, which could conceivably have been acquired by the Sun from interstellar material, assume the nebula to have been in slow rotation about the Sun; collisions between its component particles would cause it to become disc-shaped. The earliest theory was that of Kant, who supposed that the matter in the disc would collect in local aggregations and ultimately form planets. Another early theory, by Laplace, suggested that the nebula gradually contracted, and in doing so speeded up its rotation to conserve **angular momentum** until successive rings of matter, later to form the planets, were thrown from the equator. On the basis of the theories of Kant and

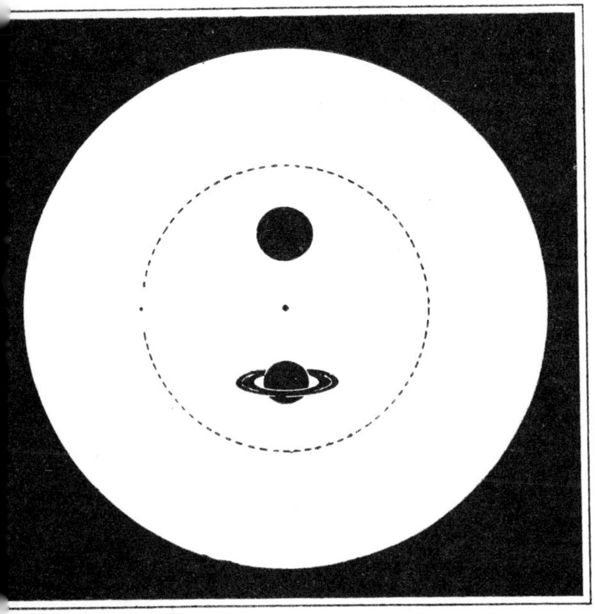

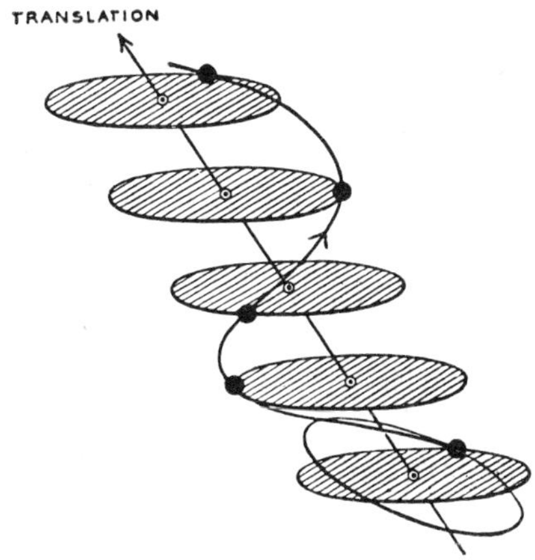

THE SCALE OF THE SOLAR SYSTEM. At top left, the large white disc is the Sun, with Jupiter, Saturn, the Earth and the Moon in its orbit round the Earth drawn to the same scale. At top right, the curved line joining the small black circles shows the Earth's motion round the Sun, while the latter, together with the entire solar system, moves in the direction of the arrow. At right bottom, a medium-sized red giant star is drawn to scale with the Sun (central dot) and the Earth's orbit within it.

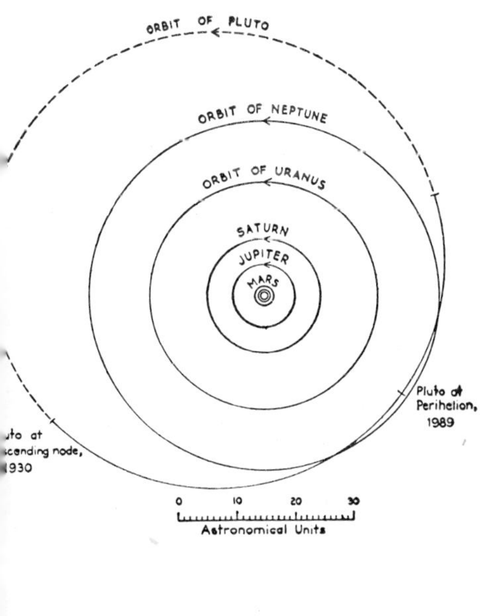

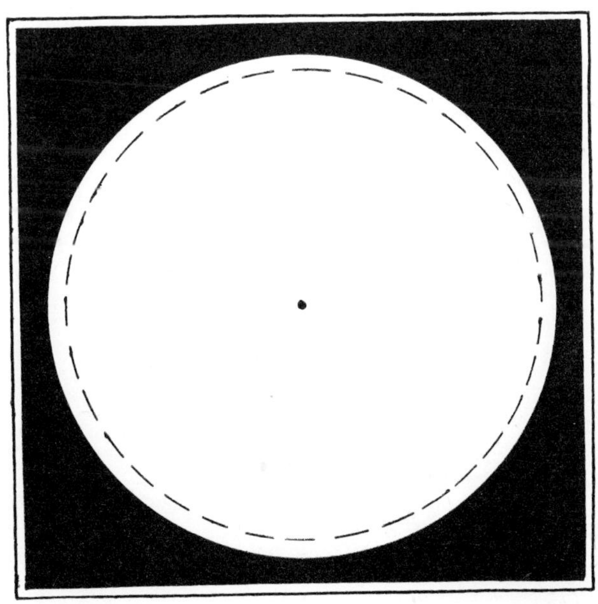

Laplace, the Sun is expected to possess most of the angular momentum of the solar system. The theories are untenable because the planets, possessing little more than a thousandth of the total mass, have 98 % of the angular momentum and the Sun only 2 %.

The most satisfactory theory so far, initially put forward by von Weizsäcker, involves the formation of vortices in the nebular disc. Plausible reasons are given for expecting the vortices to occur in rings, with five vortices in each ring. Between successive rings secondary eddies, rather like roller-bearings, develop gradually. Favourable conditions for condensation and accretion of planetary material exist in these roller-bearings; it is, however, difficult to find a mechanism for the ultimate coalescence of the ten eddies in each ring to yield the planets. Detailed mathematical analysis shows that the theory can give a law such as Bode's relating the radii of the planetary orbits to each other, and can probably also explain the differences between the inner and outer planets; the angular momentum difficulty too can be overcome. Thus, although no theory is completely satisfactory yet, there is hope that further work on Weizsäcker's may yield an acceptable solution to the problem. (R.G.)

SOLAR WIND. A continuous flow of gas outwards from the solar corona, mostly in the form of protons, electrons and complete hydrogen atoms moving with velocities of hundreds of km. per second. The motion of the charged particles is partly due to an interplanetary magnetic field and superimposes the field generated by their motion on it.

SOLSTICE. One of the two occasions during the year when the Sun, in its passage round the **ecliptic,** reaches a maximum distance from the celestial equator, i.e. when its declination is at its greatest. This happens about June 22, the *summer solstice* for the northern hemisphere, and about December 22, the *winter solstice*. At the summer solstice daylight lasts for a longer, and at the winter solstice a shorter, time than at any other period of the year.

SOYUZ. Name of a series of Russian, manned **artificial satellites.** (For *Soyuz 5*, see table on p. 38.)

SPACE LAW. Since the launching of the first space vehicle in 1957, a very considerable number of satellites and space craft have been placed in orbit and, as a result, it has become a question of some practical importance to attempt to determine the height up to which a State might exercise sovereignty over its airspace. Under the municipal law of at least 50 countries, and under the terms of the international conventions concluded in Paris in 1919 and Chicago in 1944, it is recognized that every State has complete and exclusive sovereignty over the airspace above its territory. This recognition has now matured into a rule of customary international law. However, the difficulty is to determine the meaning of the term 'airspace' and there are as many theories as there are theorists. In an attempt to present a down-to-earth point of view on the sovereignty question, special emphasis must be placed on the plain facts of State practice rather than on the speculations of writers and, of course, such practice includes both what States have done as well as the views enunciated by their representatives.

At one time or another, the space above the territory of almost every State has been traversed by some sort of space craft. However, not one single State is known ever to have protested against the launching of such space craft and alleged that they violated its territorial airspace. Neither Russia nor the United States have ever considered it necessary to seek permission from those States over whose territory the space craft was scheduled to pass. Furthermore, not one State has taken the opportunity even to reserve its position in the matter and, on the contrary, most States appear consistently to have acknowledged that outer space is free for the use of all States and cannot be appropriated by any one State.

From such uniformity of practice, it is suggested that it follows that State sovereignty in the airspace above its territory does not presently extend beyond the height at which it is possible to place a space craft in orbit and that, therefore, any State may lawfully place a space craft in orbit and expect other States to acquiesce in such a practice. It is concluded, therefore, that a State under customary international law, international convention and its own municipal legislation exercises exclusive sovereignty in the airspace above its territory but that the exact height to which such sovereignty extends is still undetermined. However, the discussion is deprived of much importance by the conclusion of the Treaty on Outer Space in 1967.

LEGAL STATUS OF OUTER SPACE. The Treaty on Outer Space was signed at Washington, London and Moscow on January 27, 1967, by representatives of more than a hundred countries, has been ratified by the majority including the United States, the United Kingdom and the U.S.S.R., and came into force on December 3, 1968. Its purpose is to ensure that the exploration and use of outer space are conducted for the benefit of mankind as a whole and that it is to be used on a basis of equality. The Treaty provides that outer space, including the Moon and other celestial bodies, is not subject to national appropriation by claim of sovereignty, by means of use or occupation or by any other means, and that States shall carry on their activities of exploration and use peaceably. No State is to place in orbit around the Earth any objects carrying nuclear weapons, and the Moon and other celestial bodies are to be used only for peaceful purposes. Astronauts are to be regarded as envoys of mankind in outer space and all parties to the Treaty are to assist them in every possible way in the event of their being involved in an accident in the territory of any State or on the high seas.

There is to be total reciprocity in the exchange of information about space exploration and use, and States are intended to inform the Secretary-General of the United Nations as well as the public and the international scientific community to the greatest extent feasible and practicable of the nature, conduct, locations and results of their activities; the Secretary-General of the United Nations, on receiving this information, is to disseminate it immediately and effectively.

States on whose registry an object launched into outer space is carried are to retain jurisdiction and control over space objects and their personnel while in outer space or on a celestial body, and ownership of objects launched into outer space, including objects landed or constructed on a celestial body, and of their component parts, is not affected by their presence in outer space or on a celestial body or by their return to Earth.

Naturally, one of the most important and difficult problems is that of liability for space 'accidents' and it seems clear that some measure of liability must be assumed by a State inflicting damage or injury by means of its space craft. Who is to be liable and how that liability is to be enforced is set out in article 7 of the Treaty which provides that each State that launches or procures the launching of an object into outer space is internationally liable for damage to another State or to its natural or juridical persons by such object or its component parts on the Earth, in airspace or in outer space including the Moon and other celestial bodies. However, the problems of enforcement, jurisdiction and relevant law will have to be dealt with when the first incident arises, and this may well pose almost insuperable practical problems. (P.E.M.)

SPACE MEDICINE. The new field of medical science which studies the human and general biological factors involved in space flight, and which provides for the first time a link between medicine and those branches of science which deal with matters of an extra-terrestrial nature. As the method of propulsion on which space flight is based is the

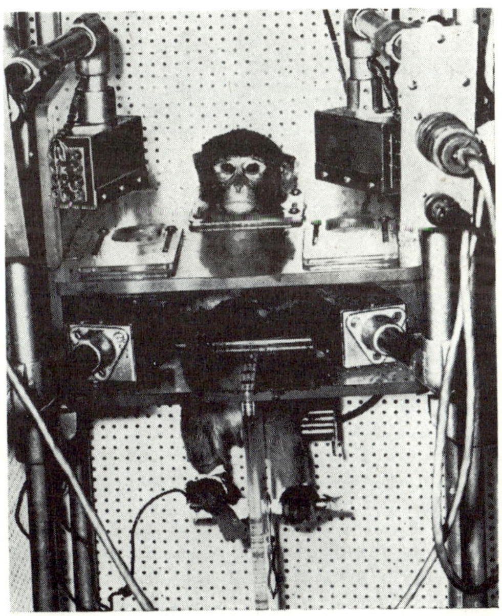

rocket, space medicine is essentially the physiology of rocket flight.

Medical problems in space flight stem, in the first place, from the environment of space *per se* and from the process of movement through this environment. Of special interest to us are the altitudes at which the characteristics of space flight begin, and what protective measures must be taken; one of the most important is the climatization of the cabin. Other problems encountered are the state of weightlessness, high accelerations during launching and during re-entry into the atmosphere, visual problems, the lack of day and night, psychological stresses, and survival outside the cabin in space and on lunar or planetary surfaces.

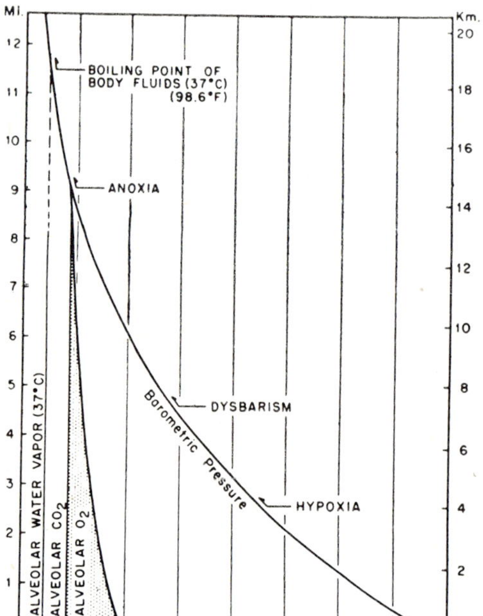

DANGERS OF FALLING AIR PRESSURE. At a height of 2½ miles, the blood no longer carries the normal amount of oxygen. At 4–5 miles, mountain sickness or *dysbarism* sets in: breathing becomes laboured and interrupted, the slightest exertion feels like hard work, and curious mental symptoms begin to show, such as partial loss of memory, extreme stubbornness, dullness and lethargy. At 9 miles, the pressure of the oxygen in the air is lower than in the lungs, and none of the still plentiful oxygen can be absorbed by the body. Pressurized flying suits with breathing equipment can overcome all these difficulties.

PROTECTIVE FUNCTIONS OF THE ATMOSPHERE. At an altitude of 600 miles the atmosphere ceases to be a continuous medium because collisions of the air particles become very rare. This is the physical border between the atmosphere and space. The various functions of the atmosphere for manned flight, however, come to an end at much lower altitudes, some even within the stratosphere. These altitudes are called the functional limits of the atmosphere.

The functions of the atmosphere can be divided into three principal categories: life-sustaining pressure functions, life-protecting filter functions, and flight-supporting aerodynamic functions. In the following, we shall subdivide these further, and shall use ten of the most important functions as a basis for differentiating between atmosphere flight and space flight:

(1) The atmosphere supplies us with oxygen for respiration; in space there is no oxygen. This atmospheric function comes to an end at 50,000 feet. At first glance this seems strange, because the atmosphere contains free oxygen at much greater heights than this. The reason is that the air in our lungs is constantly maintained at a rather high pressure by carbon dioxide and water vapour, both issuing from the body itself. The pressure of carbon dioxide is 40 mm. of mercury, of water vapour 47 – a total of 87 mm. of mercury. The air pressure of 87 mm. corresponds to an altitude of about 50,000 feet. Above this altitude no air can enter the lungs, for the pressure is greater inside than out; the contribution of the atmosphere to respiration is zero, just as if we were surrounded by no oxygen at all, as in space. This is the first of the most important functional limits of the atmosphere, or *space equivalent levels* within the atmosphere.

(2) The atmosphere exerts upon us sufficient barometric pressure to keep our body fluids from boiling. The water vapour pressure of our body fluids is about 47 mm. of mercury; as soon as the pressure falls below this our body fluids will 'boil'. Such a pressure is found at 63,000 feet. This is the second functional border of the atmosphere or space equivalent level within the atmosphere.

(3) In the denser zones of the atmosphere the outside air is compressed to pressurize the cabin; in space, there is no outside air to be compressed. In space we need a new type of cabin which is pressurized from within, a sealed cabin in which a climatically adequate atmosphere for the occupants must be artificially maintained.

(4) In the lower altitudes we are protected from **cosmic rays** by the atmosphere's filter function – in space no such natural protection exists. Below 120,000 feet, the rays lose their original power in collisions with the molecules of the air. Above 120,000 feet, however, we are beyond the protecting shield of the atmosphere, as in space.

(5) In the lower layers of the atmosphere we are protected from the sunburn-producing **ultraviolet** of solar radiation. In space, there is no protecting medium. It is the ozone layer between 70,000 and 140,000 feet altitude that forms a kind of umbrella against ultraviolet by absorbing the larger portion of these rays. Beyond 140,000 feet ultraviolet is effective in its full range as in space. (See also **Space Pollution.**)

(6) In the denser layers of the atmosphere light is scattered by the air molecules producing the blue daylight, against which the stars fade into invisibility; in space the stars are visible against a dark background at all times to the dark-adapted eye. The transition zone from the atmospheric optics to space optics lies at a height of about 80 miles.

(7) From ground level we sometimes see **meteors,** which are vaporized by friction with the air while still at an altitude of 50 to 70 miles. Above this level we are beyond the meteor-safe wall of the atmosphere, as in space.

(8) The lower atmosphere transmits sound waves; at higher altitudes sound propagation becomes impossible, as the air molecules do not collide often enough to transmit the disturbances. The region where this occurs lies between 50 and 100 miles.

(9) The atmosphere provides aerodynamic support or lift for a moving craft; space cannot. The dynamic support from the air ceases at 120 miles, for any speed. This then is the aerodynamical border between atmosphere and space.

(10) The atmosphere contains and transmits heat energy; in space, heat is transmitted by radiation exclusively. In the lower and middle altitudes friction with the air molecules causes high temperatures at the surface of a fast-flying vehicle. Above 120 miles this cannot occur, owing to the low density of the air.

This consideration, based upon ten atmospheric functions, reveals that the larger portion of the atmosphere is equivalent to free space. The region in which we encounter some, but not all, factors typical of space must be considered as partially space equivalent. This begins at 50,000 feet. The region above 120 miles is distinguished by total space equivalence when we ignore some variations caused by the bulk of the Earth itself, its magnetic field, its speed, and its own and reflected radiation. The 120-mile level is, therefore, the final functional limit of the atmosphere.

CLIMATIZATION OF THE SPACE CABIN. This is one of the most important space-medical problems.

Under the conditions of space flight a man may consume in respiration about 1½ lbs. of oxygen per day. This must be replaced in such a way that the pressure of oxygen does not fall below 100 mm. of mercury, since this is about the minimum permissible limit for efficiency; nor exceed 350 mm., because concentrations above this level are toxic. The oxygen used in respiration is converted into carbon dioxide, which is poisonous in concentrations above 3 %.

There is a natural process in our atmosphere which produces oxygen and consumes carbon dioxide. This is the photosynthesis of green plants. It may be possible to utilize this process for cabin climatization; it has been found that 5 lbs. of a certain green alga can meet the respiratory requirements of one man.

Engineering considerations dictate that the lowest total pressures concomitant with unimpaired function are used in space cabins, since this reduces the weight of the pressure vessel necessary to contain the atmosphere. This imposes the use of raised partial pressures of oxygen (pO_2) with the consequent danger of oxygen toxicity. High pO_2 also increases the incidence of lung collapse, which may result in inefficient oxygenation of the blood. Recent research has indicated that exposure to high pO_2 may increase the susceptibility of biological tissues to damage from ionizing radiation.

The use of the alternative of high total pressures with low pO_2 eliminates these difficulties but imposes the high weight penalty already mentioned. It also leads to considerable difficulties in the event of sudden cabin decompression which may result, for instance, from a meteorite penetration of the cabin walls. If the decompression occurred over a period of a few minutes there would be severe physiological effects. (If the decompression were instantaneous it would be fatal, whilst a slow loss of pressure would allow emergency procedures to be initiated.)

In atmospheres with a low pO_2 there is a considerable amount of the inert diluent gas dissolved in the body tissues. At constant environmental pressures there is an equilibrium maintained between the amount of this gas which enters and leaves the tissue. When the pressure is suddenly lowered, however, a smaller quantity of gas can remain in the tissue and the excess is released in the form of bubbles. This gives rise to the painful symptoms of *bends*, caused by the concentration of bubbles at joints. More serious is the formation of bubbles in the blood (where they impair cardiac function), or in nervous tissue (where the emboli may cause permanent incapacitation).

Such difficulties are inherent in space cabins which use diluent gases. Present research is attempting to define the aetiology of the symptoms characteristic of decompression sickness, and to find substitute gases for nitrogen. Helium seems the most useful of the alternatives – neon, argon, krypton and xenon – because much less of this gas is dissolved in tissues exposed to the gas. It has not yet been used exclusively as the diluent in artificial atmospheres.

The inclusion of inert gases in the atmosphere is very effective at reducing the danger of flash fires, in which there is an instantaneous propagation of flames to the entire enclosed volume. The fire in Apollo capsule 204 at Cape Kennedy in 1967 demonstrates how serious the problem can be. The Apollo craft now uses a diluent gas during the launch period.

The advantage of using low total pressure systems in the design and construction of space suits is discussed elsewhere. (See **Space Suit.**) However, it should be mentioned here that the flexibility of American space suits compared with those worn by Russian astronauts is demonstrated by the variety of tasks undertaken in periods of extra-vehicular activity. This is directly attributable to the use in the American space

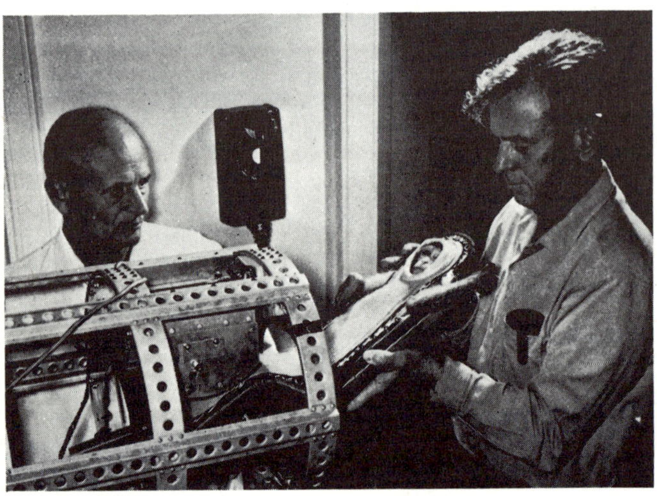

SPACE COUCH for a monkey being inserted into a biopack container before launching in a *Mercury* test. The monkey has already been accustomed to the biopack in ground tests, and is safely recovered later on. Monkeys have co-operated even more actively in space-medical tests by first learning to perform such operations as pressing a certain button upon a signal from a buzzer or a flashing bulb, and then carrying out this task in orbit on receipt of the signal from a monitoring device in the satellite or on the ground.

craft – Mercury, Gemini and Apollo – of 100 % oxygen at about 275 mm. Hg compared with 20 % oxygen and 80 % nitrogen by volume in Russian craft – Vostok, Voskhod and Soyuz.

The removal of carbon dioxide from the space cabin is essential if the effects of increased partial pressures of carbon dioxide in the blood are to be avoided. These effects range from headache, gasping and nausea induced by concentrations of about 3 %, through impaired psychomotor performance at 6 % to 7 % to the production of narcosis, coma and death at concentrations in excess of 10 %. For missions of longer than a few days concentrations of less than 1 % are the more desirable.

The system designed to remove the excess carbon dioxide (resulting from normal metabolic activity) must, therefore, be very efficient. The hydroxides of lithium, sodium and potassium are useful since they absorb carbon dioxide to form their respective carbonates. All the American space craft have used the lithium salt because this absorbs more of the gas per unit weight. There must be a constant supply of fresh absorbing material, which obviously incurs a high weight penalty for long missions. Research is being directed, therefore, to new methods of carbon dioxide removal from enclosed atmospheres. The method used in submarines of physically absorbing the gas in water by passing the atmosphere through a scrubber is prohibitive in terms of weight for space craft use. Of the alternative methods available the biological systems are the most important. Green plants grown in **hydroponic** culture can be used to absorb carbon dioxide by virtue of the photosynthetic reaction

$$6CO_2 + 6H_2O = C_6H_{12}O_6 + 6O_2.$$

The absorbed gas is used in the synthesis of new carbohydrates for cell growth. Incidentally gaseous oxygen is released and may be used to supplement the supply from gaseous or liquid stores, or from chemical reactions.

It is also theoretically possible to use enzymic reactions to control carbon dioxide levels. For instance, the enzyme carbonic anhydrase catalyses the reaction

$$H_2O + CO_2 = H_2CO_3 = 2H^+ + CO_3^{--}$$

and has been tested under laboratory conditions in a closed ecological system. However, much more work is necessary before enzymic systems have practical application.

The use of biological methods of atmosphere control also offers a partial answer to the problem of the disposal of solid faecal waste. At present this is stored until return to Earth. However, it could provide a source of the nitrogenous material essential for plant growth. Meticulous control of the physical conditions of the culture would be required to prevent the toxic effects of any additions.

The disposal of liquid waste also presents a problem. The American practice is to dump liquid overboard, but this is obviously wasteful of water, which carries a high weight penalty. About 4 litres of water per man-day is desirable, but to provide this from a store would impose severe problems for long-duration missions. In these circumstances it is essential to recover waste water. Physical processes such as distillation, freezing and dialysis all produce drinkable water, but the control of smell and taste is difficult. Recovered water may, therefore, be used solely for purposes other than drinking.

Water vapour is released into the atmosphere by respiration and perspiration. These combine to raise the humidity of the cabin. For maximum comfort humidity should be maintained between 5 and 16 mm. Hg. This is effected by an extension of the normal methods used for humidity control.

Control of the space-cabin environment is not limited to these variables. There must, for instance, be efficient removal of solid materials, odours and other trace contaminants (possibly toxic). At present, with missions of short duration, the problem is not a serious one and passage of the atmosphere through activated charcoal removes trace contaminants and odours.

Thermal control of the space cabin is essential if body temperature of 37° C. is to be maintained. There is a constant generation of heat within the space craft from the electrical apparatus and from the bodies of the astronauts themselves. Temperature control is thus a problem of maintaining a balance between the production of heat and heat loss by radiation to the space environment. This balance is achieved by passing the atmosphere through a heat exchanger which is in contact with the outside.

ACCELERATION STRESSES. Lift-off, re-entry and landing impose considerable stresses on the body. Physiologically there is no point in distinguishing between acceleration and deceleration. The important factors are the direction of the force relative to the body, its rate of onset, its magnitude, peak duration, rate of decline and total duration.

It is usual to refer to a head-first acceleration as being along the $+gz$ axis, feet-first as $-gz$, front-first as $+gx$ and back-first as $-gx$. The first two are longitudinal accelerations, the latter transverse.

The body is better able to withstand a transverse acceleration than a longitudinal one. This is because a major influence is exerted by acceleration in the + or − gz axis upon the cardiovascular system. In $+gz$ acceleration the blood is drained away from the thorax and pools in the veins of the lower trunk and limbs. This represents a decrease in blood pressure to the blood pressure sensors located in the upper thorax. The effect of the stimulus is to remove an inhibition from the heart which allows it to beat faster. In normal conditions this would raise the blood pressure. However, under the acceleration stress the blood is pooling in the lower veins and cannot be returned to the heart. It cannot, therefore, maintain the normal output of blood, and it beats even faster in the vain attempt to do so. This has an obvious detrimental effect.

The visual symptoms of *grey-out* and *black-out* are associated with $+gz$ acceleration.

A $-gz$ acceleration tends to pool blood in the veins of the thorax, which represents to the baroreceptors an increase in blood pressure. There is a reflex decrease in heart rate (bradycardia) which tends to allow a decline in blood pressure. There is also evidence that a dilation of the blood vessels (vasodilation) complements the bradycardia, just as a vasoconstriction accompanies the tachycardia (increased heart rate) of $+gz$ acceleration.

A continuation of the $-gz$ acceleration may lead to chronic cerebral congestion with

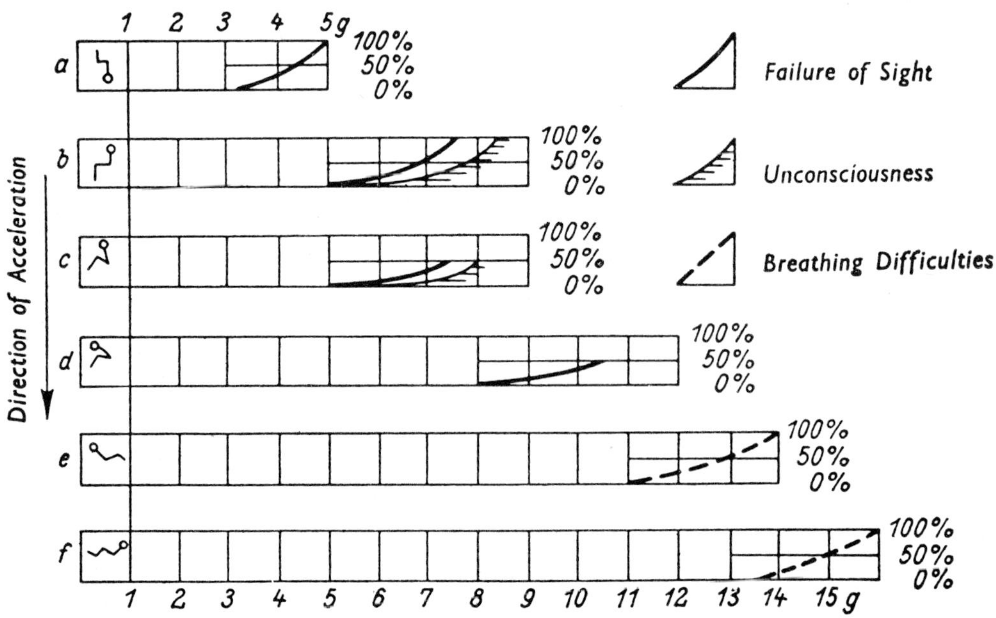

THE EFFECTS OF ACCELERATION FOR VARIOUS BODY POSTURES. Postures are indicated on the left. The percentages apply to the number of tests and not to the degree of the disturbance.
(*H. von Diringshofen*)

GIANT CENTRIFUGE for rotational studies by North American Aviation. Men have lived and worked for thirty days inside the 30-foot cabin while it rotated at the end of the 150-foot beam.

haemorrhage and brain damage. The visual disturbance of *red-out* also ensues.

Transverse acceleration is tolerated much better because there is no major effect upon the cardiovascular system – the accelerating force being applied perpendicularly to the axis of the large blood vessels. The exact effect experienced depends upon the position of the head relative to the trunk. The more inclined the thorax, the greater the increase in heart rate – as would be expected since a 90° inclination represents a $+gz$ acceleration to the thorax. Bradycardia results from $+gx$ in the fully horizontal position.

There are significant effects too upon the respiratory system – the tidal volume (amount inspired at a normal breath) decreasing as the acceleration force increases. The rate of respiration rises simultaneously. This can lead to hypoxia and unconsciousness.

These relatively minor disturbances, together with the absence of major visual effects, favour acceleration in this direction compared with the longitudinal axis. All American and Russian space craft are designed to position the astronaut to accept acceleration and deceleration in the transverse direction. Launch accelerations of 5 to 7 *g*. are routine, and these levels are easily tolerated because there is a careful consideration of the support provided to the astronaut by the launch and operations couch. American astronauts exhibit the common characteristic symptoms of tachycardia and increased respiration rate at launch.

The correct positioning and restraint is also important at times of very large acceleration forces, for instance at a time of on-pad abort, or at an emergency impact on the Earth. If these forces act for less than about 1 second the body reacts as a rigid mass, and in these circumstances restraint and impact-absorbing measures are vital.

During the non-powered phase of space flights **weightlessness** is experienced and this has several physiological effects, the most important of which are on the cardiovascular system. The hydrostatic columns of blood, against which the heart must work when at normal gravity states, are removed. The heart is, therefore, working only against the frictional forces generated by the movement of blood in the vessels. In these circumstances the muscles which control the diameter of the blood vessels and thus assist in the maintenance of blood pressure in response to postural changes are not used. They become flaccid, just as would any other muscle in similar circumstances. As a consequence of the changes (cardiovascular deconditioning) there is a reduced ability to control blood pressure on return to conditions in which postural changes normally induce vascular reaction. Thus at re-entry or on return to Earth a severe incapacitation could result from the poorly controlled movement of blood. Such ortho-static intolerance has been a feature of many manned space flights, and was thought to impose limits to the duration of tolerable weightlessness.

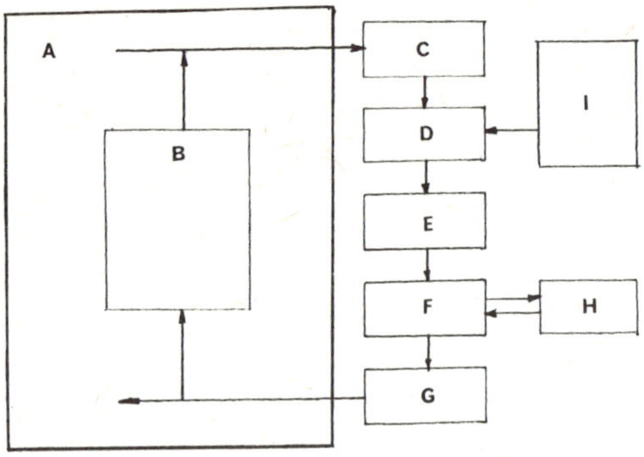

Schematic diagram of a typical environmental control system. The atmosphere from the space cabin (A) and the space suit (B) is passed through a débris trap (C), odour remover (D), carbon dioxide absorber (E), heat exchanger (F), which is in contact with the external radiator (H), and through the humidity control unit (G) back into the cabin and suit. Oxygen is added from the supply (I).

However, as a result of experiments carried out in the GT-VII mission it has been shown that a programme of exercise designed to stimulate the appropriate muscles causes a significant limit to the deconditioning problem. All American manned space flights now carry exercisers, and the intolerance suffered seems to be at a maximum after 96 hours of weightlessness.

The removal of the normal hydrostatic columns and the onset of the deconditioning tend to cause a pooling of blood in the thorax. The effect of this is to cause an increase in urine excretion after about 24 to 48 hours of weightlessness. This is a direct result of hormonal changes induced by distension of the right atrium of the heart, a stimulus which is equivalent to an increase in blood volume. The reflex changes following result in the kidneys excreting more urine. The overall effect of these changes is to reduce plasma volume.

The number of circulating red blood corpuscles is also decreased. Changes both in the fluid volume and in the corpuscular count are mitigated by an exercise programme.

The absence of physical stresses in the weightless state induces characteristic changes in bone. In normal gravity conditions there is a balance between the overall rate of bone deposition (osteoblastic activity) and the rate of bone resorption (osteoclastic activity), which is influenced by mechanical stress amongst other stimuli. Removal of one of the influencing factors leads to a dominance of the resorptive process, probably because of the removal of tiny electrical disturbances in the bone which are essential for the deposition of calcium. The loss of calcium continues as the weightlessness continues, and losses of total bone mass of some 6 to 10 % are common in space flights lasting more than a few days. The bone mass is recovered after return to normal gravity conditions at a rate of about 50 % in 6 weeks. Exercise is known to retard the loss of bone mass.

Weightlessness results in a reduction in stimulus level of the gravity-sensitive otolith organs of the inner ear. These are used, together with the eyes and pressure- and tension-sensitive nerve endings, to maintain a stable spatial orientation. A reduction in overall sensory information regarding orientation may cause abnormal righting reactions, and it is possible to suffer severe symptoms. However, only Titov has experienced any vestibular disturbances of this sort during a space flight, and it is generally considered to be a very subjective response.

RADIATION HAZARDS. The sources of danger are cosmic rays, which are powerful, and corpuscles in the solar radiation and the van Allen belts, which are plentiful.

Cellular damage by radiation is a result of ionization of cellular components, with the subsequent disruption of intracellular reactions. Radiation is thus more dangerous to rapidly dividing cells (since there will be a reaction soon after radiation damage) or to cells which are highly specialized, for example nerve cells or corneal cells. The effects can be apparent in the irradiated individual (*somatic effects*) or in his offspring (*genetic effects*). The latter are of little consequence to astronautics.

Somatic effects depend upon dose, small doses giving rise to delayed effects which may not be apparent until many years after exposure. The early effects of a larger dose of radiation are, however, exceedingly important, since they can cause severe incapacitation.

The *immediate* result of a large dose of radiation is a period of lethargy, loss of appetite and vomiting, all of which are characteristic of radiation sickness. There follows a period of apparent recovery which can last from a few hours after exposure to many weeks. During this time lethargy continues but there is little or no apparent damage. This is followed by the terminal phase of gradual decline in which physical condition becomes markedly worse until death follows. During the decline phase there are major changes in blood morphology with a decrease in the numbers of red blood corpuscles (leading to acute lethargy and fatigue, as in anaemia); a decrease in the number of white blood cells (giving an increased disposition to disease); and a disturbance in the blood-clotting mechanism which results in chronic haemorrhage. The recipient of a lethal dose of radiation often succumbs to disease rather than to a direct disturbance of any body mechanism. Sub-lethal doses of radiation can cause the initial effects of radiation sickness with subsequent recovery.

No astronaut has yet received more than a few tens of millirads of radiation, which is well below the hundreds of rads regarded as lethal to man. However, no space flight has yet lasted for more than 14 days, and it may well prove that for long missions to the planets or in Earth orbit greater attention must be paid to the problem. This may result in a specially shielded area within a space craft to which radiation presents no hazard. At present radiation protection is often vested in the space craft structure, which provides inherent shielding.

THE METEORITE DANGER is not as serious as at first thought. The danger lies in the possibility of a thermal or mechanical penetration of the space cabin by a particle of sufficient size to cause a sudden decompression. The probability of this is small, but greater protection may be afforded by constructing the craft on a multi-walled principle. The energy of the meteorite could be dissipated by penetration of one or more of the walls without rupture of the pressure vessel. This construction would also add significantly to radiation protection.

A number of other stresses are also present in manned space flight – for instance, isolation and sensory impoverishment, task complexity, work load, monotony of diet and so on. Here the subjective response must be considered, and this is an important reason for carefully selecting and training astronauts.

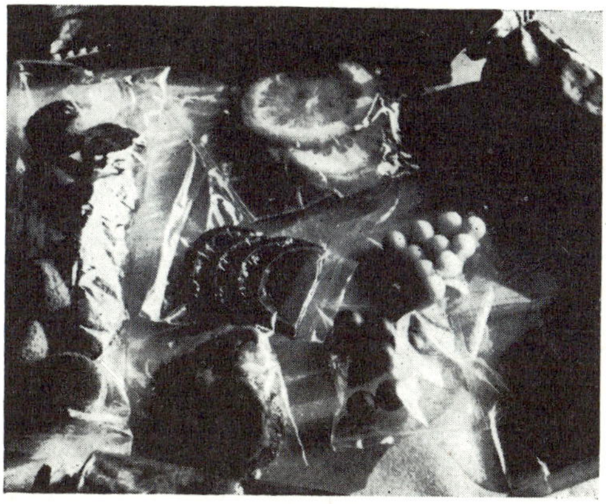

FOOD OF THE SOVIET COSMONAUTS. Apart from the inevitable tubes of pastes, Swiss roll, jellied fruit, cheese balls and biscuits are in evidence.

TRAINING OF ASTRONAUTS. A thorough physical testing programme serves the treble purpose of furthering research in simulated conditions, helping in the selection of astronauts and teaching them to cope with those difficulties which can be lessened by practice. Typical tests include the following:

Altitude Tests: the subject is placed in an altitude chamber dressed in a pressure suit and taken 'up' gradually or by explosive decompression. His reactions, heart rate, blood pressure, etc., are measured and he is in telephonic communication with the outside. He must remain capable of obeying instructions and of manipulating controls.

Gravity Tests: the subject is submitted to high accelerations in the gondola of a centrifuge, and to short periods of **free fall** in aircraft.

Heat Tests: subjects are submitted to short periods of intense heat and to prolonged periods of moderate heat with simulated partial failure of the refrigerated suits.

Equilibrium Test: the subject is placed in a chair which rotates simultaneously on two axes, and must keep the chair on an even keel by handling controls, normally and blindfolded, with or without vibration.

Cold Pressor Test: parts of the body are plunged into ice water, while changes in the blood circulation are observed.

Sensory Impoverishment Tests: curious disturbances can arise if a mammal is deprived of all external stimulation for long intervals. Subjects are placed on a very soft, moulded cushion in a dark, soundproof, odourless cabin with smooth, featureless walls, and must remain in this state for many hours without experiencing hallucinations or other mental or physical symptoms.

Space medicine is concerned not only with the immediate effects upon the body of the various stresses imposed by space flight, but also with the continuing effects. For this reason post-flight medical care is just as intensive as the pre-flight checks. The medical de-briefing of an American astronaut immediately tests his balance and postural reflexes, his extent of cardiovascular de-conditioning, the extent of changes induced by weightlessness on blood volume and cell content, and on bone demineralization, the changes in urinary and plasma concentrations of certain ions, and the amount of radiation received and its possible effects.

SPACE MIRROR. A large mirror in a satellite orbit, capable of focusing sunlight upon the Earth. Unfounded rumours that such a mirror or 'Sun-gun' was planned by the Germans to turn upon cities circulated at the end of the second world war. In fact a mirror very many miles in diameter would be needed to have any appreciable effect. However, even a large, plastic mirror would weigh relatively little and could be inflated in orbit and its attitude controlled by sensors. It may find a role as a source of energy in space, and research work is proceeding.

SPACE POLLUTION. It is not too early to draw attention to a problem which, if it is neglected, may have serious effects for man on Earth in general, and for the pursuit of astronomical observations from Earth in particular. There is a real danger that investigations may be hindered and stopped, before the end of the century, by the very devices which make them possible, and that appreciable damage to our environment will accrue. This article mentions, without comment, some of the events that have already given rise to concern.

In April 1964 a navigation satellite powered by a nuclear battery failed to achieve orbit and burnt up on falling back into the atmosphere, and 17 kilo-curies of plutonium 238 were disseminated. After extensive searches in balloons and aircraft, 88 % of the radioactive material has been accounted for in the atmosphere, mostly in the Southern Hemisphere. Plutonium 238 has a long half-life, and repetitions of such accidents could build up a particularly pernicious form of pollution.

During the first ten years following the launching of Sputnik I, some 136 out of 614 space vehicles failed to attain their orbits, approximately half of them carrying military payloads. Over 2 940 catalogued objects are at present in space, many of which are still transmitting, together with a considerable amount of uncharted débris resulting from accidents – usually on re-ignition in orbit. Occasionally a fragment falls on an inhabited area, but a graver danger is that arising out of mis-identification. In October 1962, during the Cuban crisis, a Soviet rocket carrying a Mars probe exploded in its parking orbit, and a cloud of débris was sent streaming to Alaska. An official American release of 1967 stated: 'This cloud came within range of the BEMEWS Radar Defence which saw what might for a moment have looked like a massive ICBM attack. The computer must have quickly revealed that it was not, but the potentialities for misunderstanding were there.'

The dangers to radio astronomy are equally real. The sensitivity of a large radio telescope is such that the signals collected from the remote parts of the Universe can be swamped by the ignition noise from a moped on the Moon or by echoes from the Moon of transmissions originating on Earth. It has proved very difficult to have narrow frequency bands reserved for radio astronomy. Against this background the American Defence Department's attempt to place 350 million minute metal needles in a circle round the Earth constituted a serious threat to sensitive observations of the remote parts of the Universe, particularly if several such belts were to be placed in orbit. The function of

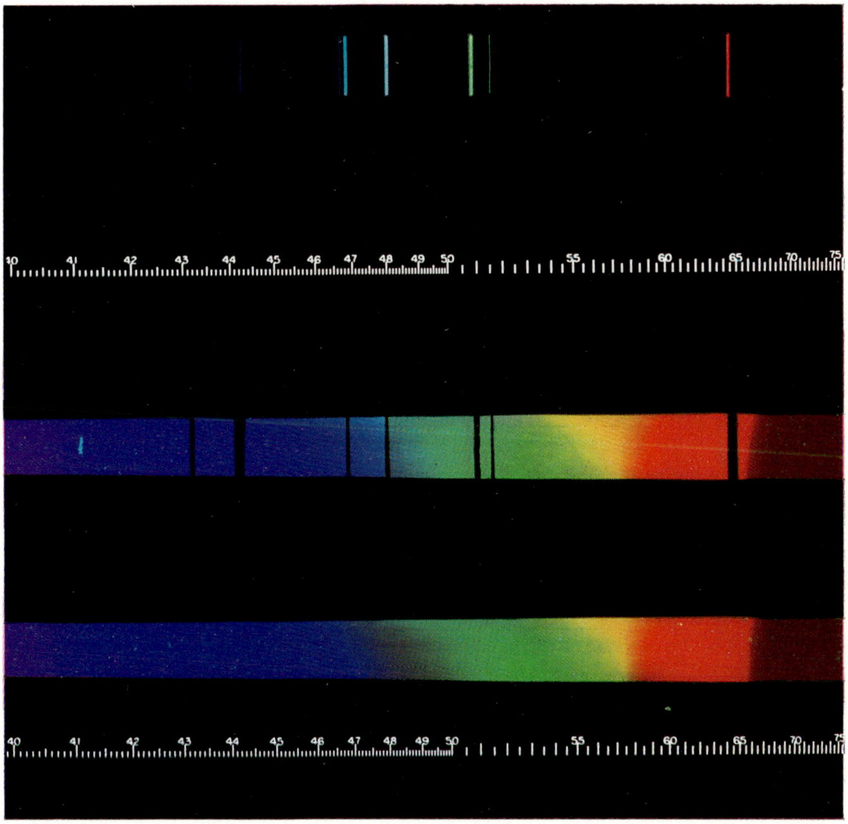

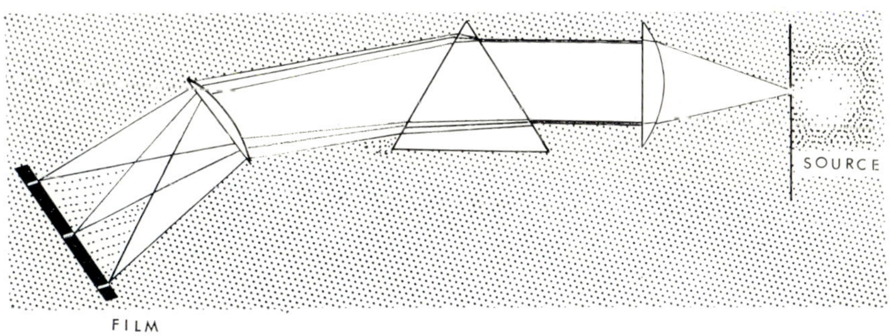

Top: The line emission spectrum of a cadmium source with the most prominent lines at 6438Å, 5086Å, 4800Å and 4678Å. *Middle:* The corresponding absorption lines, slightly exaggerated for clarity. *Bottom:* The continuous emission spectrum of a white light source between 4000Å and 7000Å, and a ray diagram of a simple spectrograph. (*Photographs by Physics Dept., Queen Mary College, University of London.*)

Above:

THE HYGINUS RILLE, two miles wide and over 120 miles long. Geologists have hailed this Apollo 10 photograph as one of the best ever taken of a lunar canyon. It is centred on the crater Hyginus near the N–E margin of Sinus Medii.

Opposite:

APOLLO 10 COMMAND MODULE separating from the lunar module 70 miles above the Moon. The lunar horizon is about 350 miles away and the Sun is almost directly overhead, so that no pronounced shadows are cast.

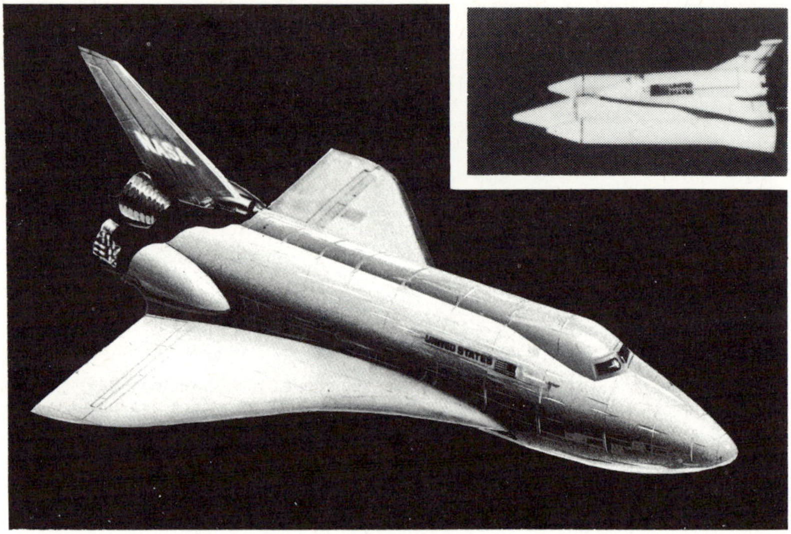

DESIGN FOR A SPACE SHUTTLE for payloads up to 300 tonnes. It does not need massive launch complexes, can go on over 100 missions, and will cut the cost of placing 1 kg of payload into orbit from $1 400 to $200. The inset shows it astride a launch booster.

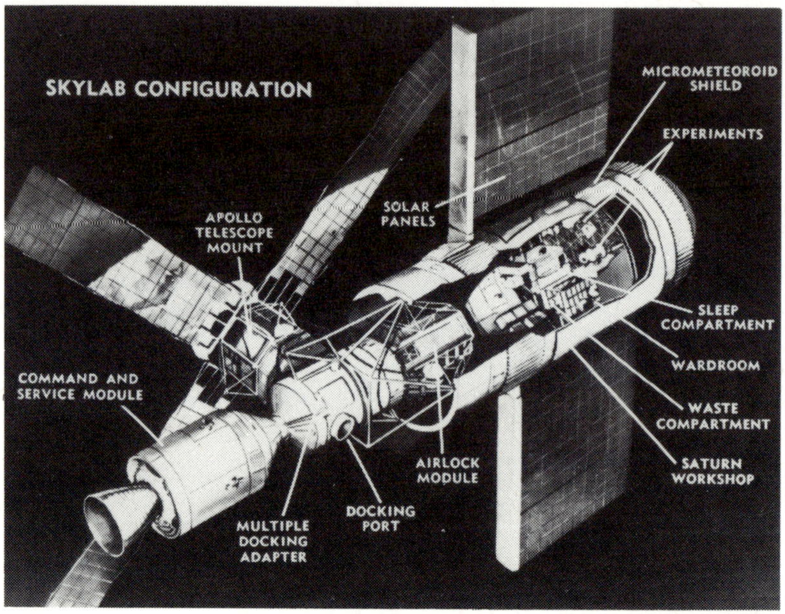

SKYLAB, the first manned space laboratory. During eight months it receives three crews, who conduct experiments on physiology, solar astronomy, Earth resources, human psychology, metallurgy and a variety of other topics.

these rings was to reflect military communications round the globe, and it would *ipso facto* scatter unwanted signals into radio telescopes. In the first attempt the package of needles did not disperse. In the second the needles dispersed as intended but their orbit had been adjusted so that after a few years the material entered the atmosphere and burned up. It is fortunate that other means of secure communication have been developed and as far as can be ascertained no further launching of needle packages is intended.

In 1958 Professor Van Allen discovered the existence of the now famous radiation zones around the Earth in which charged particles, mainly electrons and protons, are trapped by the Earth's magnetic field. The study of the structure of the inner and outer **Van Allen belts** and of the intricate movements of particles within them, their origin and the exact linkage with events on the Sun are of fundamental significance to geophysics and to our understanding of interplanetary space. In 1958 the United States exploded three *Argus* hydrogen bombs at a height of 500 km., and although the resulting aurorae were as bright as the full moon, there was no great change of particle concentrations in the Van Allen zones. Then, in 1962, a far larger hydrogen weapon was to be set off at 400 km. height, and the confident prediction was made that this, too, would have no lasting effect on the Earth's environment.

This prediction turned out to be entirely wrong. The notorious *Starfish* explosion over Johnson Island on July 2, 1962, produced 10^{27} fission particles including 10^{24} electrons which, instead of descending into the atmosphere, went outwards and became trapped in a zone underneath the inner belt. There were two long-lasting effects: certain information about the inner zone which we urgently need is still submerged in the effects arising from the artificial zone, entirely confusing the natural phenomena; and the electrons have exactly the same motion as the high-speed electrons of such objects of extreme interest as the **Crab Nebula.** A number of types of measurement are still severely handicapped today. A similar Russian experiment fortunately had less adverse consequences.

The International Committee for Space Research and other bodies are trying to secure binding agreements which will limit such risks as we can foresee, for instance that of contaminating Mars, the only likely seat of extraterrestrial forms of life that will come within our reach in a reasonable time, with organisms carried there by probes from the Earth. There is much, however, to erode one's faith in international agreements and resolutions, and they can prosper only if there is alertness and timely awareness of the latent dangers. (A.C.B.L.)

SPACE SHUTTLE. A re-usable space vehicle which is launched with the aid of an expendable booster and returns to Earth aerodynamically like an aeroplane. The first U.S. shuttle is scheduled for 1978.

SPACE STATION. Both the Russians and the Americans have expressed interest in more or less permanent orbital bases assembled in orbit from components launched separately. Space stations of this kind have a variety of possible uses, the most obvious of which are perhaps the assembly and fuelling of vehicles for lunar and planetary landing missions. For a time it seemed that the Russian plans for lunar landing missions included launch from Earth orbit of a vehicle large enough to land directly on the Moon and lift off again on a direct course for Earth without the lunar orbit rendezvous which is essential to the *Apollo*-type mission, and which has come in for a good deal of Russian criticism.

Other uses for manned space stations have been proposed, and the U.S. Apollo Applications programme has plans for a manned **orbiting astronomical observatory** equipped with a large telescope which would enjoy perfect seeing conditions.

SPACE SUIT is a generic name describing a variety of clothing designed for use in space. A space suit may be a simple one-piece garment of soft material intended for use in a shirt-sleeve environment, or it may be the complex, multi-component article designed for use in extra-vehicular activity which must perform all the functions of the space cabin.

For routine wear a space suit should be comfortable and functional. In Apollo missions the astronauts wear such a garment at all times of the flight which are regarded as non-critical – such as in the trans-lunar or trans-Earth coast phase. By so doing

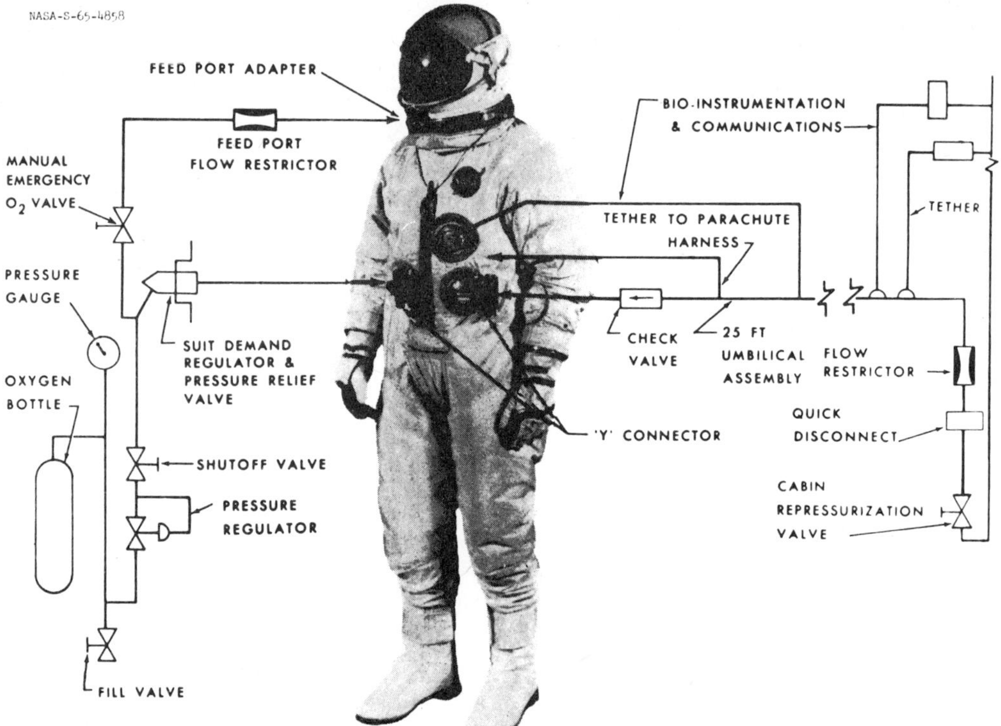

EXTRA-VEHICULAR SUIT and its diagrammatic connections to the back-pack (*left*) and through the umbilical assembly to other life-support systems (*right*).

physical and mental discomforts are minimized, and full operational efficiency is maintained. At lift-off and re-entry a fully protective suit is worn which could maintain viability in the event of emergency decompressions.

This suit, and that designed for use in extra-vehicular activity, must provide full life-support facilities. They achieve this by the use of space craft sub-systems in the first case, and by a fully independent, portable system in the latter case. The components of the life-support system have been described elsewhere (see **Space Medicine**), and the internal pressure suit can be regarded as an extension of this cabin environmental control system in all respects.

The suit designed for external use is different, however, in that it is completely independent of the space craft itself. The Apollo suit contains the miniature equivalent of the environmental control system in the back pack. It is obviously more difficult to control the variables in a space suit than in a space cabin, and the back-pack is correspondingly complex. For instance, heat exchange is effected through a radiator which is connected with a system of tubes attached to the layer next to the skin. The anthropomorphic pressure suit, meteorite protective garment and the aluminized thermal control suit complete the assembly.

The suit is designed for use with internal pressures comparable with those of the space cabin (270 mm. Hg). This low pressure has the advantage of least restraint upon flexibility of the suit as a whole, because the pressure gradient between the suit and the ambient pressure is low. Suits designed for use with higher internal pressures lack this flexibility and thus violate a fundamental design criterion. The consequences of changing pressures from space cabin to space suit have been described in the section on space medicine.

SPECIFIC HEAT of a substance is the number of **calories** required to raise the temperature of one gram of it through 1° C. Specific heats vary considerably, e.g. that of water = 1, and that of lead = 0·03. This means that the same amount of heat which will raise the temperature of water by 1° C. will raise an equal weight of lead through about 33° C. Thus bodies of low specific heat may be at very high temperatures and yet contain relatively little heat.

SPECIFIC IMPULSE. A measure of the effectiveness of a rocket engine, defined as its thrust (in pounds) per pound of fuel and oxidant consumed per second. This is equal to the **exhaust velocity** divided by **g**.

SPECTRAL CLASSIFICATION OF STARS. Most of our knowledge of stars is derived from analyses of their spectra. There are so many stars, even in our own Galaxy, that to study them all individually would be a quite hopeless task. A classification has therefore been adopted: stars of similar spectra are placed in the same spectral class, and only typical members of the class need then be investigated in detail.

The classes of spectra grade into one another, so that they can be arranged in a continuous sequence. In order of the spectral sequence, the classes are rather unsystematically named

O, B, A, F, G, K, M.

Each class is divided into subclasses numbered 0 to 9; thus the Sun is class G2; Algol, B8.

The science of **spectroscopy** has shown the sequence to be essentially a *temperature sequence*, with O stars at the hot end and M stars relatively cool. The luminosities of stars vary in a recognizable way according to their spectra: this is the basis of the **Hertzsprung–Russell Diagram.**

The spectrum of a normal star consists of a bright background crossed by dark lines, and these lines determine the spectral class. There is a smooth and progressive change along the sequence, a line at a particular wavelength appearing at a certain point, rising to a maximum intensity, and fading away through the succeeding classes. An astronomer classifies a spectrum simply by looking at the lines and comparing them by eye with a series of standards, some of which are reproduced below.

The chief characteristics of the spectral types are as follows:

O: essential feature is presence of lines of ionized helium.

B: neutral helium, strong hydrogen lines. No ionized helium.

A: hydrogen lines dominate the spectrum. Maximum at A0, later decrease as many lines due to metals appear; ionized calcium conspicuous among these.

F: hydrogen lines continue to weaken. Prominent lines of ionized calcium, the so-called H and K lines.

G: hydrogen lines weaker, metals stronger, surpassing hydrogen in later G spectra. First traces of molecules. H and K lines strong.

K: H and K lines declining. Strong metallic lines.

M: many molecular bands, especially those of titanium oxide. Very strong line of neutral calcium.

There are a few other classes of spectra which have received letter designations. Certain very hot stars, the **Wolf-Rayet stars,** have been placed in a class W allied to the type O. Classes R, N and S are similar in temperature to the M stars but seem to differ chemically: instead of titanium oxide the R and N stars contain much carbon, while S stars contain the exotic element zirconium.

A few stars do not fit in any of these classes. They are termed ‘peculiar’ in contrast to the ‘normal’ stars belonging to the recognized types. Among peculiar stars may be mentioned:

i. **White dwarfs,** which show very broad hydrogen lines and scarcely anything else.

ii. **Shell stars** – O, B and A stars surrounded by a tenuous shell of gas which superimposes emission lines on their spectra.

iii. Many **variable stars.**

iv. All **novae** have very peculiar spectra, with bright forbidden lines and other curious features.

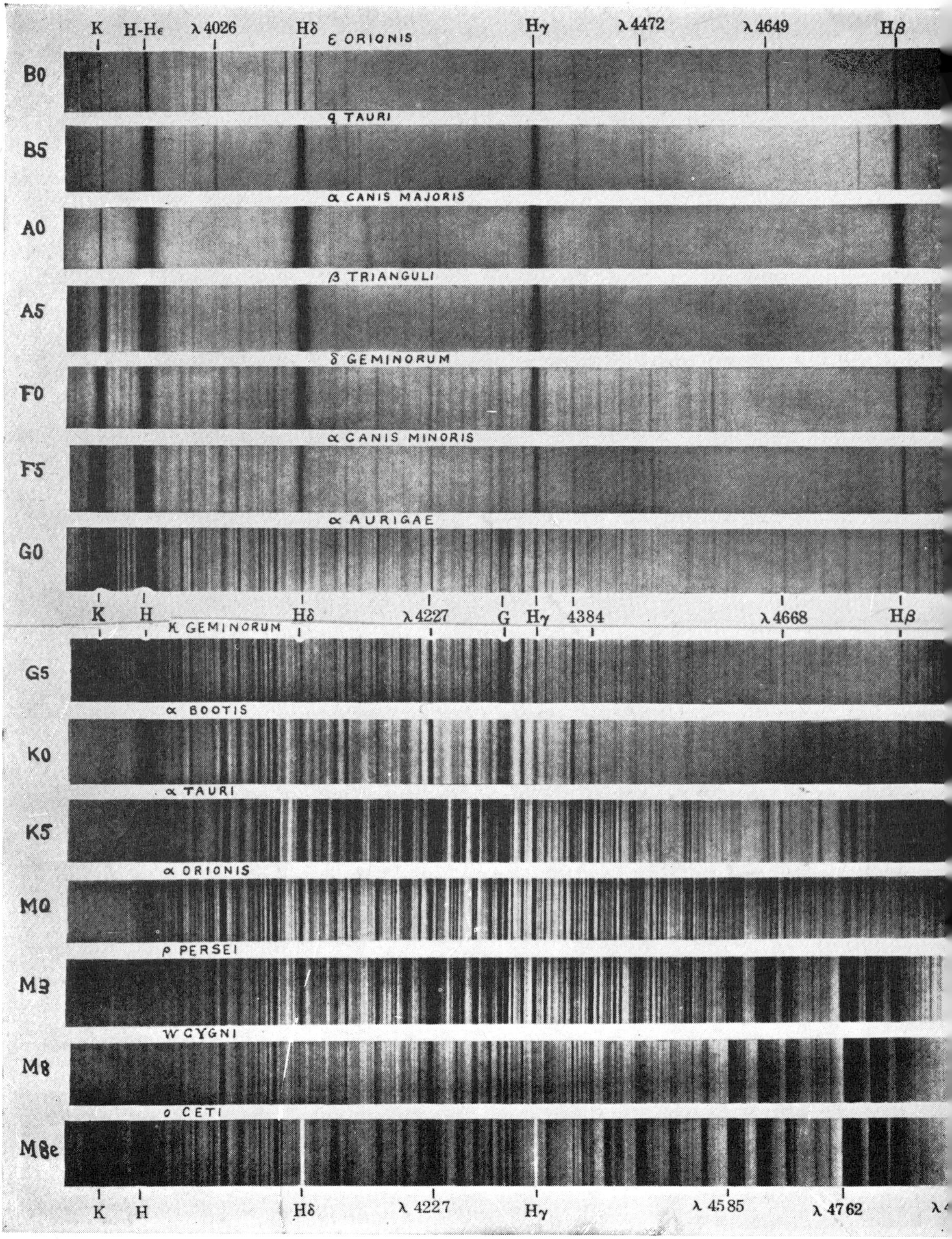
K
H-Hε
λ 4026
Hδ
Hγ
λ 4472
λ 4649
Hβ
ε ORIONIS
B0
q TAURI
B5
α CANIS MAJORIS
A0
β TRIANGULI
A5
δ GEMINORUM
F0
α CANIS MINORIS
F5
α AURIGAE
G0
K
H
Hδ
λ 4227
G
Hγ
4384
λ 4668
Hβ
κ GEMINORUM
G5
α BOOTIS
K0
α TAURI
K5
α ORIONIS
M0
ρ PERSEI
M3
W CYGNI
M8
o CETI
M8e
K
H
Hδ
λ 4227
Hγ
λ 4585
λ 4762

Opposite: SPECTRA OF STARS of different spectral classification. The spectral type appears on the left, and the name of the star above its own spectrum, while the numbers refer to wavelengths in Ångström units. The Balmer absorption lines Hβ, Hγ, Hδ, Hε are prominent in the hotter A and B stars, the G stars show strongly the H and K lines of ionized calcium, and the cool M stars have typical molecular band spectra.

---

v. Certain stars show greater intensities of certain spectral lines than normal stars of similar temperature. The chief group is that of the peculiar A stars. (R.H.G.)

SPECTROSCOPY is the science relating the nature of luminous sources to the characteristics of their emitted light. In order to determine the chemical composition, temperature, pressure and size of a star (an object seen only as a speck of light), we need to record and analyse in great detail all the evidence contained in the radiation emitted by the star.

Spectroscopy began in 1666, when Newton made his classic discovery that white light is made up of many colours. A narrow beam of sunlight was allowed to fall on a triangular glass **prism.** The beam emerged as a band of colours, comprising violet, blue, green, yellow, orange and red, due to **refraction** of the light by the prism. This band is called a *spectrum*; a rainbow is a natural example. Newton showed that the colours did not originate in the glass, but were contained in the original beam of light. Later it was noticed that the spectrum was crossed by a number of narrow, dark lines, some darker and broader than others. In 1817 Fraunhofer examined the spectrum of a lamp, and saw how different it was from that of sunlight. He observed over 750 lines in the solar spectrum and mapped out over 300, ascribing letters to the principal lines. Some of these letters are still used by astronomers, and the lines in the solar spectrum are commonly known as **Fraunhofer lines.** Some stars, for example Capella and Procyron, have spectra crossed by dark lines in exactly the same positions as the solar lines, but others, like Sirius and Castor, produce only a few intense black lines which have no obvious connection with the solar spectral lines.

The true interpretation of these observations was found in 1859 by two German chemists, Kirchoff and Bunsen, who realized that some of the dark lines seen in the solar spectrum had the same positions as certain bright lines in the spectra of laboratory sources. When light from a bright source was refracted through a prism and subsequently passed through a flame containing sodium, a pair of dark lines appeared in the yellow region of the spectrum of the source. Removal of the flame caused these lines to disappear; thus they were obviously due to absorption by sodium in the flame. Further, if the sodium flame was used as the original source, the prism did not produce a continuous spectrum of colours; instead it gave an emission spectrum of yellow lines on a dark background, the reverse of the absorption spectrum. The use of other substances instead of sodium produced different sets of lines, and Kirchoff and Bunsen showed that these sets of lines, characteristic of particular elements, could be used to identify these elements in sources of spectra. This was the beginning of *spectrochemistry*, the science of chemical analysis by spectroscopic methods. Kirchoff enunciated his famous law, that the ratio of the absorbing power of a substance to its emitting power for light of a particular colour is the same for all substances at a given temperature. He also determined that the Sun must be surrounded by layers of cool gases absorbing light emitted by the hotter layers below, the absorption lines being characteristic of the atoms present in the outermost layers of the Sun. Thus a method had been discovered which permitted the study of the chemical composition of the Sun and stars; this can be said to be the beginning of modern astrophysics. (See colour plate facing page 256.)

COLOUR AND WAVELENGTH. Light is a form of **electromagnetic radiation,** and consists of a wave motion. The distance from the crest of one wave to that of the next is called the **wavelength,** which is often conveniently measured in **Ångström Units** (abbreviated Å).

Visible light has wavelengths between 4,000 Å and 7,000 Å. When light of a particular wavelength falls on the eye, a sensation of colour is produced, different colours corresponding to different wavelengths. Light of wavelength 4,000 Å is violet in colour, and as the wavelength increases the colour changes continuously through blue, green, yellow, orange and red until, for a wavelength of 7,000 Å, a very deep red sensation is produced. The spectrum of visible (white) light is illustrated in the colour plate facing page 256. For longer wavelengths, no sensation of colour is produced, but the radiation can be recorded on photographic plates up to 14,000 Å; such radiation is called infra-red. Radiation below 4,000 Å is ultraviolet and is easily recorded photographically.

Most astronomical work is concerned with the range from 3,000 Å to 10,000 Å. Outside this range the **atmosphere of the earth** absorbs all the ultraviolet and much of the infra-red, preventing such radiation from reaching us. Recent experiments with rockets have, however, enabled the ultraviolet spectrum of the Sun to be obtained, and new developments in infra-red observational techniques (see **infra-red**) have produced further information on both stellar and planetary objects. Many observations have also been recorded from **orbiting astronomical observatories.**

TYPES OF SPECTRA. It is possible for an element to produce both emission spectra and absorption spectra, each of which can exist as continuous, line, or band spectra. *Emission* spectra are seen when light from a hot source is examined directly by a spectroscope. *Absorption* spectra are formed when a source giving a continuous emission spectrum is viewed through absorbing material; gaps are seen in the emission spectrum of the source at wavelengths corresponding to the absorption lines of the material. A *continuous spectrum* is one showing an uninterrupted band of wavelengths, while a *line* spectrum consists of a number of relatively sharp emission (or absorption) lines, separated by regions of wavelength where there is no emission (or absorption); the line emission spectrum produced by a cadmium source is shown in the colour plate facing page 256. Finally, a *band* spectrum consists of a very large number of closely spaced lines, often overlapping to a considerable extent.

Hot glowing solids and liquids produce continuous emission spectra, and hot gases may produce continuous emission spectra at high densities. Gases at lower densities, or in smaller quantities, give line or band spectra in emission. Cool layers of gas in front of hotter sources of continuous emission produce line, band, or continuous absorption spectra.

INSTRUMENTS. The *spectroscope* is the instrument which enables the astronomer to examine spectra by eye. For most purposes, however, it is more convenient to obtain a photographic record, and when the instrument incorporates a camera it is called a *spectrograph*.

A *diffraction grating* is often used, instead of a prism, in a spectroscope. This consists of a very large number of closely spaced parallel grooves ruled with a diamond, on either a polished glass surface (*transmission grating*), or a metallic mirror (*reflection grating*); there may be over 15,000 grooves per inch. When light falls on the grating, the different constituent colours are refracted or reflected at different angles, and form a spectrum, which can be made sharper and better drawn out than one formed by a prism.

MEASUREMENTS. In order to measure the wavelengths of lines on a spectrum and to identify them, some standard of comparison is required. This is obtained by photographing the known spectrum of some laboratory source side by side with the unknown spectrum. The illustration opposite shows the line emission spectrum of an iron arc, and also part of the spectrum of a star of **spectral classification** K5 between two iron comparison spectra.

DOPPLER EFFECT. The whole spectrum of a celestial body is often noticeably displaced relative to the comparison spectrum. This displacement is caused by the **Doppler effect,** and is a measure of the **radial velocity** of the body. A spectrum showing the Doppler effect in the star Zeta Herculis is shown on the opposite page.

It is important to remember that observations of the Doppler effect give the velocity of the object relative to an observer on the Earth. It is usual to correct the measurements for the orbital motion of the Earth; radial velocities of stars and nebulae are given *relative to the Sun.*

LINE EMISSION SPECTRUM of an iron arc.

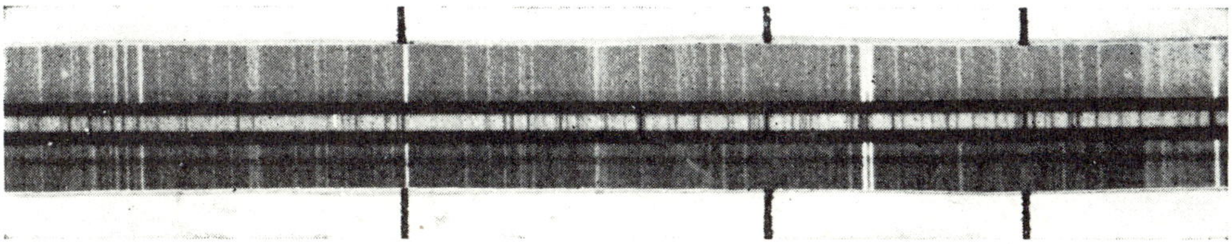

PART OF STAR SPECTRUM of spectral type K5 between two iron comparison spectra. The stellar spectrum shows dark absorption lines against a continuous emission background. Two prominent calcium lines are marked in the right half. (*R. H. Garstang*)

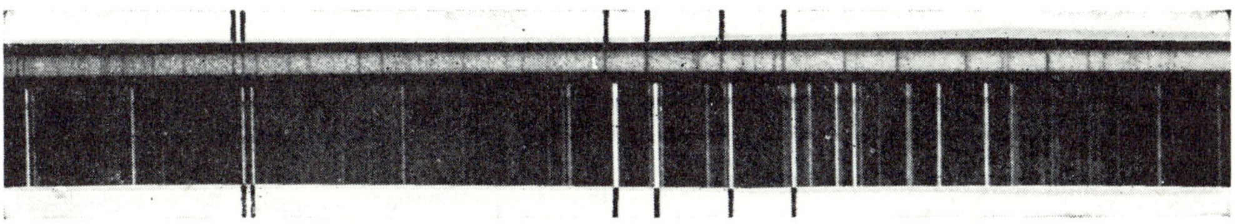

THE DOPPLER EFFECT in the spectrum of the star Zeta Herculis. Six iron lines are marked in the star and the comparison spectrum. (*R. H. Garstang*)

ATOMIC SPECTRA. Every **atom** has its own characteristic pattern of spectral lines or bands; for many atoms these patterns are extremely complicated, but nevertheless almost all observed stellar spectral lines can be explained.

The simplest spectrum is that of hydrogen, which shows a series of obviously related lines extending from the red to the ultra-violet, and converging to a well-defined limit. This series is seen in the spectra of hot stars like Sirius; more lines are seen in the stars than in laboratory spectra. In 1885 Balmer discovered empirically a simple mathematical law, which predicted correctly the wave-lengths λ of the successive hydrogen lines Hα (λ = 6562 Å), Hβ (4861 Å), Hγ (4340 Å), Hδ (4101 Å):

$$\frac{1}{\lambda} = R\left(\frac{1}{2^2} - \frac{1}{n^2}\right)$$

where R = 109,677 cm.$^{-1}$ is a constant, and n = 3, 4, 5, 6 for the respective lines. It was evident that this formula reflected a fundamental property of the hydrogen atom, but it was not until the **Bohr theory** in 1913 that a satisfactory theoretical account of the emission and absorption of radiation was achieved. In the Bohr model (see **Atom**) the hydrogen atom may be represented schematically by a **proton** nucleus, with an **electron** moving round it in any one of a series of discrete, circular orbits of fixed size. The orbits are labelled by the numbers n = 1, 2, 3, . . . , and in each particular orbit the electron has a certain energy. Thus each orbit corresponds to a particular **energy level** of the system; the smallest orbit ($n = 1$) represents the level of lowest energy, or the *ground state*. The electron may jump, or make a transition, from one orbit to another; this involves a definite change in energy, the difference between the energies of the two levels involved. The transition corresponds to the absorption or emission of radiation of a particular **frequency**, the frequency being proportional to the energy gained or lost by the electron as a result of the transition, that is, the energy difference between the two levels. Radiation is absorbed or emitted only when the electron jumps from one orbit to another, and an electron can change its state only by absorption or emission of radiation of the correct frequency.

Brackett series in the infra-red. These are exhibited schematically in the diagram opposite.

The ideas outlined above can be applied to elements other than hydrogen, but complications arise in determining the energy levels and resulting spectral lines, due to the multiplicity of electrons involved. Many absorption lines of atoms such as sodium, magnesium and iron, as well as the strong hydrogen lines, occur in the Fraunhofer spectrum of the Sun, and the H and K lines of ionized calcium, Ca II, at 3933 Å and 3968 Å are particularly prominent. Molecular spectra differ greatly from atomic spectra, because molecules can possess energy due to vibration between the constituent atoms, and due to rotation. A typical molecular spectrum contains a number of

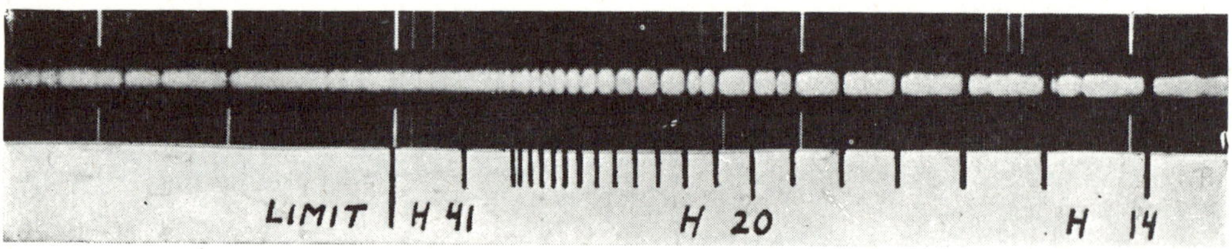

THE BALMER SERIES of hydrogen in the spectrum of the shell star HD 193182. The lines are distinguished by the quantum number of the upper energy level of the corresponding electron jump. The lines from H 14 to H 33 are marked, and on the original plate lines up to H 41 can be seen. The position of the Balmer limit is also shown. Above and below the star spectrum are iron comparison spectra.

Thus every spectral line represents a transition between particular energy levels in the atom. In the case of hydrogen, the detailed mathematics of the Bohr theory show that the Balmer formula represents the wavelengths of absorption (emission) lines caused by electron transitions from (to) the level $n = 2$ to (from) higher levels $n = 3$, 4, 5, 6. As n is increased, the lines of the Balmer series will crowd closer together and ultimately converge on the series limit at 3646 Å. The illustration above shows a number of these Balmer lines with $n = 14$ to 41 and indicates the series limit. Other hydrogen spectral line series behave in a similar manner, for example, the Lyman series in the far ultraviolet which involves transitions from (to) the level $n = 1$ to (from) higher levels, and the Paschen and

bands or *flutings* instead of sharp lines. The illustration on page 262 shows a series of typical spectra obtained from stars of differing **spectral classification**; the Balmer series of absorption lines in hydrogen is very strong in the hot A and B stars, the calcium H and K lines are dominant features in G stars, while the M stars exhibit many molecular bands.

Even in the largest possible orbit, an electron of an atom possesses only a certain energy. If it acquires, by absorption of radiation, more than this critical amount, then it is no longer bound to the nucleus, and can possess any energy above the critical amount. This process of removal of electrons from an atom is called ionization, and since the detachment of an electron entirely alters the electric field inside an atom, the resulting

ion possesses electrons with different possible orbits, and consequently has a different spectrum from that of the parent atom. In atoms with more than one electron, multiple ionization is possible, and in the far ultra-violet spectrum of the solar corona lines of eight to sixteen times ionized iron (Fe IX to Fe XVII) are observed.

If ions and electrons move freely in a volume of space, the electrons must frequently collide with the slower-moving ions, and it is possible that an electron may jump back into one of the empty orbits of an atom, emitting its surplus energy as radiation. This process is called *recombination*, and, since the free electrons have a continuous range of energies, the recombination process provides a continuous spectrum. A particular form of this process, called dielectronic recombination, is prevalent in the solar corona, and leads to a temperature estimate for the corona of about 1,000,000° C.

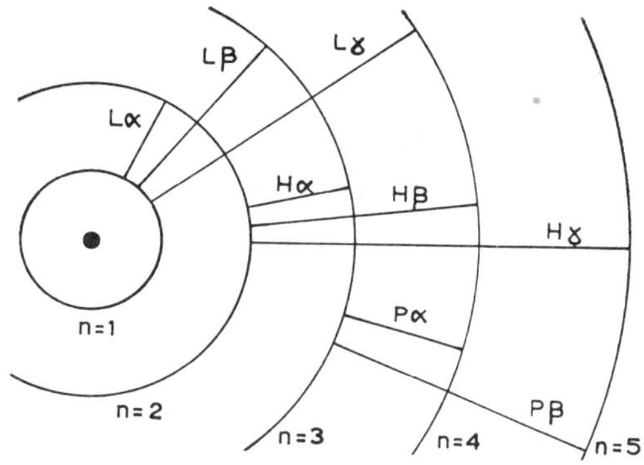

Whenever an electron jumps from a higher into the lowest (innermost) orbit, the atom gives out radiation at a wavelength corresponding to a spectral line of the *Lyman* series. Jumps down into the second lowest level contribute to the *Balmer* series (*see overleaf*). The greater the jump, the closer the emitted radiation is to the limit of the series, which is reached when an electron enters from outside the atom. Outward jumps involve the absorption of energy and give rise to absorption lines.

FORBIDDEN LINES. These are spectral lines of certain elements which, while theoretically possible, i.e. corresponding to the difference of energy between two states, are not observed in laboratory spectra, although they have been found in the spectra of various astronomical sources. Their importance lies in the information they yield about the nature of the sources. Under laboratory conditions atoms undergo frequent collisions with other atoms, ions and electrons. Collisions can excite atoms from one state to a higher state, and can de-excite them to a lower state, without any spectral line being absorbed or emitted. Forbidden lines occur only in emission. The atoms concerned must be dropping from higher to lower states by emission of radiation.

In the case of a normal or 'permitted' line, the excited atom radiates and so falls to a lower energy level in about one hundred-millionth of a second. The weakness of forbidden lines is explained by assuming that an atom spends anything from a second to an hour before emitting them. If atoms are in a level from which downward, permitted lines are possible, such lines will be strong. If no such lines are possible, forbidden lines will be emitted if the atoms have the requisite time to do this. Collisions must be infrequent, and this will be the case only if the pressure and density are low enough. This is true in the outer parts of the atmospheres of the Earth and Sun, in the gaseous nebulae and in interstellar space. From the intensities of forbidden lines we can estimate the temperatures and densities of these sources.

THE 21 CM. WAVE. Many spectral lines, originally thought to be single, have been found to consist of groups of lines exceedingly close to each other. This has been attributed to a spin of the atomic nucleus; for most atoms it may be ignored, but there is one outstanding exception. The smallest orbit or *ground state* of hydrogen is a single energy level on the usual quantum theory. When account is taken of the spin of the nucleus the ground state is seen really to be a *doublet* – a pair of energy levels extremely close together. The difference between the levels corresponds to a wavelength of 21 centimetres. This wavelength is in the radio region of the **electromagnetic spectrum,** and has been detected from interstellar hydrogen in the Galaxy (see **Radio Astronomy**). Similar lines in the radio region are observed due to OH molecules.

BREADTH OF SPECTRAL LINES. If atoms and molecules behaved exactly as has been described, absorption and emission lines would be of one precise wavelength only. Examination of the spectra reproduced on page 262 shows that whilst some absorption lines are narrow and dark, others are broad and diffuse. There are many factors which contribute to the breadth of a spectral line:

(a) *Natural breadth.* The simple picture of an atom assumes that each energy level corresponds to one and only one precise energy. This idea has been modified in the light of detailed treatments of the problem, and each energy level is now thought of as spreading over a narrow range of energies. A transition between any two levels therefore has a small but distinct range of energies, and the corresponding spectral line has a small but distinct spread. This natural breadth is a property of the atom, and does not depend on its surroundings.

(b) *Collisional broadening.* If an atom which is in the process of radiating collides with another atom, the train of light waves emitted by the atom is interrupted; the interruption has the effect of widening the spectral line. Collision broadening is clearly of most importance in dense sources where collisions are frequent.

(c) *Doppler broadening.* The **Doppler effect** applies to all moving sources of radiation. The individual atoms in a gas are moving rapidly in random directions, and these motions give rise to shifts in wavelength on either side of the mean wavelength for the gas as a whole. The effect is again to broaden the spectral lines. The broadening depends upon the velocities of the atoms and increases with temperature, so that it is greatest in hot stars.

(d) *Turbulence.* In many stellar atmospheres large-scale motions take place. Lines from the various moving clouds of gas suffer additional Doppler shifts, and the net effect on the spectrum is to broaden the lines still further.

(e) *Stark broadening.* The countless ions in a star's atmosphere cause strong electrical fields between the atoms. These slightly disturb the electrical fields *inside* the atoms, in which the electrons move. The changes cause slight differences in wavelength between lines emitted by different atoms, and the overall effect is to broaden the spectral lines.

(f) **Zeeman effect.**

(g) *Rotation.* Many stars are in rapid rotation. Different parts of the surface of a star may have different velocities towards or away from the observer owing to rotation of the star: light from these parts therefore shows Doppler shifts corresponding to all these velocities. The spectrograph necessarily photographs the spectra of both **limbs** simultaneously, as the star appears just as a point source of light; the lines are consequently broadened by the rotation of the star.

(h) *Instrumental broadening.* All the broadening factors so far listed arise in the star itself. But various factors combine to make even the most perfect astronomical instrument produce some blurring effect.

The observed breadths of spectral lines are caused by a combination of some or all of the above factors, and one of the principal problems in astronomical spectroscopy is to distinguish as many of the above effects as possible in a given spectrum. The figure shows a detailed intensity, or brightness, trace of part of the spectrum of Arcturus (α Boötis); the absorption lines of iron, nickel and calcium are fairly narrow, indicating little broadening (*cf.* spectrum of α Boötis facing page 263).

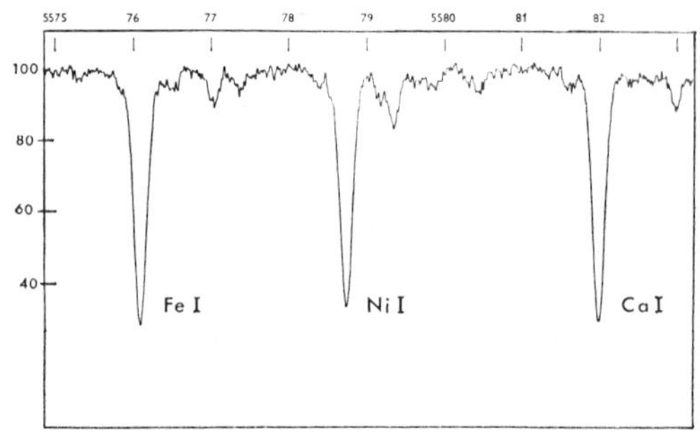

Part of an intensity trace of the spectrum of Arcturus, showing absorption lines of Fe I (λ 5576·1 Å), Ni I (λ 5578·7 Å), and Ca I (λ 5582·0 Å).
(*From 'A photometric atlas of the spectrum of Arcturus' by R. F. Griffin*)

SPECTROSCOPY IN ASTROPHYSICS. When a large number of lines in a stellar spectrum have been identified and their strengths measured, it is possible to study the temperature and pressure and chemical composition of the outer layers of a star.

The degree of ionization of the atoms of an element depends on the temperature and pressure. Electrons are removed from an atom by collision with another atom or ion, or by absorption of radiation; both the number and violence of the collisions, and the amount of radiation, are increased by increasing the temperature, and the number of collisions is increased by raising the pressure. On the other hand, increase of temperature or pressure entails more frequent collisions between ions and electrons, and therefore more frequent opportunities of recombination; this process reduces the degree of ionization. An equilibrium is set up, in which the number of electrons removed from atoms by collision or radiation is just balanced by the recombination of ions and electrons. The results of detailed investigation on the influence of pressure and temperature on ionization are embodied in an important quantitative law, known as *Saha's Equation.*

Translated into terms of spectra, this means that at low temperatures an element shows the lines of its neutral atoms, and with rising temperature the lines of the singly ionized atoms become progressively stronger at the expense of those of the neutral atoms, only to be themselves superseded by the spectrum of the doubly ionized atoms. Different elements ionize with different ease: thus, while ionized calcium is already important at 3,000° C., helium scarcely begins to ionize until 10,000° C. is reached.

Thus we are able to explain the observed *sequence of stellar spectra as a temperature sequence*, the classes from O to M being classes of decreasing temperature.

The relation between temperature and the colour predominant in the light emitted from a hot source is beautifully exemplified in the colour photograph of the Ring Nebula between pages 120–121. This shell of gas fluoresces under irradiation by invisible ultraviolet light from the very hot blue star at its centre. The temperature of the glowing gas decreases outwards from the central star, and the colour varies accordingly from blue in the middle to red at the outer fringe.

The effect of pressure is slight, but it affords us a useful way of distinguishing between diffuse *giant stars* and relatively dense *dwarf stars.* We have already seen that higher pressure leads to lower ionization: it follows that a dwarf star shows rather lower ionization than a giant of the same temperature. Different elements respond to changes in pressure to different extents, and the relative strengths of nearby spectral lines of elements of differing pressure sensitivities are measured to give an indication of the size of a star.

STELLAR ATMOSPHERES. The light which is received from a star comes from near the surface, since the material of stars is quite opaque; this is evident from the apparent sharpness of the solar limb. Radiation produced in the deeper, opaque layers of a star may be absorbed by atoms in the outermost layers, or stellar atmosphere; this absorption is of two different forms. Firstly, continuous absorption of radiation by atoms or ions, leading to ionization; for example, in hot stars, the hydrogen in the outermost layers may be ionized by radiation from the interior, while in cooler stars, such as the Sun, the negative ion of hydrogen, H^-, provides a strong source of continuous absorption (see **ion**). Secondly, radiation may be absorbed by the atmospheric constituents to produce discrete line spectra.

The results of detailed analyses are surprising; nearly all stars have very similar atmospheres, and the differences between the spectral classes, so great at first sight, are caused almost entirely by differences of atmospheric temperature and pressure. Hydrogen and helium, in the ratio of about five to one, are easily the most abundant elements. The following typical figures for the commonest of the other elements show that, at least as far as the atmospheres of stars are concerned, these may all be considered as trace elements. (See also **Abundance of Elements.**)

Element	*Percentage*
Oxygen	0·0680
Neon	0·0640
Nitrogen	0·0200
Carbon	0·0130
Iron	0·0003

(R.H.G., V.P.M.)

SPECTRUM. See **Spectroscopy.**

SPEED OF LIGHT. The fact that light does not travel infinitely fast, and arrive at its destination instantaneously, was first deduced in the 17th century. Occultations, and other phenomena, of Jupiter's satellites were found to occur systematically earlier than predicted at some times of the year, and later at others: this could be explained on the assumption that light took an appreciable time – about 15 minutes – to cross the Earth's orbit. A number of ingenious systems have been devised to determine the very short time that light takes to traverse a measured path, either inside a laboratory or outside (sometimes between mountain tops many miles apart). The most recent and accurate measurements yield a value of 299,793 kilometres per second (about 186,000 miles per second).

SPHERICAL ABERRATION is a defect of mirrors and lenses with spherical surfaces, light from the inner and outer parts of the lens or mirror converging to slightly different foci and so preventing the formation of a sharp image. It is reduced by altering the shapes of the surfaces concerned.

SPIRAL NEBULAE, SPIRAL GALAXIES. See **Galaxies.**

SPUTNIK. The name given to Russian artificial Earth satellites.

STANDARD TIME. A system of time measurement used by most countries in the world. It would make life very difficult if every place on the Earth kept strictly to its local mean time based on the position of the Mean Sun. Under the standard or *zone time* system, the same time is adopted throughout the area between two lines of longitude 15° apart, with adjustments to avoid cutting across national boundaries. The time in each zone differs from that in the adjacent zones by one hour.

STAR. A glowing sphere of gas. Unlike the planets, the stars shine by their own light: they are indeed burning fiery furnaces, where matter is pent up at inconceivable pressures and heated to inconceivable temperatures. They are not, however, 'burning' in the ordinary sense – **stellar energy** is produced, not by chemical but by *nuclear* reactions similar to that of the hydrogen bomb.

The nearest star to the Earth is the **Sun,** which is quite a typical example. Many stars are bigger and brighter than the Sun, yet so vast is the space between us and them that they appear as mere glittering specks in the night sky, their rays overpowered by the light of day.

Every star is a gigantic globe containing enough matter to make something like a million Earths – and even our puny Earth 'weighs' 6,000,000,000,000,000,000,000 tons! Most of this stupendous amount of material we cannot see; it is hidden deep below the star's surface. **Spectroscopy** gives us a surprisingly detailed insight into the surface conditions, and from these we can deduce a little about the stellar interiors. Stars are found to be formed largely of hydrogen, which seems to be the essential element in the make-up of our Universe.

Despite the tremendous size and mass of a star, our own local star system, the Galaxy, contains some 100,000 million of them; and even a galaxy pales into insignificance on the cosmic scale, the Universe certainly containing many thousands of millions of them.

The physical characteristics of stars – their temperatures, sizes, masses, etc. – and the methods used to determine these are described below. It is apparent that a great deal of our knowledge is based upon observations of **binary stars.**

TEMPERATURE. A number of possible ways of measuring temperature may be enumerated:

1. From the spectral class.

2. Kinetic Temperature, from the Doppler broadening of spectral lines (see **Spectroscopy**). The velocities of the random motions of atoms are proportional to the square root of the absolute temperature. This method of estimating temperature is not readily applicable to absorption spectra, but the random velocities can be found from line-widths in some emission spectra of objects such as gaseous nebulae and the solar corona.

3. Effective Temperature, from the total radiation of the star (**Bolometric Magnitude**)

and apparent diameter, using **Stefan's Law.** Methods of measuring star diameters are described in the next section.

4. Colour Temperature, found from Colour Index, described under **Magnitude.** This is a useful method.

The temperatures of stars are found to vary from about 50,000° K. to about 3,000° K. A few stars are thought to be hotter than 50,000° K.; for example the **Wolf-Rayet stars.** The surface temperature of the Sun is about 5,800° K.

RADIUS. Stellar radii are all too small to measure directly through a telescope, but other methods are available:

1. Eclipsing binary stars: this is the most important method.

2. The Stellar **Interferometer.**

3. The method of estimating the Effective Temperature, given in the last section, may be used in reverse to supply the apparent diameters of stars, using assumed effective temperatures based on alternative methods of temperature estimation.

The radii of stars are found to differ prodigiously. The smallest turn out to be about the size of the Earth; they belong to the class known as *white dwarfs.* On the other hand, some stars are so gigantic that they would extend to the orbit of Saturn if placed with their centres in the position of our Sun. The larger component of the eclipsing binary star system ε Aurigae must be about two thousand million miles in diameter.

MASS. The masses of stars can only be satisfactorily determined from observations of binary stars. A few stars are known to show a red shift in their spectral lines which is caused by the formation of the lines in a strong gravitational field. The Theory of Relativity predicts such a shift to be proportional to the mass and inversely proportional to the radius of the star. To find the mass we must therefore first find the radius, as well as disentangle the observed slight red shift from the ordinary Doppler effect. The uncertainties of this method are so great as to render the results of doubtful value.

The masses of stars vary far less than the radii; most masses are about that of the Sun, and few lie outside the range 0·15 to 15 Sun's masses as far as we know. It is an observational fact that the mass of a star is very closely related to its luminosity, but this must be regarded only as a relationship for stars on the Main Sequence of the Hertzsprung–Russell Diagram and near the Sun; it is not necessarily true of all stars everywhere.

DENSITY. This follows directly from the mass and radius.

SURFACE GRAVITY. This is an important quantity in the investigation of stellar atmospheres. It, too, is obtained from mass and radius.

ROTATION. The rates of rotation of a few eclipsing binary stars are measurable directly; otherwise they may sometimes be deduced from a study of the shapes of absorption lines. (See **Spectroscopy.**)

LUMINOSITY. This is a very important quantity, and is stated in terms of the absolute **magnitude** of the star. The faintest known star is of absolute magnitude 19; the brightest stars, and the ordinary novae, are around −8, and the brightest supernovae about −17. The Sun's absolute magnitude is 4·7. A relationship between mass and luminosity for stars near the Sun has been given. The period-luminosity law for Cepheid and RR Lyrae variable stars is described in **Stars, Distances and Motions.**

THE HERTZSPRUNG–RUSSELL DIAGRAM. We have seen that it is possible to discover many of the things we should like to know about the stars, albeit only approximately and for only certain of them. As at least some of the important characteristics of many stars are known, it is natural for us to ask if we can see any system in the make-up of stars, and if they conform to any general plan. The answer to such questions is that general relationships do seem to hold between the characteristics of most stars. The gradation of spectral classes and the mass-luminosity relationship have already been noted; but perhaps the most informative pattern which can be seen in the make-up of stars is that shown by the **Hertzsprung–Russell Diagram.**

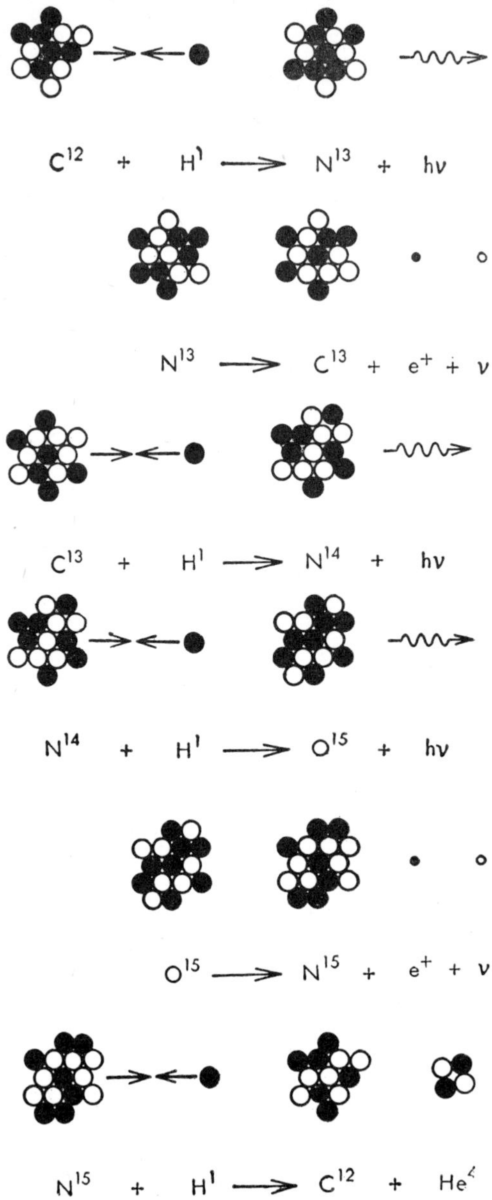

THE CARBON-NITROGEN CYCLE. It begins with a collision between a proton and a carbon atom containing 12 protons (C 12), yielding nitrogen 13 and a gamma ray. The nitrogen decays spontaneously into carbon 13, a positive electron and a neutrino. A series of further reactions ultimately reforms carbon 12, together with an alpha particle; meanwhile, there have been three releases of energy.

STARS, DISTANCES AND MOTIONS. The motions of stars relative to ourselves are measured in two directions: observations of **proper motion** show movements perpendicular to the line of sight, and those of **radial velocity** reveal motions in the line of sight. One conclusion from these measurements is that the Sun itself has a proper motion relative to nearby stars (see **Solar Apex**).

Remote stars may have their distances determined if their apparent and absolute magnitudes are known. The Cepheid and RR Lyrae **variable stars** are extremely valuable, and have been called the 'yardsticks of the Universe'. Their mean luminosities are very closely related to their periods of variation. While the relationship was correctly determined in the first place, absolute magnitudes of all Cepheids and RR Lyrae stars have until recently been in doubt, and may still be slightly in error, as none of these stars has a measurable parallax. A re-determination of the magnitudes of RR Lyrae stars in 1952 resulted in our ideas of the size of the Universe being roughly doubled. It is fortunate that Cepheids are so bright that they can be recognized in the nearest extragalactic nebulae. RR Lyrae stars are too faint to see or photograph in any extragalactic systems other than the Magellanic Clouds. (See also **Parallax** and **Distances.**)

STEFAN'S LAW. The law describing the radiation of heat from a **black body.** The heat radiated per square inch of the body's surface is proportional to the fourth power of the **absolute temperature.**

STELLAR ENERGY AND INTERIORS. The Sun, despite its relatively unimpressive absolute magnitude of 4·7, radiates as much energy every second as would be released by the explosion of several billion atomic bombs. The only conceivable sources of such a stupendous quantity of energy are nuclear reactions.

The temperatures of stellar interiors may be estimated from the physical conditions at the surface and a knowledge of the laws of physics. It is concluded that the centres of main sequence stars, at least, are all at fairly similar temperatures, of the order of 15,000,000° C. As analysis of star atmospheres suggests that the Universe is endowed with, to say the least, a goodly proportion of hydrogen, it is natural that we should look for a nuclear reaction which consumes hydrogen and goes at a reasonable rate at 15,000,000° C. It turns out that there are two such plausible reactions.

The first is the *carbon-nitrogen cycle*. In this, four successive hydrogen nuclei combine with a carbon nucleus, energy being liberated at each state. The carbon is transmuted into nitrogen during the first three additions, but the fourth addition regenerates the original carbon nucleus and gives also a helium nucleus. Helium is stable at stellar temperatures. It is worth noting that further carbon cannot be formed at these temperatures. The carbon-nitrogen cycle is thought to be the principal source of energy of the stars in the earlier spectral classes.

The second reaction is the *proton-proton reaction*. A proton is the nucleus of a hydrogen atom; such a nucleus combines successively with three more to give a helium nucleus with the liberation of energy. This reaction is more likely to occur in the stars of later spectral type.

Thus the net effect of both reactions is to transmute four hydrogen nuclei into one helium nucleus. When this occurs about half of 1 % of the mass of the hydrogen disappears, being converted into energy.

The carbon-nitrogen cycle and the proton-proton reaction are the most important energy-producing reactions which occur in stellar interiors. Many other nuclear reactions can occur at different stages of a star's evolution; these reactions do produce energy but their main importance is that they synthesize elements. (See **Stellar Evolution** and **Nucleogenesis.**)

Nuclear reaction rates are very sensitive to temperature, hence the energy-producing region of a star is confined to its very hot central zone. The energy produced has to be transported from the centre to the surface of the star, where it can be radiated away. Except in those regions containing **degenerate matter,** the energy is transported by either *convection* or *radiation*. (See **Heat.**)

Convection occurs either when the temperature is rather low and radiated energy is easily absorbed, or when there is such a large amount of energy to be transported that the material itself has to move and carry the energy. Radiation occurs when the temperature

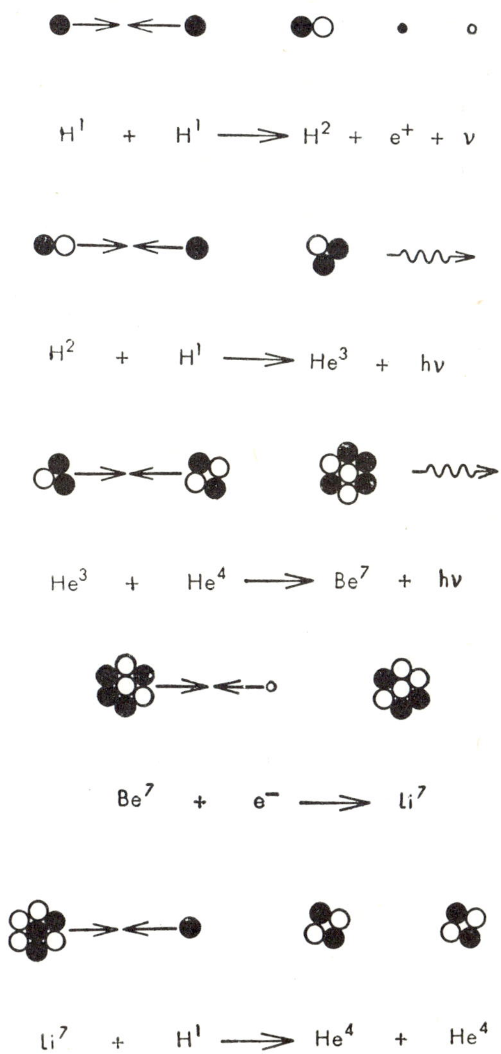

THE PROTON-PROTON REACTION. The nucleus of a hydrogen atom is a single proton (black circles). By successive collisions and the release of some energy these protons can coalesce to form helium nuclei (i.e. alpha particles) each containing two protons and two neutrons (white circles). As this process continues in stellar interiors, hydrogen is used up, helium becomes more abundant, and energy is released.

is rather high and absorption is therefore small and when there is not too much energy to be transported. The early spectral type main sequence stars have high luminosities, so there is a large amount of energy to be transported from the centre and they have a convective interior zone; but since they have high surface temperatures they have a radiative outer zone. Late spectral type main sequence stars are not as luminous and are not producing as much energy in their centres, therefore they have a radiative interior zone; but because of their low surface temperatures they have a convective outer zone. Mixing of layers of the star takes place in convective regions but not in radiative ones. Since the interior convective zone is usually small, the helium, formed by transmutation of hydrogen, is not mixed with the rest of the star, but remains where it was produced.

Observations of binary stars imply the mass-luminosity relation for main sequence stars: the larger the mass, the greater the luminosity. Thus bright stars of high mass and early spectral type need to produce energy much faster than the fainter, less massive stars of late spectral type, and so the brighter stars exhaust their central hydrogen sooner. The Theory of Relativity shows that the rate at which hydrogen must be used up to account for the observed energy release of stars like the Sun is such that these stars could continue to radiate at their present intensities for a period several times as great as their probable ages (about 5,000 million years). On the other hand, very luminous giant stars would exhaust their hydrogen in a comparatively few million years, suggesting that they must be very young on the cosmic time scale.

What happens to a star when it exhausts its central hydrogen is part of the study of **stellar evolution.** (R.G., J.A.J.W.)

STELLAR EVOLUTION. The formation and development of stars.

There is some uncertainty about the early evolution of a star. The process probably begins with the streaming of local, randomly formed cloudy condensations in **interstellar matter** towards what may loosely be called their common centre of gravity. The farther an incoming particle has fallen towards this centre, the later it arrives and the greater the attracting mass that has already accumulated near the centre, so that after a while particles will arrive with considerable velocities. Their energy will tend to compact the central mass and to heat it, while the central mass will itself tend increasingly to contract under its own gravity, and to radiate away energy when it begins to glow. A star of large luminosity, large radius and low density and surface temperature will have been formed. The large luminosity at this stage may play an important part in the theories of formation of a **solar system.**

A few nuclear reactions involving the light elements lithium, beryllium and boron now occur, but most of the energy balancing the luminosity is obtained by gravitational contraction. The contraction stops when the centre of the star is hot enough for hydrogen to ignite; the star is then a **main sequence** star. A balance of various opposing forces has been set up which can remain in near-equilibrium for long periods; but as certain processes exhaust themselves the balance shifts, sometimes catastrophically, but for most of the time quietly along the main sequence of the **Hertzsprung–Russell Diagram.**

Our understanding of stellar evolution after initial formation has made notable advances during the past two decades for three main reasons. Firstly, the structure of stars and their energy-producing mechanisms are now fairly well understood. Secondly, the significance for stellar evolution of the Hertzsprung–Russell Diagrams of star clusters is now appreciated. Thirdly, high-speed computers have made possible the solving of the complicated mathematical equations which describe the evolution of a star.

Stars must evolve because they are continually radiating energy from their surfaces. For a main sequence star this energy derives from the conversion near the centre of hydrogen into helium. (See **Stellar Energy.**)

When most of the hydrogen near the centre of a main sequence star has been converted to helium, the remaining nuclear reactions are not able to produce as much energy as the central region, or *core*, is radiating. To restore the balance of energy, the star releases gravitational energy by contracting. The contraction raises the temperature and soon a shell of hydrogen, surrounding the helium core, is ignited. The hydrogen-burning shell stops the star's contraction and the helium

MESSIER 16, a galactic nebula of intermingling hot and cooler gases in violent motion. The small, compact, dark *Globules of Bok* are believed to be stars in an early stage of formation. The larger globules may give rise to an entire star cluster, such as the Pleiades. Their final condensation into stars may be very sudden, each globule collapsing violently under the pull of its own gravitational field.
(*Mount Wilson – Palomar*)

produced in it is added to the core. No energy is now flowing in the helium core, so all parts of it have the same temperature: it is an *isothermal core*. When the core reaches a critical mass of about one-tenth of the mass of the star, the pressure on it from the rest of the star is so large that it collapses rapidly; at the same time the rest of the star expands rapidly. This very fast phase of evolution continues until the radius is very large and the surface temperature is so low that the star appears red (on the main sequence the surface temperature was higher and the star appeared blue). It is a **red giant.** Slow contraction of the *core* continues until its temperature becomes high enough for helium to burn.

The brighter stars use their hydrogen faster than fainter ones and therefore leave the main sequence sooner. Hertzsprung–Russell diagrams of star clusters confirm this: stars at the top part of the main sequence have evolved and the number of stars expected to be there is about the same as the number of red giants. The rapid evolutionary transition from the main sequence to the red giant region accounts for the existence of the *Hertzsprung gap*. The differences between the Hertzsprung–Russell diagrams of **globular clusters** and **galactic clusters** are also accounted for by the theory of stellar evolution. The diagrams show that the top of the main sequence ends at a smaller luminosity in globular clusters. So, in globular clusters, fainter main sequence stars have had time to evolve and become red giants. Globular clusters are thus older than galactic clusters in general, which agrees with our knowledge of *stellar populations*.

Evolution beyond the red giant stage is very complicated. It seems that higher-mass stars burn helium and later carbon in their cores. The lower-mass stars have cores of **degenerate matter,** and helium ignition produces an explosion, the *helium flash*, with uncertain results. It is possible that stars after the helium flash may form the *horizontal branch* of the Hertzsprung–Russell diagrams of globular clusters.

The timescale of stellar evolution is very long. For example the Sun took a few million years to contract from an interstellar gas cloud and become a main sequence star. Its lifetime as a main sequence star is about ten thousand million years. It has been a main sequence star already for five thousand million years, so in five thousand million more years it will have exhausted its central hydrogen and will become a red giant. Its expansion may be so great that for some time the Earth will orbit the centre within its tenuous outer reaches. Its lifetime as a red giant will be a few million years.

There are many uncertainties in the theory, especially concerning the possibility of mass loss from red giants and the amount of energy which may be carried away by **neutrinos.** (See **Neutrino Astronomy.**) Nor is it completely understood how **white dwarfs, novae** and **supernovae** fit into the scheme of evolution. **Quasars** and **pulsars** add to the types whose formation must be explained, but may themselves give us new insights into stellar processes. (J.A.J.W.)

SUN. The Sun is by far the closest star to the Earth, its nearest rival being more than a quarter of a million times farther away. It is the only star which presents an appreciable disc in a telescope, and consequently our knowledge of the Sun is much more detailed than that of any other star.

The Sun presents a disc a little over half a degree of arc in diameter. Its distance is 93,000,000 miles on the average; its diameter is some 860,000 miles. The solar **mass** is just about a third of a million times that of the Earth, or 2×10^{33} grams. The surface gravity on the Sun is about 28 times that on the Earth, and this causes great compression of the material comprising the Sun, so that, although it is gaseous, its average density is about 1·4 times that of water. The high surface gravity is also partly responsible for the sharpness of the Sun's **limb** as seen in a telescope. Our luminary is a very undistinguished star, a vast sphere of glowing gas; it is a very typical dwarf star of **spectral class** G2, and is represented by a point lying neatly on the Main Sequence of the **Hertzsprung–Russell Diagram.** Its apparent stellar magnitude is a little fainter than -27, and the absolute magnitude of $+4{\cdot}7$ follows from this. The Sun's internal structure and source of heat are discussed under the heading of **Stellar Energy.**

When we observe the Sun through a telescope (with, of course, suitable arrangements for dimming its light to a comfortable value) we see no solid surface but just the outer layers of a sphere of gas. The gas

THE CORONA OF THE SUN IN TOTAL ECLIPSE. Coronal streamers following the pattern of magnetic fields can be discerned everywhere and form an arc over a strong local disturbance just to the right of the Sun's top in this picture. There are many bright prominences, and irregularities in the edge of the Moon's profile are clearly visible. (*Science Museum*)

contains many **ions** which make it opaque when seen in any great thickness, so we cannot see regions of the Sun many miles below the 'surface'. In fact, this part of the Sun may be likened to a thin fog which is self-luminous by virtue of being white hot. The surface layers of the Sun are called the *photosphere*. Above the photosphere lies a tenuous, and for the most part transparent, atmosphere. The lower part of the solar atmosphere is the *chromosphere*, about 10,000 miles deep and merging into the higher *corona*, which extends several millions of miles from the Sun. There is no hard-and-fast line of demarcation between the photosphere and the chromosphere, or between

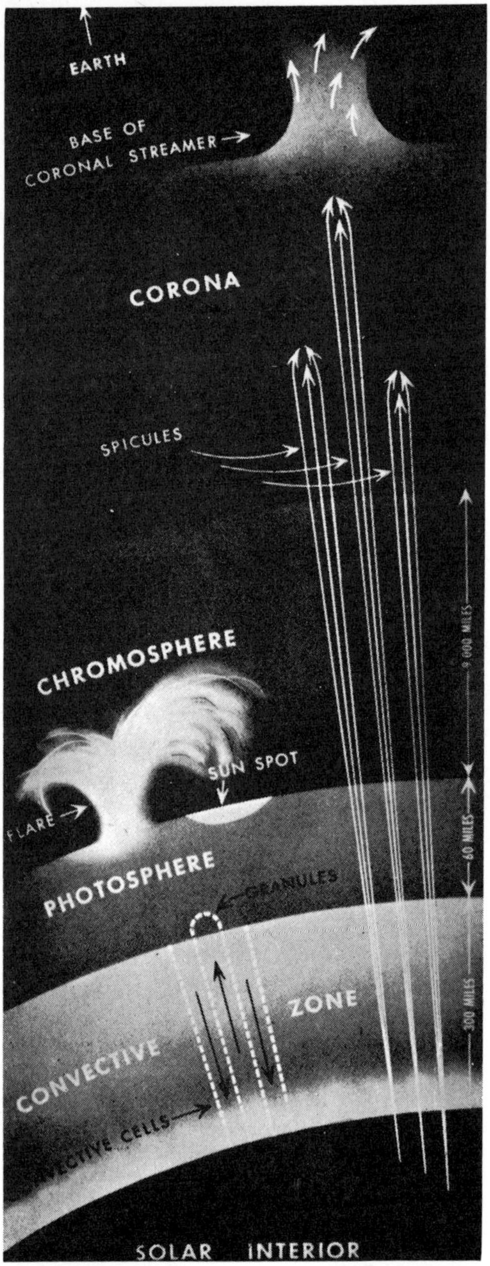

the chromosphere and the corona; neither does the corona possess any definite outer boundary.

The photosphere presents a picture of ceaseless activity and motion. It often has areas of intense disturbance centred around dark markings known as *sunspots*.

SUNSPOTS are areas where the temperature of the photosphere is abnormally low. The average photospheric temperature is around 5,500° C., while sunspot temperatures are about 4,000° C. Spots are of all sizes up to (exceptionally) 100,000 miles or more across. The smallest are called *pores*.

Typical sunspots have two well-defined areas: a dark *umbra*, surrounded by a less dark *penumbra*, which shows a great deal of fine striation and other structures. Spots often occur in groups which are extended in longitude more than in latitude, and there are often two large spots, one at either end of the group, which are named the leader and trailer spots; between these, there may be as many as a few dozen smaller spots, often enveloped by the same penumbra. The lifetime of sunspots is very variable: small spots and pores last a few days, or even less than a day, while larger spots and groups last up to 100 days or so.

THE SUNSPOT CYCLE. The spottedness of the Sun varies markedly. The spottedness is sometimes given in millionths of the visible hemisphere. The largest group on record was seen in April 1947, and had an area of 6,100-millionths. But the customary system uses WOLF RELATIVE NUMBERS, which take account of the number of spots and groups, and the observer's instrument. When the mean sunspot number for each year is plotted on a graph, the latter shows a very obvious cycle of about 11 years; the length of time between successive maxima varies from 7½ to 16 years. The rise to sunspot maximum is normally more rapid than the decline. The annual sunspot numbers from 1750 to 1948 are shown opposite.

SUNSPOT LATITUDES. Sunspots are only found within 40° of the solar equator, and the mean latitude varies during the sunspot cycle. Shortly after minimum the spots occur in the higher latitudes on both sides of the equator, and as the cycle progresses they move gradually into lower latitudes, nearly reaching the equator by the time sunspot minimum is again reached, and new spots break out in high latitudes. This is pleasingly portrayed in the 'Butterfly Diagram'.

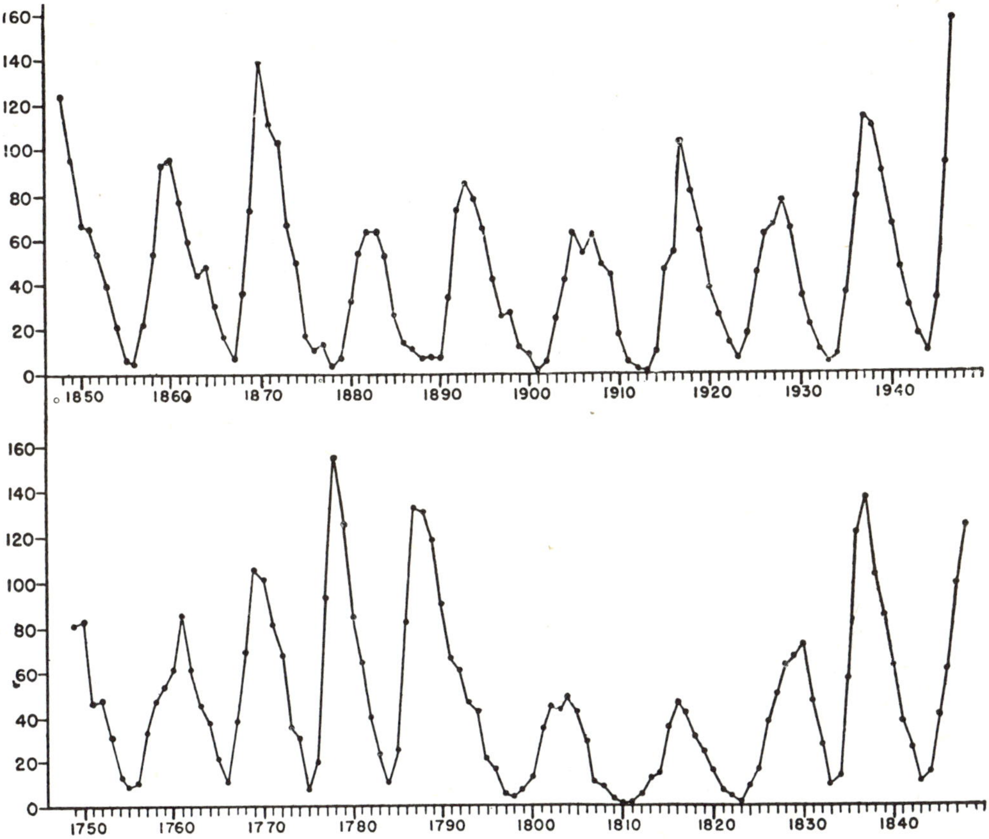

THE SUNSPOT CYCLE, 1749–1948. The average interval between maxima is 11·2 years. (*Menzel*)

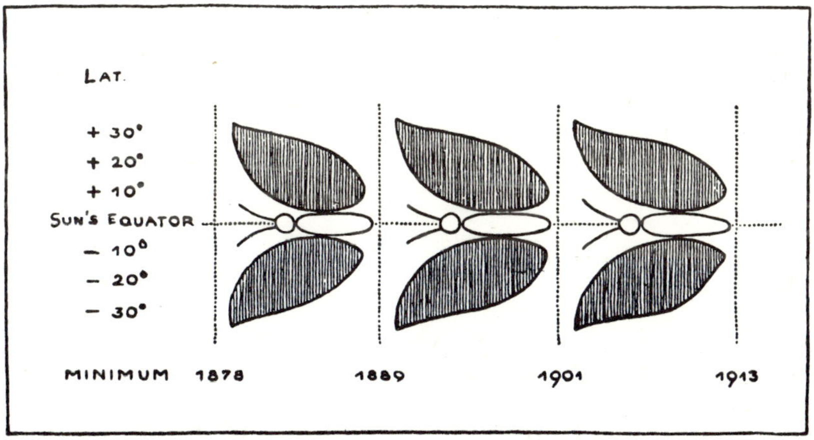

THE BUTTERFLY DIAGRAM. The shaded areas extend over the range of latitude in which sunspots were observed at any given time. For instance, in 1881 all spots were between + 15° and + 28° and between − 16° and − 31°. The dotted vertical lines indicate the sunspot minimum at the end of each 11-year cycle. The body and feelers of the 'butterfly' have been added only to complete the pattern.

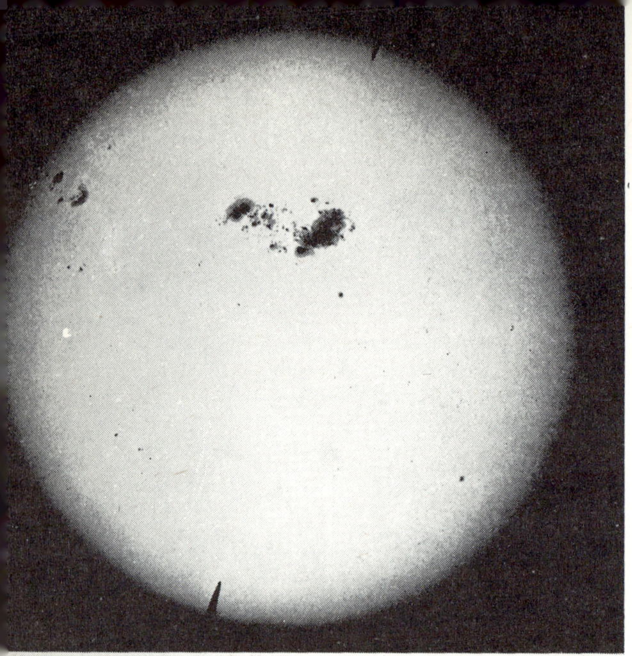

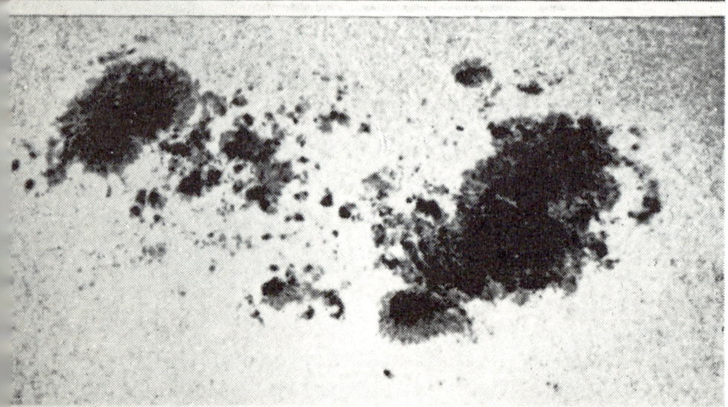

The Sunspot of April 7, 1947

GRANULES. The photosphere shows structure not only in and around sunspots, but everywhere except close to the limb. It is covered with a system of *granules*, which give the surface a mottled effect. Individual granules are only about one second of arc across, and are difficult to see, let alone photograph. In the 1880s the French astronomer Janssen secured some amazingly good photographs of granules, which have only recently been surpassed by the expedient of sending a photographic telescope to an altitude of 80,000 feet by balloon, to get well above the layers of the Earth's atmosphere which are responsible for bad seeing. The bright granules seen in the photograph are probably the tops of convection currents bringing up hotter material from below, while in the dark intervening lanes cooler matter is subsiding. Any particular granule is visible for only a few minutes.

FACULAE. These are brighter patches of the photosphere usually associated with sunspots. They sometimes appear on the solar surface before a spot in the same region, and often persist after the disappearance of spots.

ROTATION. From the movement of sunspots, we know that the Sun rotates on an axis inclined at 83° to the ecliptic in a period of about a month. The exact time varies with the latitude, being shortest at the equator. At latitudes above 40° there are no sunspots or other long-lived features to act as fixed marks on the solar disc, and the **Doppler effect** caused by the approach and recession of opposite limbs has to be used to find the rotation period. The table gives the period at different latitudes.

PERIODS OF SOLAR ROTATION

Latitude (*degrees*)	*Period* (*days*)
0	24·65
20	25·19
40	27·48
60	30·93
90	34

The rotation is in the same direction as the movement of the Earth round the Sun. Consequently the period appears about two days longer from the Earth. Spots take a fortnight to cross the disc, and if very long-lived they reappear at the east limb a fortnight after passing round the west limb.

THE SOLAR SPECTRUM. The upper layers of the photosphere are largely responsible for the Sun's absorption spectrum. The lower part of the photosphere may be considered to radiate nearly as a **black body** and to give a continuous spectrum; **atoms** in the higher layers absorb specific wavelengths of this light and in doing so give rise to the many thousands of dark absorption or Fraunhofer lines which cross the solar spectrum (see **Spectroscopy**). The lines are all identifiable with elements found on Earth; more than sixty elements have been recognized in the Sun. Hydrogen is much the most abundant of the constituents. The region of the solar surface in which the absorption or 'reversed' lines are found used to be termed the *reversing layer*; this name has been dropped as it is now known that not all the lines are formed at the same levels. As a general rule, the stronger an absorption line appears the higher is the layer which causes it. Some of the strongest absorption lines in the spectrum of the Sun arise in the chromosphere. The chromosphere has in addition a rich but feeble emission spectrum. Very few of the emission lines are revealed even by powerful spectroscopes in normal conditions owing to the overpowering brightness of the photospheric spectrum against which the chromospheric lines are seen. However, at the time of a total eclipse, the chromosphere alone is seen at the initial and final moments of totality, and at these times the so-called **flash spectrum** can be observed. The flash spectrum is very similar to the absorption spectrum in reverse, but surprisingly enough shows the temperature of the chromosphere to increase outwards until it becomes considerably higher than that of the photosphere. There is, therefore, a particular level in the solar atmosphere at which the temperature is lower than in the layers below and above it. It is the region above this temperature minimum which is primarily responsible for the emission spectrum.

The general fogginess or *opacity* of the Sun's atmosphere is not the same at all wavelengths, so the level to which we can look 'down' into the Sun also varies with wavelength. In the visible part of the spectrum the variation is not very large; but in the far ultraviolet, shortward of a wavelength of about 1,700 Ångströms – only observable from above the Earth's atmosphere, from rockets or satellites – the opacity becomes so high that we cannot see down to the photosphere at all but only to the regions of the chromosphere above the temperature mini-

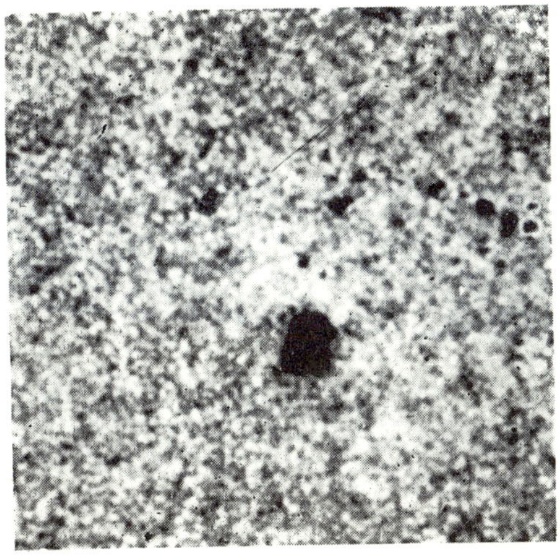

SUNSPOTS AND SOLAR GRANULATION. The fine mosaic-like pattern is difficult to reproduce in print; it represents the tops of the convective cells. (*Janssen, Meudon*)

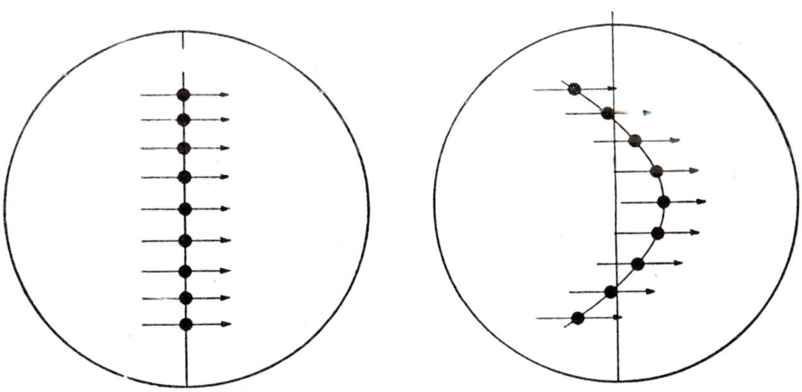

The Sun does not rotate as a solid body; the equatorial region is continually drawing ahead of the rest. As a result, sunspots ranged in a line as in the left would reappear after one rotation with those nearest to the equator in the lead. (*After Menzel*)

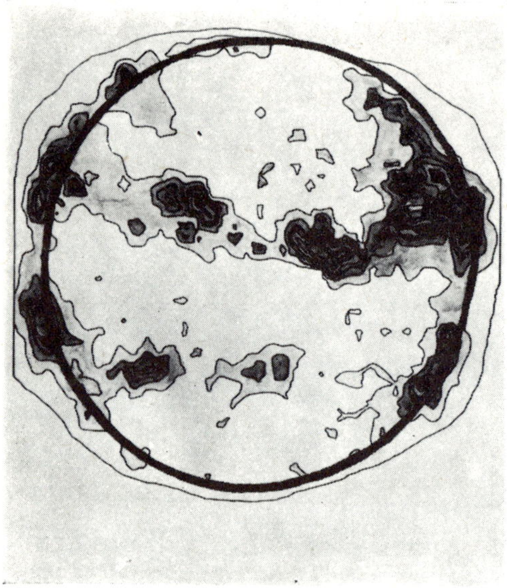

THE FIRST ULTRAVIOLET VIEW OF THE SUN was received from the Orbiting Solar Observatory OSO-4. It shows regions of UV intensity in the corona, at considerable heights above the photosphere, where the emissions from Magnesium-10 ions indicate temperatures above 850,000° C. This is one of 50 ultraviolet wavelengths which the spectrometer aboard OSO-4 can be commanded to record. The satellite is helping to build up a three-dimensional map of solar radiation in frequencies which cannot be observed from the ground because of atmospheric absorption.

Relative sizes of the Sun as seen from Mercury, Earth and Pluto. Approximately correct diameters will be seen if the book is put about 5½ feet away.

mum. Accordingly, the solar spectrum, which shows absorption lines throughout the visible and near-ultraviolet spectral regions, undergoes a drastic change near 1700 Å: at all shorter wavelengths it is an emission spectrum.

LIMB DARKENING. Near the limb of the Sun we look very obliquely into the solar atmosphere and consequently we do not see to such deep levels as when we look at the centre of the Sun's disc, where we see vertically into the atmosphere. In visible light the deeper levels are hotter and therefore brighter. Accordingly, the solar disc appears brightest in the middle and progressively dimmer towards the limb – a phenomenon called *limb darkening*. The effect is not very apparent to the eye when the Sun is viewed through a dark glass, but is always very obvious on photographs.

In the far ultraviolet part of the spectrum, where the opacity of the Sun is so great that, even in the centre of the disc, only the high chromosphere is seen, the temperature of the layers seen at the centre of the disc is less than that of the higher layers seen near the limb. At these wavelengths the centre is the faintest part of the disc – a case of *limb brightening*.

PROMINENCES. These are seen during total eclipse as purple appendages of the chromosphere. They are huge clouds of glowing gas, of diverse forms which change continuously. Some are quiescent, remaining almost the same for hours; many are arches, and are shot off the chromosphere with great velocities; there are also geyser-like eruptions and other forms. Some types are associated with sunspots. Speeds of over 100 miles per second are observed, and

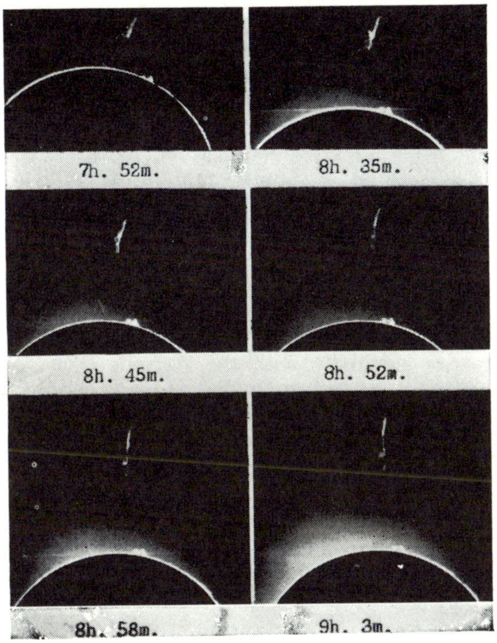

A very high eruptive prominence on the Sun.
(*Royds, 1928*)

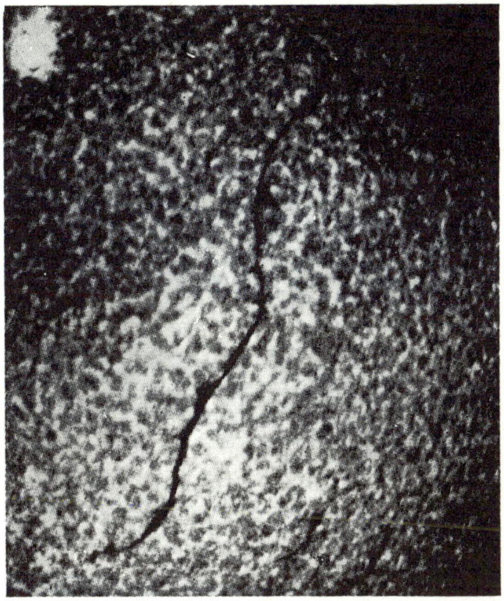

SOLAR FILAMENT of July 20, 1922 – one of the largest on record. A flocculus appears at top left.
(*d'Azambuja, Meudon*)

exceptional prominences are thrown several hundred thousand miles above the Sun. The purple colour of prominences is due to the fact that most of their light arises from the emission spectrum of hydrogen gas, whose principal lines are in the red and blue parts of the spectrum.

SPICULES are like tiny prominences – glowing tongues of gas which shoot out of the chromosphere near the Sun's poles; each lasts a few minutes only. Together they cause the chromosphere to appear like a lawn viewed horizontally, with many grass blades sticking up. They may be related to granules.

The **coronagraph** enables spicules and prominences to be observed without a total eclipse; but for its aid very little would be known about these features.

SPECTROHELIOGRAPH. This instrument enables us to observe the Sun in light of one colour only. If the colour is chosen to be exactly that of an absorption line the instrument 'sees' only the atoms of the kind which is forming the line, and a photograph showing the distribution of these atoms is obtained. The most usual absorption lines to use are the red line of hydrogen, designated Hα, and the violet line of ionized calcium. Approximately the same structures are seen whatever line is observed. The layers depicted lie in the lower parts of the chromosphere and show a mottled structure, far coarser than granules. Bright regions often occur; these are called *flocculi.* Dark markings called *filaments* are prominences seen in projection against the solar disc instead of against the sky beyond the limb. Near the centre of the disc filaments appear very narrow, showing that prominences are thin, blade-like sheets of gas.

FLARE. This is a sudden, short-lived increase in the light intensity of the chromosphere in the vicinity of a sunspot. It is normally seen in Hα light, using a spectrohelioscope, but very intense ones can be observed in white light, as was first done in 1859. Flares are bursts of *light*, not of matter. They have great influences on the Earth (see below).

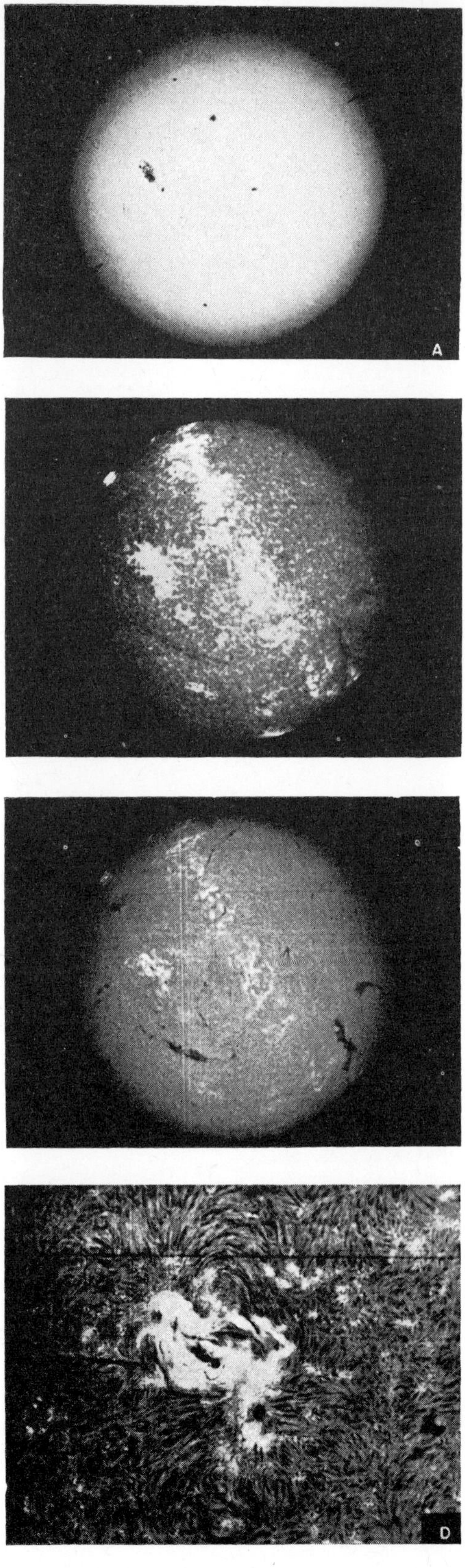

CORONA. This is the outer part of the solar atmosphere and is composed of exceedingly tenuous gas. Although the innermost parts can be revealed by the coronagraph, the corona can only be well observed during the fleeting moments of a total eclipse. Then it is seen as a shining halo of white light, decreasing in intensity away from the Sun until it is lost against the remaining brightness of the sky. The total light is comparable with that of the full Moon. The corona has on occasion been seen to extend as far as five solar diameters from the Sun. Its form varies markedly with the sunspot cycle: at sunspot minimum it has long equatorial streamers and short polar plumes, while at maximum it is much more nearly circular. Radial structures and streamers are always noticeable.

The spectrum of the corona is complicated and puzzling. It points to the existence of three sources of light:

i. An emission spectrum, whose source has been called the E corona. The bright lines were at first unidentifiable, and were attributed to a new element, 'coronium'. It transpired that they are emitted by common atoms such as iron, nickel and calcium, ten to thirteen times ionized. Such ionization could occur only at stupendous temperatures of the order of a million degrees Centigrade. The emission lines are seen only in the spectrum of the inner corona.

ii. The solar absorption spectrum; this must be seen by reflection from particles in the corona. These presumably constitute the innermost zone of the Zodiacal Light, and their light, called the F corona, dominates in the region beyond one solar diameter from the limb of the Sun.

iii. The K corona, dominating near the Sun, shows a solar-type spectrum in which the absorption lines are so immensely widened

THE SUN. From top to bottom: (a) ordinary photograph, (b) calcium spectroheliogram, (c) hydrogen spectroheliogram, (d) enlarged portion of (c) showing details of the sunspot group.
(*Mount Wilson – Palomar*)

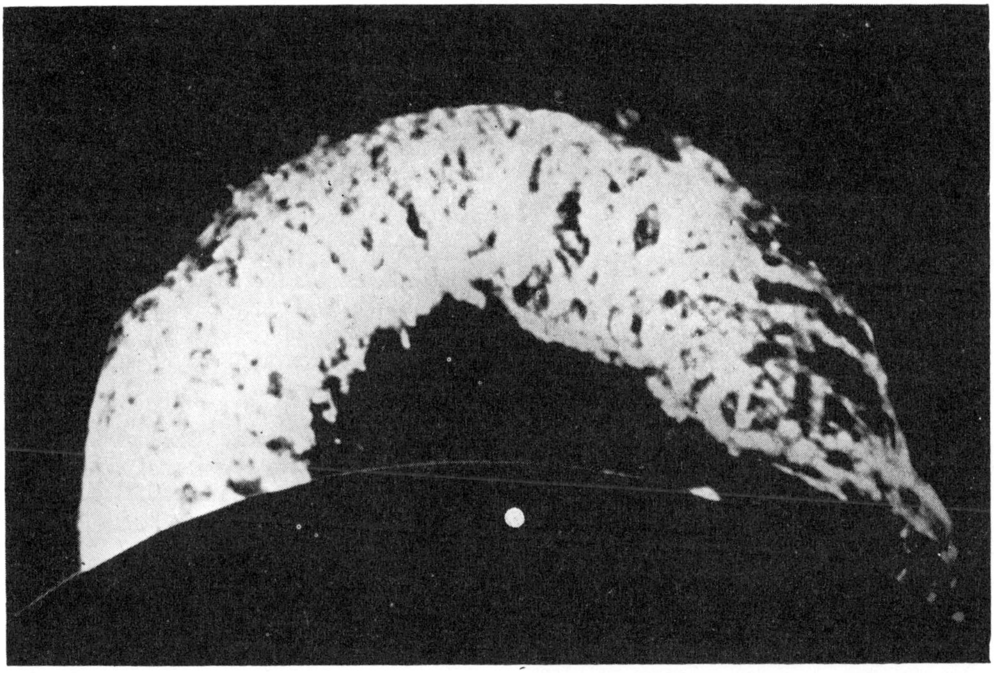

The great solar prominence of June 4, 1946. The round, white dot indicates the size of the Earth. When this picture was taken the arch of the prominence was shooting outwards with a speed of over 100 miles per second. Less than two hours later it had dissipated itself. (*Roberts, Climax*)

as to be undetectable. It is due to sunlight scattered by free electrons whose thermal motions at a temperature of 1,000,000° C. would account for the extreme Doppler broadening. (See **Spectroscopy.**)

RADIO EMISSIONS from the Sun emanate in the corona and corroborate the extreme temperature suggested by spectral observations. They can be detected at all times but are subject to enormous variation. The appearance always has the same polarity in one hemisphere and the opposite polarity in the other. In the next cycle these polarities are reversed, so the sunspot cycle is really 22, not 11, years.

Very accurate scanning of the solar surface with a **magnetograph** has now revealed that the Sun as a whole has a general magnetic field which at the surface is a few times stronger than that of the Earth. In 1959 this general field was observed to reverse its polarity.

SOLAR WIND. Many **planetary probes** have carried equipment which has shown the existence of a continuous and measurable flow of **plasma** away from the Sun. The plasma, which is known as the *solar wind*, consists of streams of charged particles, mainly protons with velocities which are typically 1,000 kilometres per second. The flow of these particles is quite irregular and consists of bursts of varying intensities and thickness, which are related to the solar surface activity.

EFFECTS OF THE SUN ON THE EARTH. Apart from the grosser effects such as the retention of the Earth in its orbit and the climatic effects of the solar radiation, other important atmospheric phenomena are associated with the Sun. Solar flares produce energetic **ions** which can travel to the Earth and be trapped in the Earth's magnetic field to form the **Van Allen belts.** When the particles reach the Earth they cause the

aurorae and **magnetic storms.** The ultraviolet light from the Sun is variable and, together with the particles, affects the **ionosphere** and hence terrestrial radio propagation.

SUN-GRAZERS. A group of spectacular comets which at perihelion pass through the outer atmosphere of the Sun. (See **Comets.**)

SUN-SEEKER, SUN-SENSOR. A device used to determine the direction from a space craft to the Sun. Once locked-on to the Sun, the device defines one axis for orientation of the space craft or its sun-sensitive apparatus.

SUPERGIANT STARS. Stars of enormous size and luminosity compared with main sequence stars of the same spectral type. (See **Hertzsprung–Russell Diagram.**)

SUPERNOVA. A star which explodes catastrophically, with a sudden liberation of most of its energy.

A supernova has a light curve superficially like that of a normal **nova** with a rather slower accession of brightness and a smoother decline, uninterrupted by irregular fluctuations. However, the scale of the phenomenon is quite different, a supernova being about ten thousand times as bright as an ordinary nova. The absolute magnitude of a supernova at maximum is around -16 or -17; it releases as much energy in one second as the Sun does in 60 years.

Unfortunately supernovae are rather rare and difficult to study. Even in a large galaxy such as our own the frequency is only a very few per millennium. The only historical supernovae observed in our Galaxy occurred in 1054, 1572 and 1604; all reached apparent magnitudes considerably brighter than zero, and that of 1572 was clearly visible in daylight. About fifty have been observed in other galaxies.

SUPERNOVA in an irregular galaxy. *Top:* 1937 Aug. 23, exposure 20 min., maximum brightness. *Middle:* 1938 Nov. 24, exposure 45 min., supernova faint. *Bottom:* 1942 Jan. 19, exposure 85 min., supernova too faint to observe. Only the last exposure was long enough to bring out the galaxy itself. (*Mount Wilson – Palomar*)

A supernova outshines the whole light of any but the largest galaxies. One photographed in the thirteenth-magnitude galaxy IC 4182 outshone the galaxy by five magnitudes – a hundredfold!

There appear to be two types of supernova. Type II seem like novae on a grand scale: they have recognizable spectra which are quite similar to those of novae, and are rather fainter than Type I. Their rates of expansion are around 5,000 km./sec. Type I have entirely unrecognizable spectra in which not a single band has been identified. The widths of the bands suggest explosive velocities in excess of 10,000 km./sec.

It seems that a supernova explosion results in the dissipation of nearly the whole mass of the star into space, and therefore it can happen only once to a star, in contrast to nova outbursts. The remains of the supernova of 1054 are still visible as an irregular, expanding nebula (the **Crab Nebula**) consisting of matter propelled outwards from the star by the explosion.

The recent discovery of a **pulsar** in the Crab Nebula, and of another pulsar in the same direction as a nebulosity which is similarly thought to be a supernova remnant, has led astronomers to suggest that pulsars are produced in these cataclysmic explosions. But the discovery of pulsars themselves is too recent to allow this suggestion to be more than a tentative association. However, if it is found to be true, then any attempt at explaining the cause of supernova outbursts must also account for the production of pulsars by these explosions.

SYMBIOTIC STARS. A class of stars with a characteristic spectrum, exhibiting features of *two* different **spectral classes.** This spectrum is frequently a combination of the classes B and M. These are normally produced under very different conditions of stellar surface temperature, and therefore in very different types of star. Hence it is thought that the stars producing these *combination spectra* are **binaries,** but so close together in their orbits that their orbital motion cannot be detected.

Symbiotic stars are frequently found to be irregular **variables,** especially when the difference in temperature between the two stars is extreme. Their range of brightness variation is similar to that of U Geminorum variables.

SYNCHROTRON RADIATION. A charged particle which experiences an **acceleration** will emit electromagnetic radiation (see **Electromagnetic Waves**). If the charged particle is moving in a magnetic field (see **Magnetism**), it will experience an acceleration due to this field as long as its motion is not parallel to the magnetic field. The radiation which such a particle emits is called synchrotron radiation or sometimes *magnetobremsstrahlung*. Since the magnetic fields in radio sources (see **Radio Astronomy**) are thought to be quite small, the charged particles are required to have velocities near the speed of light in order to produce radiation in the radio range of the **electromagnetic spectrum.**

SYNODIC PERIOD. The interval between successive oppositions of an outer planet, or between successive inferior conjunctions of an inner planet. (See **Conjunction.**)

SYRTIS MAJOR. The most conspicuous dark area on **Mars.** It is visible in a small telescope when Mars is well placed.

T

TEKTITES are objects consisting of silica-rich glass (70–80 % SiO_2) which superficially resemble obsidian, yet are different from any terrestrial obsidian, and could be fragments of liquid rock chilled to solidity during flight. The name is derived from the Greek word *tektos* – molten. They do not appear to bear any obvious geological or mineralogical affinity to the local rocks when found *in situ.* They have diverse shapes like teardrops, rods, discs, flanged buttons or dumb-bells. They range in colour from black through dark brown to bottle-green, in size from walnut or smaller to apple.

Although they have been recognized for at least two centuries, their origin is still uncertain. The two main ideas are: (*a*) that they are of terrestrial origin and might be explained by volcanic action; (*b*) that they stem from large meteoric impacts – and that tektites are some of the original impact material or terrestrial material formed as a result of the impact. The meteoric theory has rapidly gained ground, for many specimens contain small nickel-iron spherules of undoubted meteoritic origin. Recent debates on tektites have centred on the kind of meteoric impact that could have been responsible. Some opt for the Moon as a likely place of origin, while others consider that comet or asteroid impacts are more likely.

Tiny glass particles, less than 1 mm. in diameter, were discovered in sediments in the South Pacific dated at 700,000 years. These *microtektites* appear to be directly related to the Australian mainland tektites, and their deposition coincides in time with a reversal of the Earth's magnetic field. The idea that a large meteorite impact could have caused both the reversal and the tektites is being investigated.

TELEMETRY. The technique of transmitting the results of measurements or observations which were made by instruments in inaccessible positions (such as an unmanned satellite in orbit) to a point where they can be used. By definition this includes all means used to transmit the readings obtained from instruments carried on missiles and space craft, but similar techniques are widely used for monitoring the behaviour of large terrestrial systems from a central control point. Control of electricity supply networks and of many industrial processes depends heavily on telemetry. Telemetry systems consist of multiplexing devices for combining the outputs of a number of instruments into one communications link, the communications link itself, and the system at the control centre which 'unscrambles' the signals and reproduces the instrument readings. Systems can vary widely in complexity and information-handling capacity, and can operate in 'real time', transmitting the readings as they are at the moment of transmission, or can transmit batches of data stored up from one communications 'session' to the next. The latter method is ideal for satellite applications where it is desired to use only one ground station; the satellite can collect data continuously, but may only be 'visible' to the ground station (and hence in radio contact) for fifteen minutes or so five or six times a day. The penalty to be paid is that data must flow from the satellite much more rapidly during communications sessions than they are acquired, and hence a communications system with wider bandwidth (larger information capacity) and therefore needing more power is required.

TELESCOPE. (The reader is asked to refer to **lens** and **mirror** before reading this article.) A telescope is an optical instrument used for examination of distant objects. Its most fundamental component is a lens or mirror capable of gathering light from the distant object and focusing it into an image which is a miniature representation of the object. Other components are placed at or near the focus, their nature varying according to the purpose for which the telescope is being used. For instance, a photographic plate may be put at the focus to record the image; or an **eyepiece,** which is just a small magnifying glass

THE BACK OF THE 200-INCH MIRROR. It was cast in Pyrex glass, and 10 years were spent in letting it cool evenly and grinding its optical surface. The honeycomb spaces reduce the weight of the mirror by 20 tons, and enable it to follow changes in atmospheric temperature far more rapidly than a solid mirror.

or simple microscope, may be provided to enable the image to be viewed directly. Professional astronomers often mount complicated auxiliary equipment, such as photometers or spectrographs, at the foci of their telescopes.

REFRACTORS AND REFLECTORS. A telescope which employs a lens or *object-glass* to form the image is termed a *refracting telescope*, or simply a *refractor*. The object-glass is made up of two or more lenses one behind the other; this is essential to minimize **chromatic aberration,** and the multiple lens is said to be *achromatic*.

If a mirror is used to form the image instead of a lens, the result is a *reflecting telescope*, or *reflector*. As the image in a reflector is formed between the mirror and the object under observation, if one attempted to view it *in situ* one's head would obstruct the incoming light. Various systems have been devised to overcome this; most employ a small secondary mirror to deflect the light and cause the image to be formed in a more convenient place. The commonest system for small reflectors is the *Newtonian*: a small, flat mirror, always called simply the *flat*, is set at an angle of 45° in the telescope tube and reflects the light out to an eyepiece at the side.

Large reflectors often use a convex secondary mirror which reflects the light through a hole in the main mirror; this is the *Cassegrain* system. A modification of the Cassegrain arrangement is the Ritchey–Chrétien, which is much in vogue among the designers of large telescopes nowadays. In this system the primary mirror, as the large mirror which first reflects the starlight is called, is deformed, so that it would not *by itself* produce a good image at all. The convex secondary mirror is, however, correspondingly deformed to give a proper image at the Cassegrain focus, and by an appropriate choice of the shapes or *figures* of the two mirrors the **field of view** covered in good definition is made considerably larger in the Ritchey–Chrétien system than in the straightforward Cassegrain.

The auxiliary equipment used at observatories is sometimes too heavy to be carried on the telescope, or it may require to be kept in one particular attitude instead of being tipped in all directions as the telescope is moved to different parts of the sky. This difficulty is overcome by the use of one or more additional flat mirrors which are arranged so as to reflect the beam of a Cassegrain reflector in a fixed direction, out of the telescope tube itself to a stationary focus where the equipment may remain stationary and yet be able to receive the light from stars in any direction. Such an arrangement is called a *coudé* reflector, and the fixed focus is the *coudé focus* of the telescope.

The world's largest reflecting telescopes are so big that they can afford to let not just the observer's head but his whole body obstruct part of the incoming light beam and yet not lose an important fraction of the total light. They may therefore be used at the *prime focus*; a cylindrical observing station – the *prime focus cage* – is provided at the top of the telescope tube to house the observer, who, with any auxiliary equipment he may be using, is carried around by the telescope (see picture on page 196). Smaller telescopes can also be used at the prime focus if the observer does not need to be there in person; for instance photography is often performed at this focus. The **Schmidt camera** is an example of a type of photographic telescope which is always used at the prime focus.

Loading a photographic plate into the 15-inch refracting telescope at the U.S. Naval Observatory.

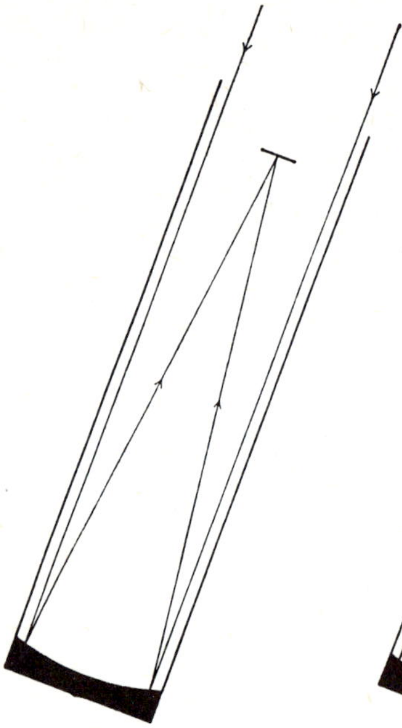

PRIME FOCUS

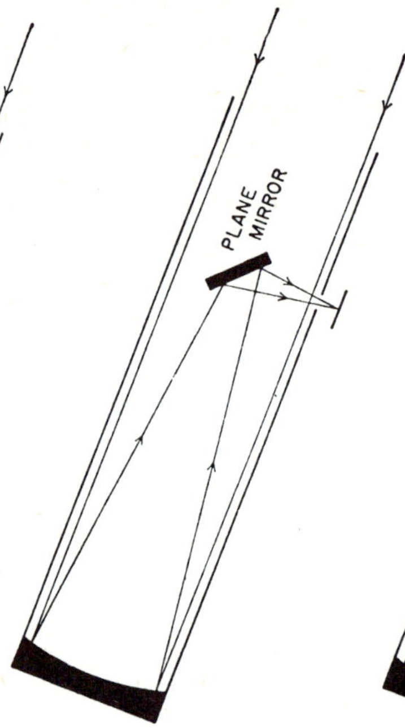

NEWTONIAN

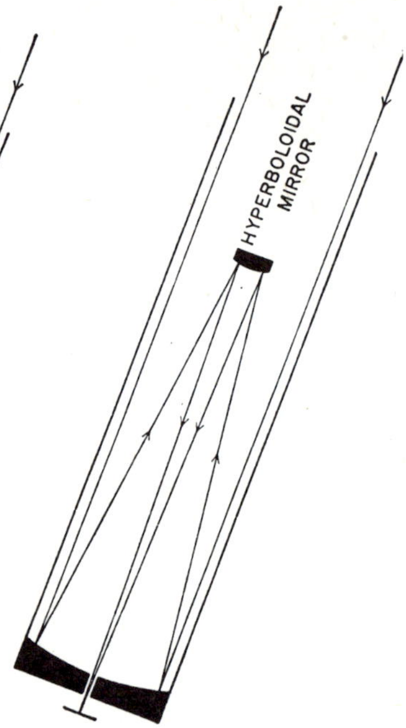

CASSEGRAIN

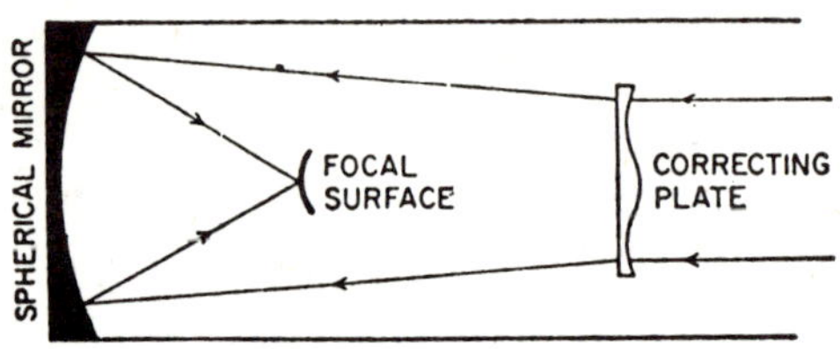

Schmidt Camera

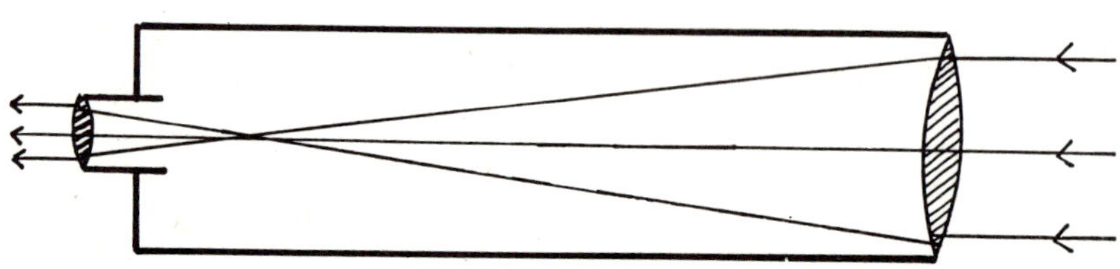

Refractor

Other focal positions are available on some large reflectors. Such telescopes usually have a number of secondary mirrors which are employed alternatively to give the image in different places. This is done in an effort to make the instrument suitable for many different research projects; unfortunately it normally means that the telescope is less effective for all these different purposes than it could be for one particular purpose if it were designed especially for that one.

APERTURE, LIGHT-GRASP AND RESOLVING POWER. The *aperture* of a telescope is the diameter of its main mirror or lens. The chief task of a telescope is to gather as much light as possible; the greater the aperture, the more light is collected, and the fainter the stars that can be observed. This power of a telescope to collect light is sometimes called its *light-grasp*. The table gives the approximate limiting **magnitudes** of stars visible to the eye on a first-class observing night with various apertures:

APERTURE (*inches*)	MAGNITUDE
1	9·5
2	11·0
3	12·0
4	12·6
6	13·5
12	15·0
30	17·0

Appreciably fainter stars may be photographed with the same apertures.

The *resolving power* of a telescope is the smallest angular distance between two stars for them to appear just separated in the telescope, in favourable circumstances. This angle is inversely proportional to the aperture of the telescope. Theory and practice agree that it is about five seconds of arc divided by the aperture in inches. However, the **seeing** in the Earth's atmosphere prevents very large telescopes from ever reaching this limit of resolving power.

The *magnifying power* of a visual telescope depends upon the eyepiece used, and can be almost any desired value. The answer to the question 'How powerful is that telescope?' cannot therefore be given in terms of magnifying power, but rather in terms of light-grasp. Great magnification is not necessarily an advantage. It is obviously useless to magnify an image beyond the point at which the telescope's limited resolution causes it to appear ill-defined; this point is reached when the power is about 50 times the aperture of the telescope in inches. Thus the best 6-inch telescope in the world cannot magnify with clarity more than $\times 300$.

Often, the seeing limits the reasonable magnification to a much lower value. In any case, a high power has the disadvantage of giving a small field of view, so that observations of extended objects such as star clusters and nebulae are generally best carried out with a low power, of 6 to 10 per inch of aperture.

TELESCOPE TUBES. A refractor has a cylindrical tube with the object-glass at one end and the eyepiece at the other. A reflector, however, must have a tube open at the 'top' to let light pass down to the mirror at the far end, and the tube design is a matter of choice. The essential function of the tube is simply to keep the mirrors and eyepiece in the correct relative positions; it need be no more than a framework.

Large reflectors housed in observatory buildings very often have lattice-work 'skeleton' tubes made of steel struts. Amateurs' telescopes used outdoors also sometimes have lattice tubes, but these allow dew to form on the mirrors, and extraneous light from nearby, or from the Moon, enters the eyepiece and spoils the image. The remedy is to make the tube of sheeting, instead of isolated bars.

Unfortunately reflectors, necessarily having one end open, are very prone to air currents which flow up and down the tube, refracting light irregularly and giving an unsteady image (see **Seeing**). If the tube is a mere framework, it has the advantage that currents cannot flow along it; in a tube made of sheeting they can be very troublesome. A large hole in the side of the tube near the main mirror prevents cold air collecting at the bottom of the tube. Making the tube out of heat-insulating material prevents rapid changes in temperature inside and helps to minimize air currents: thus wood is preferable to sheet metal, or a metal tube may be cork-lined. If the tube is of square section, the currents may tend to flow in the corners where they are least harmful; circular-section tubes have no such advantage.

Dr. Hubble, the noted American astronomer, guiding the 48-inch Schmidt camera on Mount Palomar. Because of its wide-angle vision this instrument has been able to complete in 7 years a photographic survey of the sky which would have taken the 200-inch Hale telescope 5,000 years to accomplish.

TELESCOPE MOUNTINGS. There are two systems of mounting – altazimuth and equatorial.

An *altazimuth* mounting allows the telescope to move in two directions, vertically (altitude) and horizontally (azimuth). This type of mounting is easily made by amateurs and can readily be made portable.

The *equatorial* mounting carries the telescope about an axis which points towards the celestial pole and is therefore parallel with that of the Earth. The Earth rotates one degree in four minutes, so that frequent adjustments must be made to the position of a telescope to prevent the object under observation drifting out of the field of view. An altazimuth mounting requires corrections in both altitude and azimuth, but the equatorial needs to be turned about only one axis to follow a particular star. As the Earth

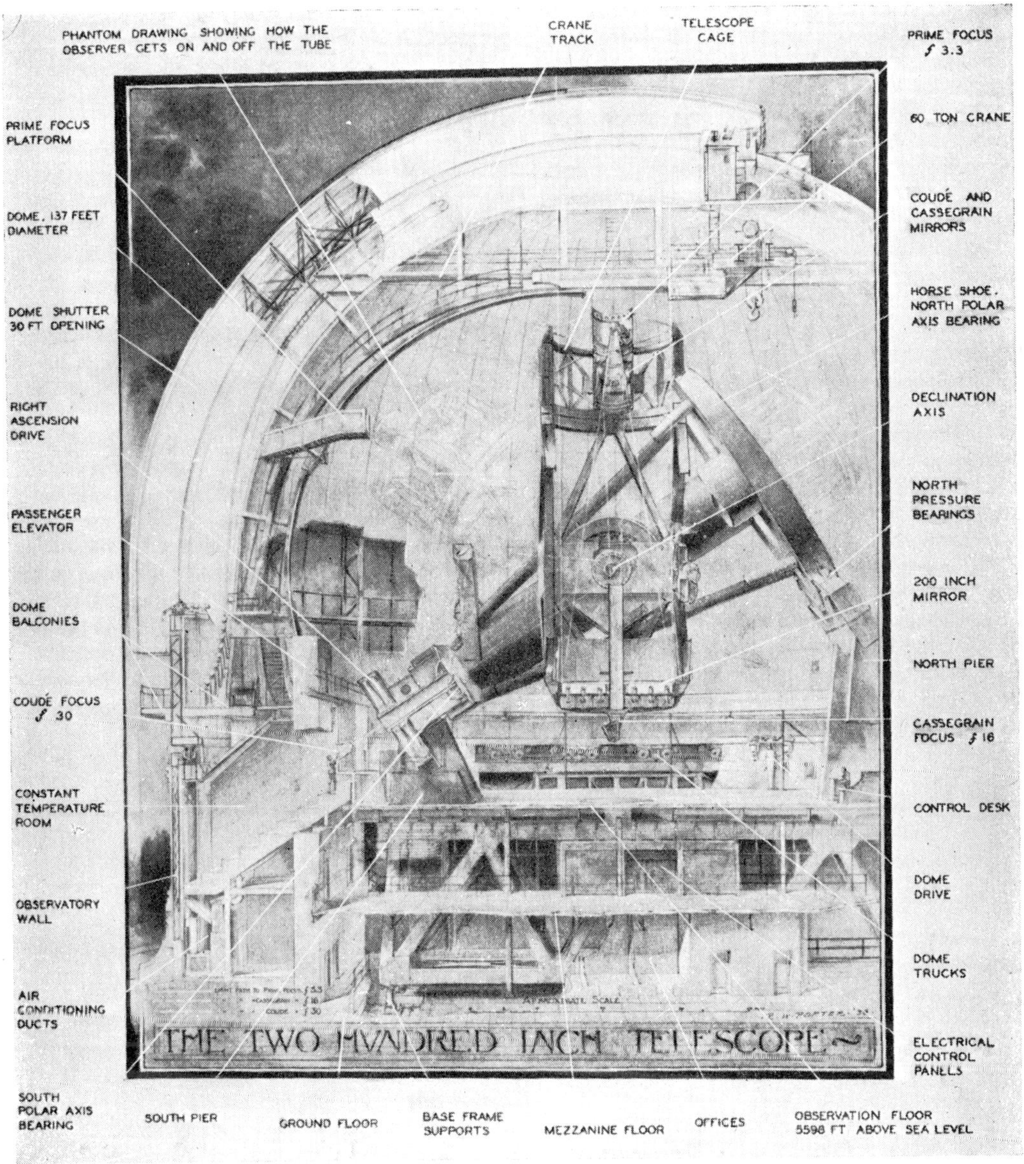

turns, the equatorial is turned about its parallel axis in the opposite direction; this can be done continuously and automatically by a mechanism called the *sidereal drive*, and the observer's hands are left free. The axis parallel to the Earth's is called the *polar axis*; there is also a *declination axis* which enables the telescope to be trained on objects at any distance from the celestial pole.

The so-called *English* type of equatorial has the telescope pivoted in declination inside a rectangular frame or yoke which is pivoted at the polar axis. The 100-inch reflector at Mount Wilson is mounted in this way (see picture on page 197). It has the disadvantage that the end of the yoke obstructs the instrument's view of the celestial pole, and the 100-inch cannot get above $+56°$ declination. For the 200-inch at Mount Palomar, this system was modified, and the upper bearing hollowed out as a horse-shoe between whose arms the telescope can swing to reach the pole. A variant of the Palomar design has been chosen for several new large telescopes.

The *fork mounting* has the polar axis bifurcated above its upper bearing, and the telescope turns in declination within the fork so formed. The 48-inch Schmidt on Mount Palomar is so mounted, and the 120-inch reflector at Lick is supported in the same way. In this design the whole weight of the telescope tube overhangs the polar axis bearings, and there is difficulty in making the fork sufficiently rigid to carry a large telescope accurately.

The *German equatorial* has a declination axis above the upper bearing of the polar axis; the telescope is mounted at one end of the declination axis, and a counterweight at the other. The world's largest refractor, the Yerkes Observatory 40-inch, is mounted in this manner.

The *cross-axis* mounting has a stout polar axle between two well-separated bearings; the declination axis intersects it half-way along, carrying the telescope on one side of the polar axle and a counterweight on the other. Both the German and the cross-axis mountings have the disadvantages of needing substantial counterweights and also of restrictions of movement such that observations have sometimes to be interrupted while the telescope is 'reversed' to the other side of the piers which support the polar axle. Nevertheless the cross-axis mounting has been used for a number of large telescopes; indeed, astronomers at the McDonald Observatory in Texas found it so satisfactory for their 82-inch reflector that they adopted it again when they came to design the 107-inch.

The types of equatorial mounting described above are by no means the only practicable forms, but they are the most common and are used for the majority of the telescopes in the world's observatories.

TELESCOPE HOUSINGS. Large telescopes are always housed in some sort of building. Occasionally, the building or its roof is removed a short distance upon rails when the telescope is used, but generally the roof is a *dome*, often of hemispherical shape, provided with a shutter; this can be moved to leave a slit through which the telescope is pointed. The dome can rotate above the circular wall of the observatory to allow the telescope to be used in any direction.

THE WORLD'S LARGEST TELESCOPES
(*operational in* 1970)

APERTURE (*inches*)	NAME	PLACE
200	Hale Telescope	Palomar Mountain, California
120		Lick Observatory, Mt. Hamilton, California
107		McDonald Observatory, Texas
102		Crimea Astrophysical Observatory
100	Hooker Telescope	Mount Wilson Observatory, California
98	Isaac Newton Telescope	Royal Greenwich Observatory, England
88		University of Hawaii, Mauna Kea
84		Kitt Peak National Observatory, Arizona
82	Struve Telescope	McDonald Observatory, Texas
79		Tautenberg Observatory, East Germany
79		Ondrejov Observatory, Czechoslovakia
77		Observatoire de Haute Provence, France
74		David Dunlap Observatory, Toronto
74		Radcliffe Observatory, South Africa
74		Mount Stromlo Observatory, Australia
74		Kottamia Observatory, Egypt
74		Okayama Astrophysical Observatory, Japan

THE LARGEST TELESCOPES. The Earl of Rosse had a 72-inch reflector, at Parsonstown in Northern Ireland, in 1845. It had a speculum-metal mirror and a primitive altazimuth mounting which restricted observations to the vicinity of the meridian.

The first large reflector of the modern era, with a glass mirror and a well-engineered mounting, was the 60-inch which was brought into use at Mount Wilson Observatory in 1908. It was surpassed in aperture ten years later when the Dominion Astrophysical Observatory of Victoria, British Columbia, completed the 72-inch; but within a few months the distinction of having the world's largest telescope returned to Mount Wilson with the inauguration of the 100-inch reflector, which remained unsurpassed for thirty years. The Palomar Mountain 200-inch reflector (see plates on pages 288 and 293) came into operation in 1948. No larger telescope has yet been used, but a 236-inch is under construction in the U.S.S.R. All the telescopes referred to in this paragraph are still in nightly use and continue to yield important observations. The Mount Wilson and Palomar instruments owe their existence to the initiative of one man, George Ellery Hale, who was earlier responsible also for the Yerkes refractor referred to above. (R.G.)

TEMPERATURE. The degree of hotness of a body, as measured by one of several arbitrary scales; it is proportional to the average kinetic **energy** of the molecules in the body.

Very tenuous matter, such as that composing the exosphere, may have a very high temperature but contain little heat. The individual atoms possess large kinetic energies and the gas is therefore by definition at a high temperature. However, the gas is so rarefied that the amount of heat contained in a given volume is relatively small. A cupful of water at normal temperature contains more heat than forty thousand cubic yards of the outer atmosphere at 2,700° C. Conduction of heat ceases to be effective in a tenuous gas as collisions between atoms become rare.

The temperature of an empty space depends upon the radiation passing through it, and is the temperature which a **black body** would reach if placed in the space. (See **Heat**.)

Temperatures are usually expressed in degrees Centigrade, or in absolute or Kelvin degrees (°A. or °K.). For the Centigrade scale, the temperature of melting ice is taken as zero, and that of boiling water at normal atmospheric pressure as = 100. The scale is then extrapolated upwards and downwards.

Since cold is merely the absence of heat, there is an absolute minimum temperature corresponding to the total absence of heat. This temperature is −273·16° C. and is taken as the **absolute zero** of the Kelvin scale, whose subdivisions are otherwise identical with those of the Centigrade scale. Different methods of measuring a temperature can lead to slightly different results, and a precise statement of a temperature implies or includes a statement of the method.

TERMINATOR. The boundary between the daylight and night hemispheres of a planet or satellite.

In the case of the Moon, the terminator appears rough and broken owing to the unevenness of the lunar surface. Irregularities have also been seen in the terminator of Venus, but are probably due to phenomena in the atmosphere of that planet.

Relief is best perceived by the eye when it casts shadows. It is therefore often easier to detect surface contours on, for instance, the Moon or in aerial photographs near the terminator (i.e. near dawn or dusk at the surface), where the shadows are long, than in the regions that are not obliquely illuminated. Full Moon is thus a disappointing time for lunar observations, particularly since those shadows that are formed away from the centre of the disc lie almost wholly out of sight behind the features which cast them.

THREE-BODY PROBLEM. Given three bodies of known masses, their positions in space at any one instant and the speed and direction of their motions, is it possible to calculate what their positions will be after a given interval of time, if the only forces acting on them are the gravitational attractions between them? The laws governing these attractions are fully known, and in the case of two bodies present no special difficulties. For three or more bodies the problem has, however, so far proved insoluble unless (1) the three bodies form an equilateral triangle as in the case of the Sun, Jupiter and the **Trojans**, or (2) one of the bodies is very small compared with one of the others, e.g. in the Earth–Moon–Sun system, or (3) the three bodies are ranged in a certain way along a straight line. In all other cases, practical answers can only be found by a laborious step-by-step method of approximation. Modern computing machines facilitate this kind of calculation, but the full general solution remains one of the outstanding problems of celestial mechanics.

THRUST. The propelling force exerted on a rocket by its exhaust jet.

TIDAL FRICTION. The force tending to slow down the rotation of one body in the gravitational field of another.

As the Earth rotates 'beneath' the Moon, water is heaped up under the latter; this bulge remains always in the same place relative to the Moon and the Earth therefore rotates under it. This effect is the cause of the ocean tides. The friction of the tides on the Earth is overcome only by the continuous expenditure of about two thousand million horsepower derived from the rotational energy of the Earth which has, however, a large store of it. The length of the day is being increased by this tidal friction by about a thousandth of a second per century.

Smaller tides are also raised on the land and within the rotating bodies even if they have no liquid on them. Work is done in deforming a body to raise tides, and this work comes from the rotation. The rotational energy of all the satellites in the solar system has, as far as is known, been exhausted by tidal friction, so that they keep one face permanently turned towards their primaries.

TIDES. The tides on the Earth are caused by the rotation of our planet in the **gravitational fields** of the Sun and Moon; the ability of the Moon to raise tides (its *tide-raising force*) is nearly three times as great as the Sun's.

The gravitational field of the Moon, as of all other bodies, weakens with increasing distance; consequently the lunar attraction upon the Earth as a whole is weaker than upon the ocean immediately 'under' the Moon. This water therefore tends to be heaped up into a *high tide* by a flow of water from neighbouring areas. The flow constitutes the tidal currents. In a similar manner, the body of the Earth is attracted more than the waters on the side farthest from the Moon and is pulled away from them, causing a second high tide on the opposite side of the Earth from the first. As the Earth turns under these two tidal bulges, the latter pass any given point at intervals of little more than twelve hours, the Moon's motion accounting for the difference from the exact half-day. Owing to

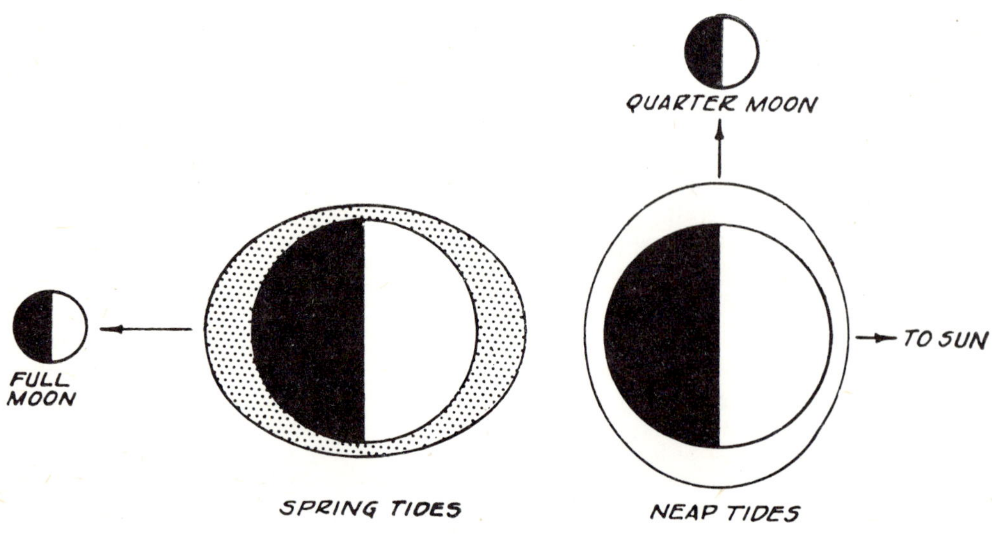

the friction of the seas upon their beds and shores, the time of high tide at a point on the coast differs from the time calculated from the position of the Moon.

Exactly similar considerations apply to the Sun, although the solar tides are less than those of the Moon. When the tidal bulges of the Sun and Moon coincide, at New and Full Moon, they have the maximum height and are called *spring tides*; at First and Last Quarters of the Moon, the two bodies are pulling at right angles to one another, and the *neap tides* which then occur have a relatively small range.

An asymmetry arises from the fact that the Moon (and consequently the axis of the tidal bulge) does not usually lie in the plane of the Earth's equator; this causes the heights of the two daily tides to be unequal.

TIME DILATATION. The slowing-down of events taking place in one system as observed from another system, when the two systems are not at rest relative to each other, or are at different gravitational potential. This effect, which was predicted by the Theory of **Relativity,** is minimal except for velocities comparable to the speed of light, but is steadily receiving experimental verification.

TIME MEASUREMENT. The fundamental standard for time measurement has always been the period of the Earth's rotation, the day. All time-measuring instruments or clocks are intended to subdivide this period, so that the time of an event may be given accurately.

The first really efficient clocks were regulated by pendula. In theory they are perfect, but in practice many factors conspire to introduce errors: temperature changes alter the length of the pendulum, air pressure fluctuations alter the air resistance, and the pendulum must be periodically interfered with in order to keep it swinging. The *Synchronome-Shortt Free Pendulum*, which until recently was the standard at most of the world's observatories, overcomes some of these difficulties: it runs in an evacuated case at constant temperature, and is accurate to within one second per year.

Quartz clocks have improved on this for short periods. They have enabled us to detect slight changes in the length of the day during the year, a phenomenon which is caused by the locking up of great quantities of water as snow and ice in the polar regions during winter in each hemisphere.

Atomic clocks are even better, and are accurate to one part in 10,000,000,000. Their frequency standard is provided by the light emitted or absorbed in the transition of an atom (e.g. caesium) between two stable states.

TRAJECTORY. The path described by a missile. (See **Ballistics.**)

TRANSDUCER. Any apparatus which converts physical quantities from one form into another. A microphone is a transducer, for example, since it changes fluctuations in air pressure (sound) into fluctuations of an electrical current; a tape recorder transduces sound into patterns of magnetization. Transducers of many kinds are used in rocket instrumentation and guidance.

TRANSFER ELLIPSE. The most economical path by which a rocket can transfer from an **orbit** about one planet into an orbit about another. It is part of an **ellipse** with the Sun in one focus. A rocket can travel in a planetary orbit without using propellents, as a planet. A relatively short burst of its motor can accelerate it into a transfer ellipse, and for the greater part of its journey the rocket moves part of the way around the Sun as an independent member of the solar system. By combining careful timing with a suitable acceleration or retardation upon reaching the second planet's orbit, the vehicle can then either land on it, follow it, or go round it and return to the first planet.

TRANSIT (*lit.* 'he crosses'). The passage of a planet across the Sun's disc, or of a star or other celestial body across an observer's meridian. (See **Meridian Passage,** and picture on following page.)

TRANSIT CIRCLE. A telescope permanently aligned for the timing of transits. (See **Meridian Passage.**)

TRITON. The larger satellite of **Neptune**. It has a diameter of perhaps 3,000 miles, and despite its distance is easily seen in a moderate telescope. It should be able to retain an atmosphere; indications of a methane mantle have been reported but not confirmed.

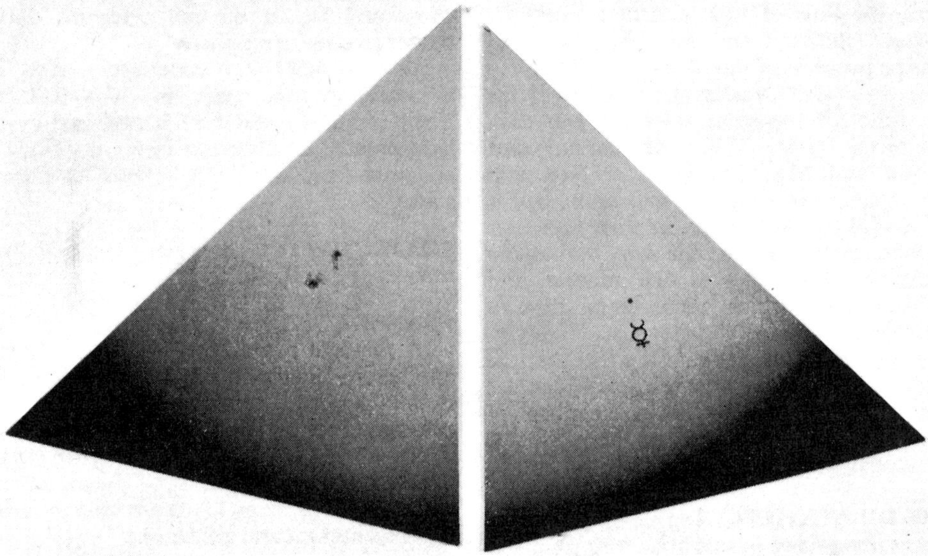

A TRANSIT OF MERCURY. The small dot in the picture on the right is Mercury crossing the Sun's disc on November 7, 1914. On the left, a relatively small group of sunspots from a different part of the same photograph. (*Royal Greenwich Observatory*)

TROJANS. Several **asteroids** revolving in the same orbit as Jupiter; one group is 60° in front of Jupiter, the other 60° behind. Achilles and Patroclus, the largest members, are over 150 miles in diameter, but their remoteness makes them difficult to observe. The orbital motion of the Trojans is an example of an interesting special case of the **Three-body Problem.**

TROPICAL YEAR. See **Year.**

TROPICS. The area of the Earth's surface over which the Sun can appear in the zenith. It is limited to the North by the *Tropic of Cancer* and to the South by the *Tropic of Capricorn*, the parallels of latitude 23° 27′ N. and 23° 27′ S. respectively.

TROPOPAUSE. The junction of the troposphere with the stratosphere. (See **Atmosphere of the Earth.**)

TROPOSPHERE. See **Atmosphere of the Earth.**

TWILIGHT. The periods after sunset, and before sunrise, when the sky is not dark. *Astronomical twilight* lasts while the Sun is less than 18° below the horizon; it lasts longer in high latitudes than near the equator because of the shallow angle at which the Sun sets and rises as seen from high latitudes. Twilight lasts all night around midsummer in Great Britain and areas farther from the equator.

TWINKLING. The rapid changes in brightness and colour of stars, especially when at low altitudes above the horizon, due to **seeing** conditions.

U

ULTRAVIOLET EXCESS of a star is the excess in its **colour index** ($U - B$) over and above that of a star of the same colour index ($B - V$) in the **Hyades.**

ULTRAVIOLET RADIATION. Electromagnetic radiation emitted by very hot sources at wavelengths slightly shorter than those of light. The Sun is a powerful source of ultraviolet rays, but only a small proportion penetrates to ground level. At high altitudes, exposure to ultraviolet radiation can cause injury ranging from sunburn to blindness and wholesale destruction of tissue, but suitable clothing and shielding for the eyes easily provide protection.

Ultraviolet rays can be registered on photographic plates; they excite fluorescent substances into emitting light, and are responsible for a good deal of the ionization in the upper atmosphere. (See **Electromagnetic Spectrum.**)

UMBRA. When a source of light casts a shadow of an object, the shadow usually consists of two portions: the inner, dark *umbra*, which receives no light from the source, and the outer half-shadow or *penumbra*, which is illuminated by light from part of the source.

When the Earth enters the shadow of the Moon cast by the Sun, observers in the penumbra see a partial or annular eclipse of the Sun, and those in the umbra see a total eclipse. (See **Eclipse.**)

The word *umbra* is also applied to the darker region in a sunspot (see under **Sun**).

UNIVERSAL TIME is the time to which astronomical data are usually referred. It is the same as **Greenwich Mean Time.**

URANIUM. A chemical **element** whose atoms each contain 92 protons; this is the greatest number possessed by any naturally occurring element, although artificial elements have been made with up to 101 protons.

Uranium is radioactive metal, and is used as a fuel in nuclear reactors.

URANUS, the third of the giant planets, was discovered in 1781 by Sir William Herschel. Herschel was not, however, the first to record it; Flamsteed, the first Astronomer Royal, saw it six times between 1690 and 1715 without realizing that it was anything but an ordinary star. It is just visible to the naked eye.

ORBIT. Uranus revolves round the Sun at a distance which varies between 1,699 and 1,867 million miles, giving a mean of 1,783 million miles. The orbital eccentricity is 0·047, about the same as that of Jupiter; the inclination 0°.8, less than that of any other planet, and the mean orbital velocity 4·2 miles per second. The sidereal period is 84 years, so that Uranus has completed rather more than two revolutions since its discovery.

DIMENSIONS AND MASS. Uranus has a diameter of 29,300 miles, less than half that of Saturn. The volume of the globe is 64 times that of the Earth, the mass 15 times; the density is 1·3 times that of water, very slightly less than that of Jupiter. The surface gravity on Uranus is greater than that of the Earth by only one-tenth, but the escape velocity is 13 miles per second.

ROTATION. The most remarkable feature of Uranus is the tilt of its axis, which lies almost in the plane of its orbit. The ‘seasons’ there are in consequence most peculiar. First much of the northern hemisphere, then much of the southern will be plunged into darkness for many years at a time, with a corresponding period of sunlight in the opposite hemisphere. Sometimes, as in 1945, the pole appears to be in the centre of the disc as seen from the Earth; in 1966, the equator was presented.

The rotation period is 10 hours 40 minutes, not much longer than that of Saturn. There is probably a difference between the equatorial and polar periods, but no certain information is available.

SURFACE FEATURES. Even a large telescope reveals very little detail on Uranus. Faint belts are sometimes visible, and occasional spots have been reported, as in 1949 and 1952; but all these are beyond the range of small or moderate instruments.

VARIANCE IN BRILLIANCE. Observations since 1951 have shown that the brightness of Uranus varies. Certain fluctuations are to be expected; when it is at its closest to the Sun the surface will be more brightly lit than when near aphelion, and owing to the polar compression a slightly larger apparent diameter will be presented when a pole appears central. The variations recorded are, however, additional to those of known cause, and it has been suggested that they are due to disturbances upon Uranus itself.

TEMPERATURE. Uranus is a bitterly cold globe. The temperature appears to be in the region of −185° C., appreciably lower than that of Saturn, and the satellite system must be equally frigid.

COMPOSITION OF THE GLOBE. In composition, Uranus is probably similar to Jupiter and Saturn. Spectroscopic research has revealed abundant methane and a trace of ammonia.

SATELLITES. The five satellites of Uranus, reckoning outwards from the planet, have been named Miranda, Ariel, Umbriel, Titania and Oberon. Estimates of their diameters are most uncertain. All are difficult objects – particularly Miranda, the closest to Uranus, which was discovered as recently as 1948. They revolve in orbits virtually in the plane of the planet's equator, and must be extremely cold worlds devoid of any trace of atmosphere.

TRAVEL TO URANUS. It seems hardly necessary to add that there is no prospect of reaching Uranus in the foreseeable future. The planet is even less welcoming than the nearer giants, and any form of life on it or its satellites seems to be out of the question, because of the low temperature (see **Life**). (P.M.)

URSIDS. A **meteor** stream with maximum display about December 22.

V

VACUUM. Space almost devoid of matter.

It used to be said that 'Nature abhors a vacuum'; but over 99·9999 % of the known Universe is given over to a vacuum far better than any that can be achieved in a laboratory.

VAN ALLEN BELTS. Zones surrounding most of the Earth in the shape of shells in which charged particles are trapped and accelerated by the Earth's magnetic field.

The existence of these belts was deduced from data received from the Explorer IV artificial satellite and from rockets fired to 450 miles. In November 1958 a nuclear explosion took place at a height of 100 miles above the central Pacific; charged particles from this explosion travelled in a loop which spanned the equator, and produced spectacular aurorae on descending again near Samoa. The radiation belt in which they moved was monitored by Pioneer III, and from the results Dr. Van Allen constructed a novel picture of the outer atmosphere. Prior to this Jacchia had inferred the existence of radiation belts from irregularities in the orbital decay of satellites. It was found that drag in the belts increased on the arrival of particles ejected by solar flares.

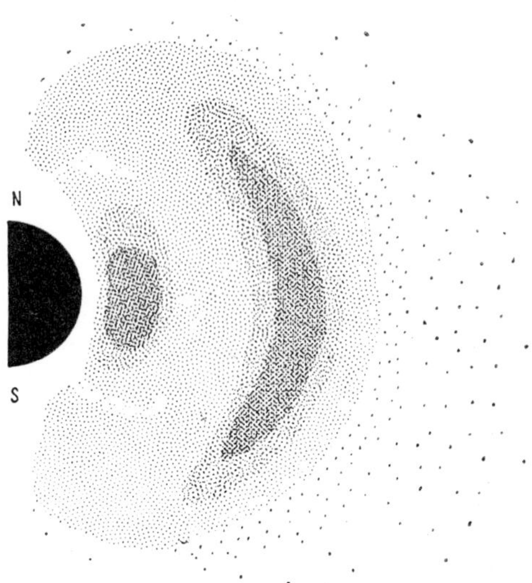

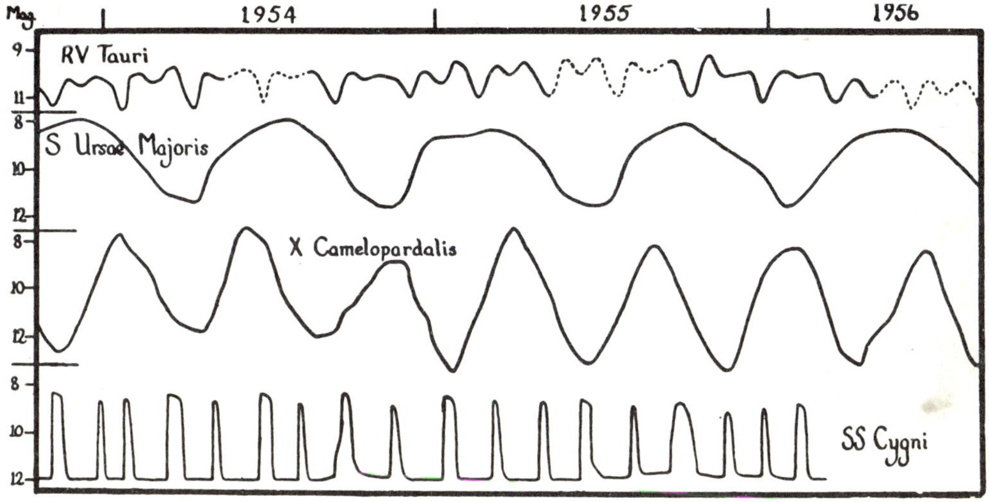

Light curves of four variable stars. The second and the third are long period variables; the curve for SS Cygni is typical of a class named after another of its members, U Geminorum. (*R. Griffin*)

Two belts of fluctuating depth have been found in the region between 200 and 30,000 miles from the Earth. The outer belt curves down above the polar regions and is clearly related to auroral displays. Ionization by cosmic rays is probably the chief source of charged particles in the inner belt. The particles move in complicated spiral paths, and their high velocity imparts temperatures of over 2,000° C. to the belts.

The diagram opposite shows a section of the belts and half the Earth; the Earth's magnetic poles are marked N. and S., and the depth of shading indicates roughly the intensity of charged particle radiation.

VARIABLE STAR. A star which varies in brightness, or **magnitude.**

NOMENCLATURE. In **constellations** where there are more naked-eye stars than could be named with the letters of the Greek alphabet, recourse was had to the small letters of the English alphabet, and finally to the capitals. The constellation richest in bright stars, Argo, used these up as far as Q. When variable stars were discovered, the first found in each constellation was named R, the next S, and so on to Z. Double-letter names were then given: RR, RS . . . RZ, SS, ST . . . SZ, TT . . . ZZ. The later stars were designated AA, AB . . . AZ, BB, etc.: at QZ this system too was exhausted, and numbers were given. Because 334 variables had received letter designations, the variable after QZ was numbered V335, then V336 and so on.

METHODS OF OBSERVATION. These are three in number:

(1) The magnitude of the variable may be estimated visually by comparison with other nearby stars of known and constant brightness; the accuracy of a careful observation is about one-tenth of a magnitude. This method is particularly suited to amateurs, and a large proportion of the observations made of variables which are not strictly periodic and have a range of variation, or *amplitude*, of the order of two magnitudes or more is made by them.

(2) A camera may be used. The diameters of star images on a photographic plate vary according to the brightnesses of the stars; measurement of the image sizes enables magnitude determinations to be made with somewhat higher accuracy than by visual methods. Photography has the advantage of giving a permanent and objective record of the magnitude. A disadvantage is that the photographic plate and the eye do not have the same colour sensitivity: of two stars having equal visual magnitudes, the bluer one is brighter photographically. A suitable filter, however, may be used to bring photographic measurements into approximate accord with visual ones: magnitudes determined in this way are called *photovisual.*

(3) A *photometer,* containing sensitive photo-electric equipment, attached to the telescope is able to compare the brightnesses of stars to considerably greater accuracy than the other methods, the error of an observation being of the order of a hundredth of a magnitude. This method of magnitude determination is often slow and requires expensive apparatus. The use of differently coloured filters in front of the photometer permits the magnitude to be determined in any one of a number of more-or-less standardized colours or *systems.*

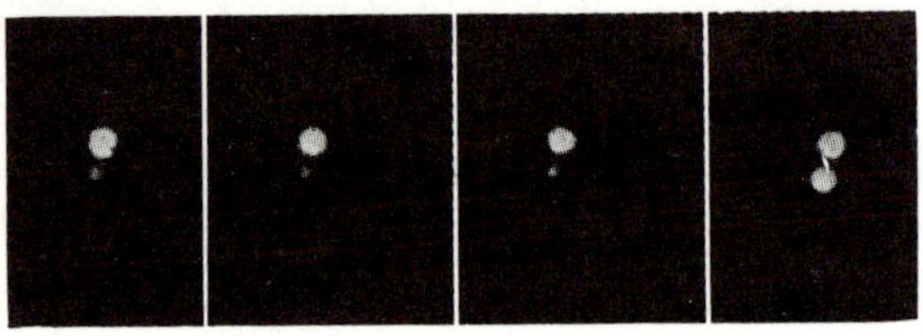

A FLARE STAR. Four photographs of the two close components of the multiple star Krüger 60, taken at intervals of about 135 seconds. There is a sudden and remarkable burst of luminosity from the normally fainter component. Flares like this are not uncommon among yellow and red dwarfs, and are basically similar to solar flares.
(*Sproul Observatory*)

CLASSES OF VARIABLE STARS. Observations of variable stars are plotted on graphs called *light curves* which show the variation of brightness with time. The shape of the light curve is of the utmost importance in deciding what type of star we are dealing with. The amplitude of variation and, in the case of a star with a repetitive light curve, the *period* are also derived by inspection of the curve.

It is evident from the diversity of light curves that there are various types and causes of stellar variability. In some cases the curve is due to the **eclipse** of one star by another; in others there seems to be a periodic, or nearly periodic, oscillation of brightness of the star itself, while in yet others the light curve shows irregular, and in some cases very sudden, fluctuations. In the following paragraphs the characteristics of the main types of variables are described; the types are often named after the first-discovered star, which is known as the *type star.*

ECLIPSING VARIABLES. The light curves of these variables, and their importance in adding to our knowledge of the stars, are discussed in the article on **binary stars.** From the observational point of view, there are two chief types:

Algol stars, named after the type star Algol or *β Persei,* which was known as a variable by the Ancients who gave it its name meaning 'The Demon', have long, approximately horizontal portions of the light curve between minima. Amplitudes of variation up to four magnitudes are known, and periods range from a few hours to twenty-seven years. This longest period belongs to the naked-eye star ε Aurigae; this star is so large that, although the relative velocity of the components is nearly twenty miles per second, the eclipse lasts for two years! The last eclipse began in 1955. The light curve of an Algol star is shown in the diagram on page 60.

β Lyrae stars have continuously changing brightness. Chiefly hot stars of O, B and A types, they are close together and tidally distorted into elliptical shapes, so that a constantly changing area of star surface is presented to us. A secondary minimum is always observed. In a sub-class named after W Ursae Majoris the minima are equal; the periods of these stars are usually short, less than one day. *β* Lyrae stars are known with periods from one-half to two hundred days.

CEPHEIDS are named after the naked-eye star *δ* Cephei. They are variables in their own right, unlike the eclipsing stars. It is not possible to draw a typical light curve, as considerable differences exist between different stars: particular shapes of light curve seem to be associated closely with definite periods. In general it may be said that the curve shows a sudden accession of brightness occupying only about a fifth of the period, followed by a slower decline and a relatively flat minimum. Periods are known from about one day (rarely) to about sixty days, and amplitudes up to 2½ magnitudes. It was early noticed that, of the many Cepheids visible in the **Magellanic Clouds** (and therefore at similar distances from us), the ones of longest period were also the brightest, and there was a definite relationship between period and luminosity. This period-luminosity law has been of the greatest value in determining the scale of the Universe (see **Stars, Distances and Motions**).

In the **globular clusters** of our **Galaxy** there are numerous Cepheid-like stars of remarkably short periods, almost all less than a day. These variables were originally known as *cluster Cepheids*, but examples were soon found outside clusters and the group is now named after one of these stars, RR Lyrae. Their light curves are generally similar to Cepheids, but there appears to be a marked break in characteristics between the two groups; very few stars of either type have periods of about one day, at which the break occurs.

Stars of both groups pulsate in the period of variation, alternately expanding and contracting. This does not by itself, however, account for the whole of the variation: the maximum light is accompanied by a higher surface temperature and an earlier **spectral class.** Spectra at maximum vary from class A for RR Lyrae stars to G for the Cepheids of longest period; at minimum they are about one class later. The reasons for, and mechanisms of, pulsation are understood in detail by astronomers, especially as regards the RR Lyrae stars.

SEMI-REGULAR AND IRREGULAR VARIABLES. Unlike the foregoing types, these stars do not have repetitive light curves. In semiregular variables a period is sometimes traceable; it is often about 100–200 days. In irregular variables the changes in brightness seem wholly capricious. Amplitudes are generally around two magnitudes. An interesting sub-group is that of RV Tauri. The light curve of the type star shows a period of about 80 days, involving two maxima and two unequal minima, superimposed on a slower oscillation of about four years.

LONG PERIOD VARIABLES. These are giant red stars with a characteristic period of about a year; only a few, mainly atypical, members of the group have periods outside the range 140 to 500 days. Amplitudes are large, commonly three to six magnitudes, with χ Cygni exceeding eight. The light curves show that deviations from the average of the order of a magnitude in amplitude and twenty days in period are commonplace. The extreme observed range of χ Cygni is well over ten magnitudes – a factor of ten thousand.

The brightest long period variable, *o* Ceti, occasionally reaches second magnitude. Its appearance and disappearance aroused considerable interest in the 17th century, before its periodic nature was recognized, and it was given the name of Mira, meaning 'The Wonderful'. Long period variables are still sometimes known as Mira stars.

U GEMINORUM stars are small, faint stars which undergo outbursts at intervals, like recurrent small-scale *novae*. They have almost constant minima, then a sudden rise lasting only about 24 hours, followed by a slower decline. The decline may set in at once or the star may remain at maximum for ten days; such short and long maxima tend to follow each other alternately. Two stars, SU Ursae Majoris and AY Lyrae, have 'supermaxima' much longer and brighter than normal maxima. The mean interval between outbursts is generally between 14 and 100 days, UV Persei having an exceptional average of 270 days; but the interval is very variable and may range from half to twice the mean. Amplitudes increase with period from two and a half to five magnitudes. A sub-class with an additional peculiarity is named after Z Camelopardalis. This star has a period of about 23 days, but occasionally it stops its variation half-way down its decline and remains at almost constant brightness for months before resuming its normal variations. In 1948–50 a standstill lasted almost two years.

Several of the U Geminorum stars, including U Geminorum itself, have been found to be binary systems showing eclipses and having periods of a few hours.

R CORONAE variables make up in peculiarity for what they lack in number. They remain at constant brightness much of the time (the type star did so for nine years in 1925–33) and then, in a few weeks, decrease in light by anything up to eight or nine magnitudes; there is then an irregular recovery lasting some months. Both the amount of the decline and the interval between successive minima are entirely random. As is the case with several types of variation, the cause is still in some doubt; but it seems to be external to the star itself as the spectrum remains singularly constant during the light changes. The stars contain much carbon, and one theory suggests that condensation of carbon in their atmospheres may obscure their light at intervals.

PULSARS were first discovered by radio astronomers to vary with remarkably short periods of the order of a second or less. Subsequently certain of them were found to show flashes of light having the same short period as the radio flashes. Special equipment is needed to observe light variations which occur so rapidly. Further information on **pulsars** will be found in a separate article.

Novae and **Supernovae** are also discussed under their own headings. (R.G.)

VELOCITY is speed *in a given direction* although the word is often loosely used in the same sense as speed. Thus a car moving due North with a speed of 30 m.p.h. has northward, eastward and southward velocities of 30, 0, and −30 m.p.h. respectively.

VENUS is the second planet in order of distance from the Sun. It is considerably larger than Mercury or Mars, and has a much higher albedo (0·59), so that it appears brighter than any object in the sky apart from the Sun and Moon. The magnitude at maximum brilliancy is −4·4, compared with −2·8 for Mars and only −1·8 for Mercury. Moreover, Venus may remain above the horizon for some hours after sunset,

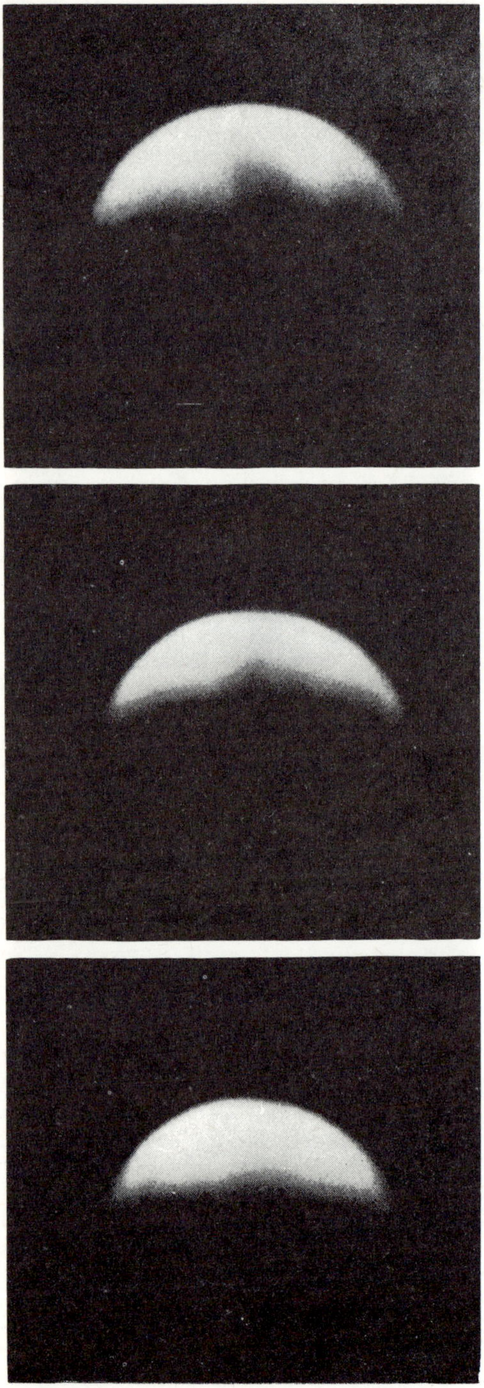

VENUS. Three 100-inch photographs showing irregularities in the edge of the terminator.
(*Mount Wilson – Palomar*)

though owing to the extreme brightness of the disc the best observational work has been carried out during daylight.

ORBIT. Venus has an orbital eccentricity of only 0·0068, which is less than that of any other planet. The almost circular orbit means that the distance from the Sun changes very little from its mean value of 67,200,000 miles; the perihelion and aphelion distances are 66,750,000 and 67,650,000 miles respectively. The orbital inclination of 3° 24′ means that transits across the Sun are very rare (see **Transits**). The sidereal period is 224 days 16 hours 48 minutes, and the mean orbital velocity is 31·7 miles per second.

PHASES. The phases of Venus were first detected by Galileo, and are obvious with any small telescope. Since Venus is comparatively remote when half or gibbous, it appears at maximum brilliance during the crescent stage, 35 days after eastern elongation and 35 days before western elongation. The moment of exact half-phase is known as *dichotomy*. Owing to the presence of an atmosphere surrounding Venus, the times of theoretical and observed dichotomy may differ appreciably.

During the crescent phase, the unilluminated hemisphere of Venus is often faintly visible. The precise cause of this *Ashen Light* is uncertain. One theory attributes it to high-altitude aurorae in the atmosphere of the planet.

DIMENSIONS AND MASS. Venus has been described as the Earth's twin, and it is true that the two globes are remarkably similar in size and mass. Venus has a diameter of 7,700 miles, while the volume is 0·92 of that of the Earth and the mass 0·81. The density is 5·0 times that of water and the escape velocity 6·3 miles per second. These figures give a value for the surface gravity of 0·87 that of the Earth. Yet despite all this, Venus must be regarded as a non-identical twin, and from our point of view it is certain to prove decidedly hostile.

VENUS THROUGH THE TELESCOPE. Telescopically, very little detail can be seen on Venus apart from the characteristic phase. This is because of the dense, all-concealing atmosphere. We can never observe the actual surface; all we can make out are vague, impermanent patches, which are certainly 'atmospheric' in nature and are not noticeably informative. Bright areas are frequently seen to cover the cusps, but are certainly not analogous to the polar caps of the Earth or Mars; neither are they permanent. It is not surprising, then, that before the launching of Venus probes our knowledge of the planet was extremely slight.

THE CRESCENT VENUS photographed in blue light with the 200-inch telescope. There is a complete absence of markings, but the softness of the terminator indicates the presence of an atmosphere. (*Mount Wilson – Palomar*)

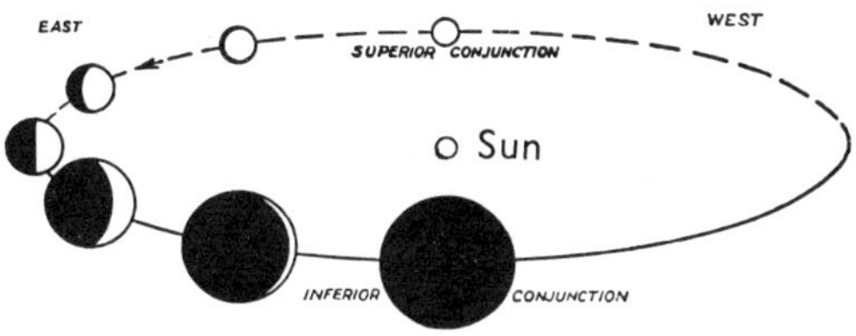

THE PHASES OF VENUS. As the planet approaches the Earth, we see more and more of its unilluminated side, while its apparent size increases.

VENUS PROBES: NEW INFORMATION. The first successful probe to Venus was Mariner II, sent out in 1962. It was followed by others, both Russian and American; in late 1967 the Soviet vehicle Venus IV achieved a soft landing, while new information was sent back by the U.S. Mariner V.

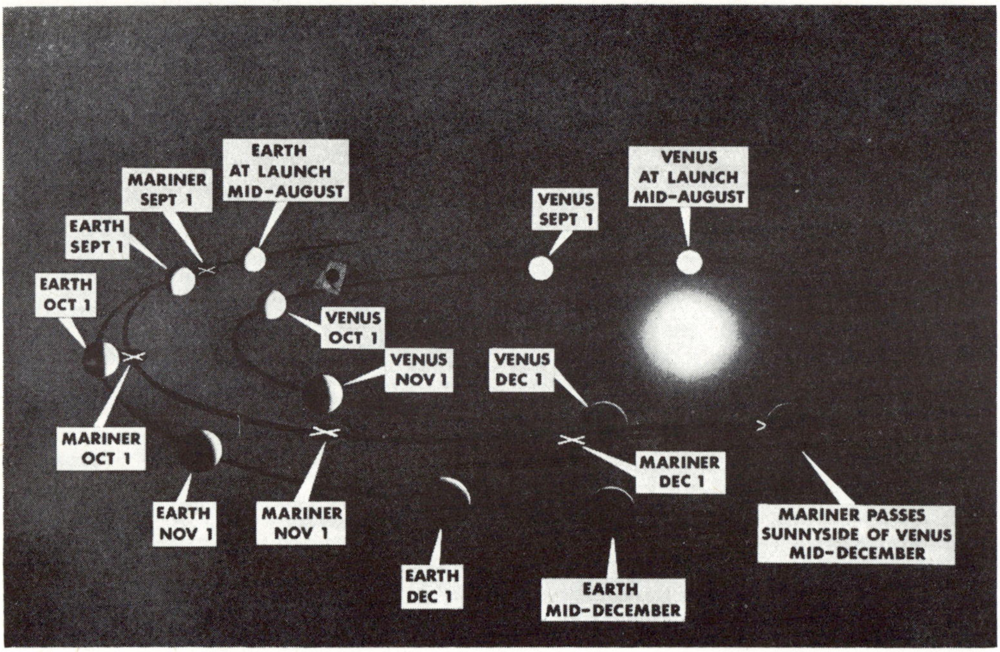

A TYPICAL VENUS PROBE TRAJECTORY, here drawn with the Sun stationary, and of course not to scale. This sketch should be compared with the path of a Venus probe illustrated under **Planetary Probe.**

All these probes have given similar data. For instance, Venus seems to lack any measurable magnetic field. More important, the surface temperature is very high, and must be over 400° C.; it is certain that this temperature relates to the actual surface, and not to any layers in the planet's atmosphere. The rotation period is extremely long (about 243 Earth-days), and the atmosphere is very dense indeed. In fact, everything that the probes have told us makes Venus appear overwhelmingly hostile.

All this has now been confirmed by the measurements transmitted by the Russian probes *Venus 5* to *Venus 8* while they descended for nearly an hour by parachute through the Venusian atmosphere. The sets of readings are all in broad agreement: atmospheric carbon dioxide amounts to over 97%, nitrogen and inert gases to about 2%, oxygen and water vapour to less than 1% each; the pressure at ground level is a little less than 90 atmospheres, and the temperature is confirmed at 470° C, roughly corresponding to a depth of 100 metres in a bath of molten zinc. The focus of attention has now shifted from Venus to Mars.

ROTATION. The rotation period of Venus was for many years a matter for debate, and estimates ranged between 24 hours and 224¾ days – the latter being the length of Venus' revolution period round the Sun. Prior to 1962, the most generally accepted value was about 30 days, due to photographic measurements by G. P. Kuiper in America. However, the measures sent back by the probes, together with radar researches carried out from the United States, have shown that the actual period is about 243 days – longer than the revolution period – and that the rotation occurs in an east-to-west or retrograde sense. The reasons for this curious state of affairs are quite unknown.

Series of photographs taken at the Pic du Midi observatory and elsewhere between 1953 and 1957 showed dark bands which moved

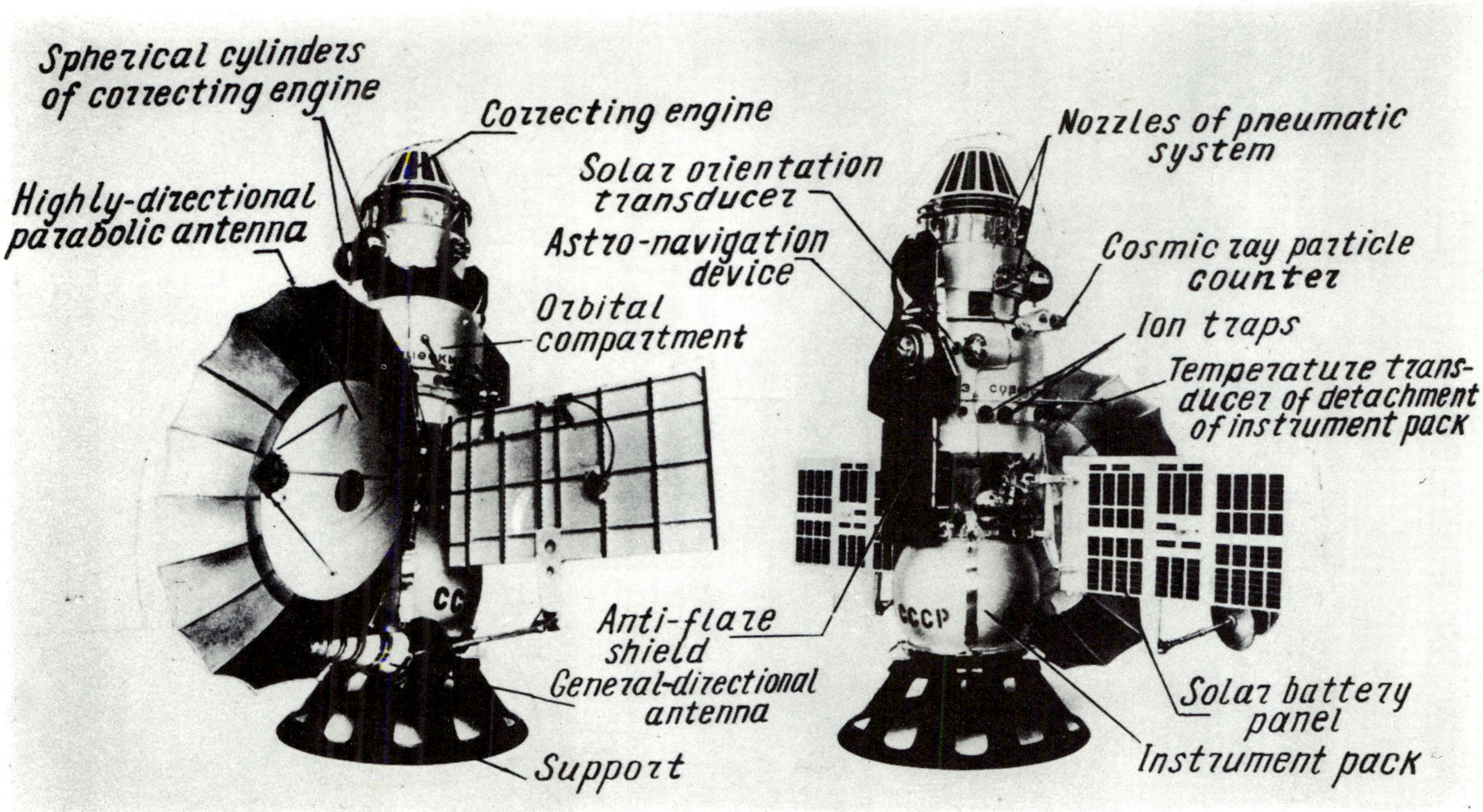

THE VENUS 5 PLANETARY PROBE

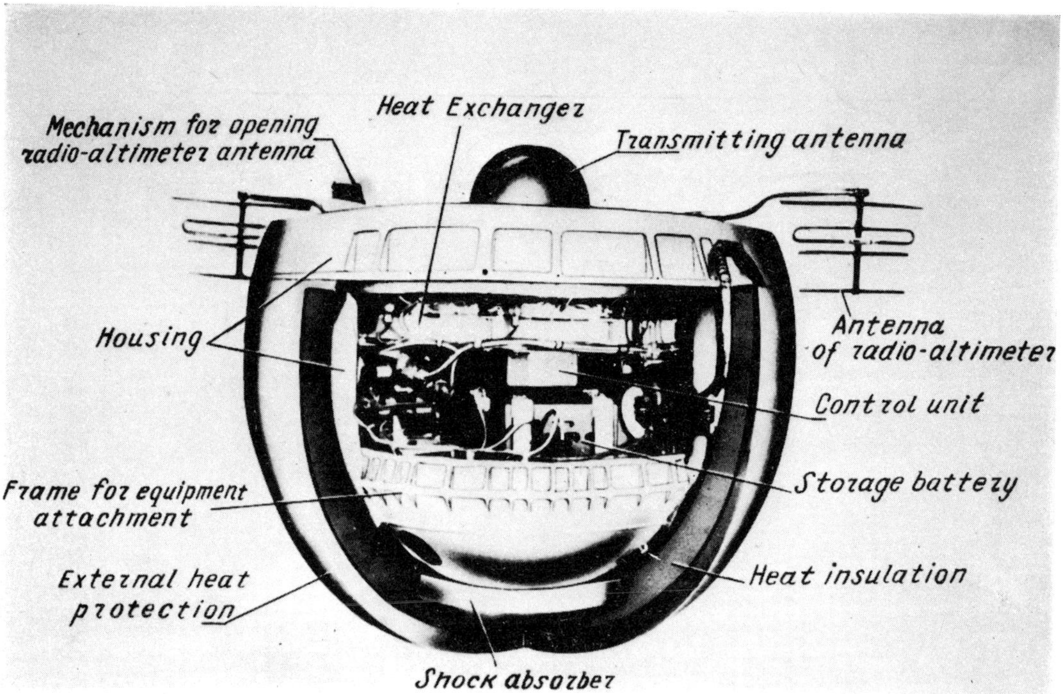

Above: VENUS 5 INSTRUMENT PACK shown in section. This is the part of the probe shown on the previous page which descends to the surface of the planet by parachute.

Below: The instrument pack being lowered into a container for testing on a centrifuge. (*Novosti*)

across the visible part of the disc and re-appeared every four days, and Boyer in 1966 obtained a further excellent series which gave the same mean period. Thus there is soundly based evidence for two entirely different rotation periods. One way to reconcile these apparently conflicting results is to accept that the upper atmosphere of Venus, in which features visible in ultraviolet light lie, turns once in four days, while the globe turns within this mantle in 243 days, both in the retrograde sense.

ATMOSPHERE. The nature of the atmosphere of Venus provides another mystery. The main constituent seems to be carbon dioxide; but it is the density which is so surprising – perhaps 40, 50 or even 100 times as great as that of the Earth's atmosphere. Astronomers were somewhat reluctant to accept these findings, but they have now been confirmed by several different means of investigation, and so must presumably be regarded as at least of the right order. Just why Venus should have an atmosphere of this kind, we do not know.

ASHEN LIGHT. This is the name given to the faint luminosity of the night side of Venus when the planet appears as a crescent. It does not seem to be a mere contrast effect, and it may be due to electrical phenomena in Venus' atmosphere – though this explanation is weakened by the discovery that there is no appreciable magnetic field.

SURFACE CONDITIONS. It used to be thought that Venus might be a world in a Carboniferous state, with swamps and luxuriant vegetation. When this attractive picture had to be given up, there were two main theories. The planet had to be regarded either as a dusty desert, or else as a world covered chiefly with water. The marine theory, due to F. L. Whipple and D. H. Menzel, was popular in the late 1950s, but the probe information seems to have disposed of it. Even under high pressure, water could hardly exist in the liquid state at so high a temperature, and we must now assume that Venus is an arid, hostile world. No direct sunlight will penetrate to ground level, and to set up any form of manned base there will be difficult in the extreme.

THE PHANTOM SATELLITE. Like Mercury, Venus has no known satellite. If there is a minor attendant, it cannot be as much as a quarter of a mile across. A satellite of Venus reported now and then up to the mid-18th century must certainly be dismissed as a telescopic ghost.

LIFE ON VENUS. From what has been said, it is evident that conditions on Venus are unlikely to favour life. Although the existence of lowly organisms cannot be ruled out, most authorities regard the planet as completely sterile; but see **Life.** (P.M.)

VENUS PROBE. See above, and under **Planetary Probe.**

VERNAL EQUINOX. The time when the Sun appears to cross the celestial equator in a northerly direction, about March 21 each year. (See **Equinox** and **Celestial Sphere.**)

VESTA. Although it has a diameter of only 214 miles, much less than that of Ceres or Pallas, Vesta is the brightest of the **asteroids.** It can reach almost 6th magnitude, and so can be glimpsed without optical aid. It has a period of 3·63 years, and was the fourth member of the swarm to be discovered (1807).

VULCAN. The name that was proposed for the non-existent **Intra-Mercurian Planet.**

W

WARGENTIN. One of the most unusual lunar craters; it appears 'full to the brim'.

WARHEAD. The explosive nose compartment of a missile.

WAVE AND WAVELENGTH. A wave in physics is a regular, oscillating disturbance which spreads through a medium. In a *transverse* wave, the particles of the medium move to and fro at right angles to the direction in which the wave is travelling. Ripples on a pond are an example of transverse wave motion; the direction of propagation is horizontal, and a small particle floating on the surface moves up and down together with the water without being carried forward by the wave. It is the *disturbance* that moves forward, not the water (as long as the wave does not break).

In *longitudinal* waves, the particles of the propagating medium oscillate forwards and backwards in the direction of the wave motion, but here too their mean position does not change. Sound is an example of a longitudinal compression wave. The particles of the medium through their oscillations form alternate regions of compression and rarefaction, and it is the location of these regions which advances, not the medium. A graph of density along a line of propagation shows the familiar wave pattern (see illustration).

Both kinds of wave can be *reflected* or *absorbed* by obstacles in their path, and can be *refracted* (i.e. their direction of advance is changed) when they pass from one medium into another in which they travel at a different speed. When they pass through an opening which is small compared to their *wavelength* (the distance between one wavecrest and the next) they can also be diffracted, i.e. they will spread out in all directions after passing through the opening. Two identical waves exactly out of step with each other can cancel

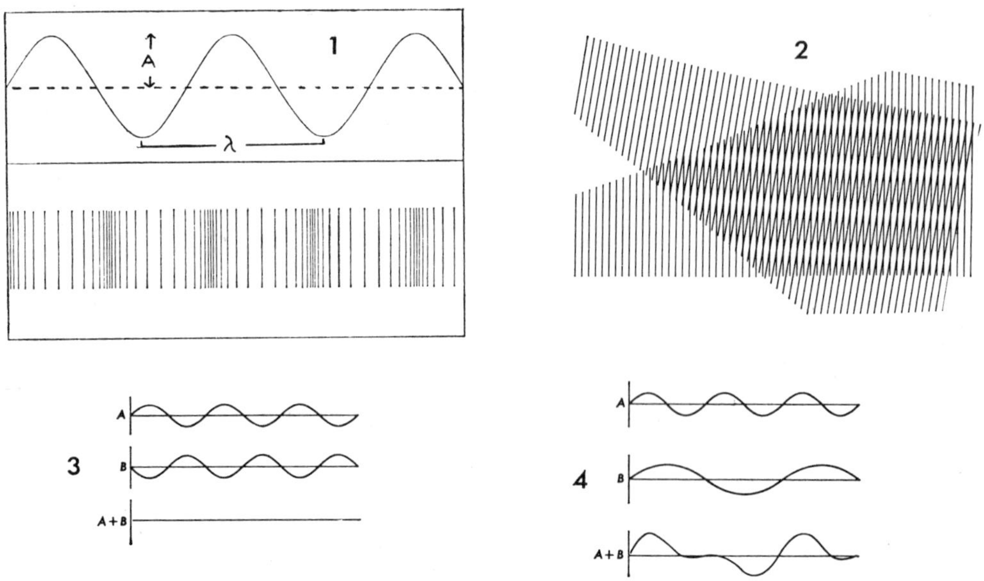

(1) A longitudinal wave, and its representation as a sinusoidal wave. λ is the wavelength and A is the amplitude. (2) Two wave trains intersecting each other at a small angle give rise to interference patterns in the form of alternate regions of weak and strong disturbance. (3) Two similar waves out of step with each other can cancel each other. (4) Two dissimilar sinusoidal waves added together result in a more complex wave form.

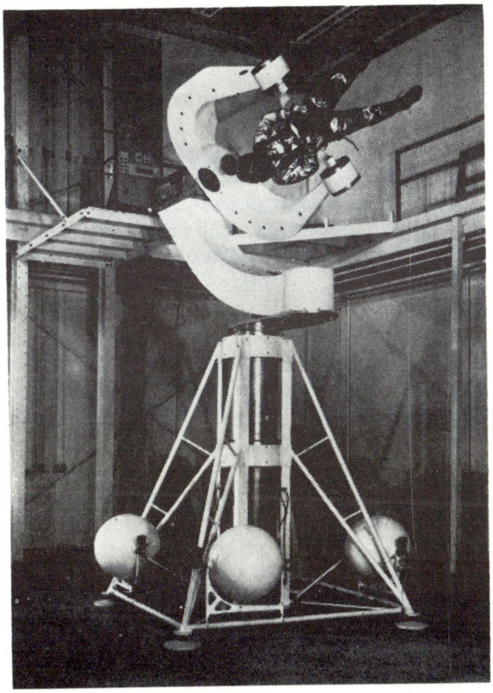

One way to simulate weightlessness to some extent during training is to place astronauts under water. Another is provided by this horseshoe agitator, which tumbles the astronaut in various orientations.

each other; two waves in step reinforce each other; and generally two or more waves travelling through the same medium at the same time give rise to a more complex, combined wave.

Physicists and mathematicians find that there are many phenomena which behave exactly *as if* they were wave motions and obey the same laws, but for which there is no material medium which 'waves'. A fundamental example is **electromagnetic radiation,** which includes light.

The **frequency** of a wave motion is the number of cycles which it completes in unit time, and for a given speed of propagation is clearly inversely proportional to the wavelength. For electromagnetic radiation in space,

$$\lambda\nu = c,$$

where $c = 300{,}000$ km. per second is the velocity of light *in vacuo*, λ is the wavelength and ν the frequency.

BUBBLE TROUBLE in a rocket fuel tank can be serious if agitation during the launch forms bubbles which cannot 'rise' out of the liquid during the weightless, coasting phase. The force of sloshing fuel can also cause a rocket to veer off course. These are some of the myriads of problems which space engineers must anticipate and overcome.

WEIGHT. The force of attraction of a gravitating body upon a mass. (See **Gravitation.**)

WEIGHTLESSNESS is the situation in which the gravitational forces acting upon a body are exactly balanced by equal and opposite inertial forces, such that the only accelerating force is that of gravity. The terms zero gravity and gravity-free state are often incorrectly used as a synonym, whilst another frequently used phrase – **free fall** – merely describes one condition in which the apparent loss of weight occurs.

The occurrence of the phenomenon has important physical and biological consequences. For instance, the contents of a partially filled liquid propellent tank of a rocket engine accumulate in a mass which may or may not be in the correct position for pumping prior to an engine start. Ullage motors are, therefore, used to impart a sufficient force to the space craft to ensure that the propellents may be pumped. An unrestrained astronaut will experience an equal reaction to any force exerted by him on any other object, and this will impart to him an uncontrolled motion in a direction opposite to the initial action. Where such powerful forces as those necessary for extra-vehicular maintenance and construction tasks are involved, the consequences could be serious. This has led to a great deal of research into the need for careful restraint and for the use of torqueless tools which do not turn the user when he tries to turn them.

The physiological effects are described under **Space Medicine.** Numerous methods have been employed to simulate the weightless condition on Earth in order to investigate these reactions. An exact reproduction of the weightless state is possible only at the point of an aircraft trajectory, at which the sum of the centrifugal and tangential forces exactly equals the weight of the aircraft (which is a very limited period at the peak of a vertical parabolic trajectory), or in a freely falling device. The former method can provide a maximum period of about 15 seconds of weightlessness, whilst the latter has severe practical limitations.

Neutral buoyancy by total body water immersion can provide some of the characteristics of the weightless state, and is widely used for simulation purposes.

Since weightlessness is experienced at all times of unpowered flight in space, such as in a planetary orbit or in a minimum energy transfer orbit, and since this state is known to produce certain adverse effects upon the body, it may be desirable to provide **artificial gravity** on long, manned space missions.

WHITE DWARF. A star of high surface temperature but very small diameter, consisting of matter in which the electrons are degenerate. (See **Degenerate Matter.**) A typical white dwarf has a mass about that of the Sun, a size about that of the Earth, a density about 1,000,000 times that of water and a surface gravity of the order of a few hundred thousands that at the surface of the Earth. Their spectra, which vary from types B to G, show marked collisional broadening (see **Spectroscopy**) caused by the high atmospheric pressure due to their enormous surface gravities. They are about eight to eleven magnitudes fainter than main sequence stars of the same spectral type and occupy a region of low luminosity and high surface temperature in the **Hertzsprung–Russell diagram.**

A white dwarf has no source of nuclear energy, and to maintain its luminosity it uses the thermal energy of the **ions.** The ions gradually cool down and the star gets fainter and fainter till it has no more energy left to radiate.

A white dwarf may arise at the end of the evolution of a normal star or it may condense directly from **interstellar matter** if nuclear reactions cannot start and halt the contraction from the interstellar gas cloud. (See **Stellar Evolution.**) Stars of mass greater than a certain critical mass cannot become white dwarfs. Thus most stars in order to become white dwarfs must lose mass, and they may do so by a **supernova** outburst. (J.A.J.W.)

WILSON CLOUD CHAMBER. The first detector of charged sub-atomic particles. It consists of a cylinder fitted with a movable piston, at least part of the cylinder being transparent. The space above the cylinder contains air saturated with water vapour; if the piston is suddenly withdrawn a little way some of the vapour condenses as tiny droplets. If, at the critical moment, a charged particle enters, it leaves a trail of ionized molecules, and the vapour condenses more easily upon the ions than elsewhere. Consequently a particle leaves a trail of small water droplets which mark its path and allow the latter to be photographed.

WIND TUNNEL. A device which produces a gas flow of known velocity and pressure in its *working section* to enable tests to be carried out to determine the aerodynamic behaviour of scale models of aerospace vehicles.

Subsonic tunnels consist normally of a system of fans and baffles upstream of a straight tube which forms the working section. The baffles are used to straighten out the flow downstream of the fans to ensure that the flow in the working section is smooth. Low-speed tunnels are often used to test models of civil engineering structures such as tall buildings and large bridges as well as such things as investigating the wind forces on a large rocket standing on the launch pad.

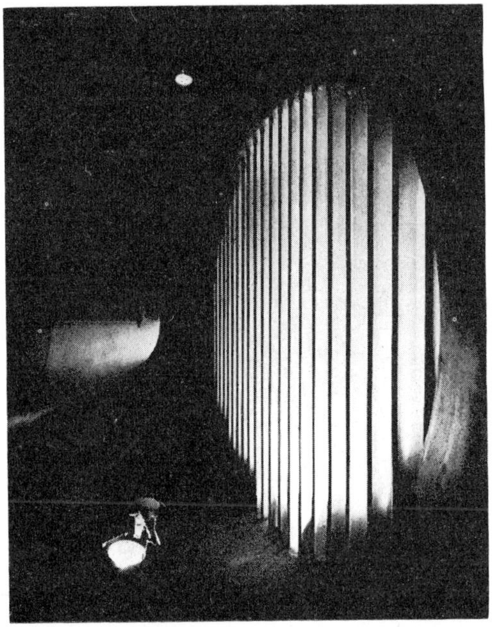

WIND TUNNEL capable of generating wind speeds of 3,000 m.p.h. for testing space vehicles. The cone-shaped nozzle injects the airstream through the baffles, which smooth and direct it, into the working section whose diameter ranges up to 62 feet.

Supersonic tunnels operate by expanding a supply of compressed gas through a convergent-divergent nozzle, and the flow speed reached is a function of the expansion ratio. The gas supply can be provided from a bank of compressors, or a single sufficiently large compressor, in a continuous flow tunnel, or from a large pressure reservoir in the 'blowdown' type of tunnel. This type has considerable cost advantages as the power input can be spread over a long period, but it can be used only for very short test runs as the pressure in the reservoir quickly falls.

Hypersonic research (Mach numbers above about 5) requires a rather different type of tunnel called a *shock tube* in which the energy source is explosive. (See also **Air Resistance**.)

WINDOWS, ATMOSPHERIC. See **Atmosphere of the Earth.**

WOLF-RAYET STAR. A star of a class related to the early main sequence stars. The hottest main sequence stars, those of types O and early B, tend to show bright emission lines in their spectra. These are known as *shell stars.* In extreme cases the spectra consist almost entirely of bright lines: such stars are placed in Class W, the *Wolf-Rayet* stars. Their bright lines are usually very broad and hazy, and it used to be thought that matter was streaming from their surfaces with high velocity, causing great Doppler broadening of the lines. The spectra certainly come from gaseous envelopes at extreme temperature, but matter does not seem to be rapidly lost from these. Ionized helium, and highly ionized carbon, nitrogen and oxygen, are present in the envelopes. Recently many of the Wolf-Rayet stars have been discovered to be spectroscopic binaries, and an appraisal of the dimensions of both envelopes and stars should soon be possible.

X

X-RAY. The region of the **electromagnetic spectrum** with wavelength between about 0·1 Å and 200 Å.

X-RAY ASTRONOMY. The observation of celestial objects in the X-ray region of the **electromagnetic spectrum.**

Although X-rays in the mean penetrate powerfully through solids (they are strongly absorbed only at particular discrete wavelengths by electrons in the atoms), they are unable to penetrate the **atmosphere of the Earth,** which at this wavelength has the stopping power of about a metre of lead. This contrasts with the fortunate transparency of the atmosphere for that select octave, the optical region. For this reason observations of X-ray emission by celestial sources can only be made using rocket or balloon platforms, or **orbiting astronomical observatories.**

The Sun has been the most closely observed source, because of its proximity and large angular size. Besides a general background of continuous thermal X-rays originating in the corona of the Sun, a great intensification of emission has been found to be correlated with solar **flares**. This enhanced production of X-rays by flares has been shown to produce sudden disturbances in the Earth's **ionosphere.** The mechanism responsible for these X-rays associated with flares is not well understood. Both thermal emission and **synchrotron radiation,** due to the acceleration of electrons in the Sun's magnetic field, are thought to contribute to their production.

Cosmic X-rays from sources outside the solar system were found for the first time in 1963. Since then, more than twenty X-ray objects have been discovered. The majority lie in the galactic plane and are generally thought to be members of our Galaxy. Three of the strongest sources have been identified with optical objects.

The first source found, Sco X-1, has been identified as a blue star whose spectrum is very similar to that of an old **nova**. It is thought that, like novae, the source may be two stars forming a close binary system in which the process of mass transfer is occurring. The acceleration of material between the two stars could give rise to X-ray emission if they are sufficiently massive. A second source, Cyg X-2, has a similar optical counterpart and this process may account for most of the observed discrete sources. The third identified source of X-rays is in the **Crab Nebula.** This source is very different from Sco X-1, with a much larger source diameter of the order of several light years and a relatively more powerful emission at short X-ray wavelengths. The cause of the X-ray emission is probably related to the mechanism producing the radio emission of the Crab Nebula. (G.T.B.)

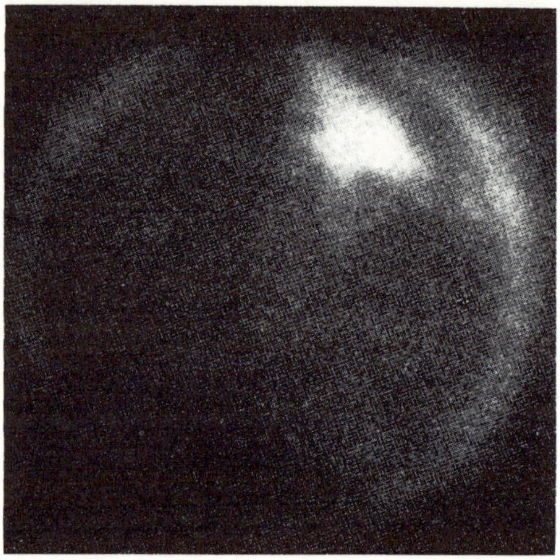

X-RAY PHOTOGRAPH OF THE SUN made during a rocket flight, using a telescope based on the fact that X-rays behave like light rays if they strike surfaces like mirrors at a low angle. The Sun is a relatively weak emitter of X-rays; a high proportion of the X-rays emerge from the corona, but the bright area at upper right corresponds to a local disturbance.

(*Giacconi/Goddard Space Centre*)

Y

YEAR. Several measurements of the year are recognized, and each gives a different length. They are enumerated below and their lengths in mean solar days (civil days) are tabulated for epoch 1960.

Tropical Year is the time between successive vernal equinoxes. This keeps in step with the seasons, and the civil year of the Gregorian calendar is based on it. Owing to precession, the equinoxes regress round the ecliptic about 50 seconds of arc per year, making a whole revolution in 26,000 years; consequently the tropical year is shorter by about one part in 26,000 than the

Sidereal Year, which is the 'true' year or the time taken by the Earth to traverse its orbit once and return to the same direction in space relative to the Sun.

Anomalistic Year is the mean interval between succeeding perihelia in the Earth's motion. The position of perihelion progresses round the Earth's orbit by an average of 11 seconds of arc per year, making the anomalistic year slightly longer than the sidereal year.

Eclipse Year is the length of time between successive returns of the Sun to the Moon's ascending node. If the Moon's orbit were fixed in space the eclipse year would be equal to the sidereal; but the Moon's nodes regress rapidly round the ecliptic, completing a revolution in about 18 years, causing the eclipse year to be much shorter than the other years. The eclipse year is so named because it is the interval between eclipses, which can only take place near a node.

Besselian Fictitious Year begins at the instant when the Sun's mean longitude is 280°; this occurs on December 31 or January 1 each calendar year. The Sun's longitude at this time is 80° less than at the vernal equinox, and therefore the beginning of the Besselian year simply marks a particular time in the tropical year.

Gregorian Year is the mean length of year on the Gregorian calendar, and

Julian Year the mean length on the Julian calendar.

The yearly course of the seasons is a result of the inclination of the Earth's axis to the plane of its orbit. The axis maintains a fixed direction in space, so that the northern hemisphere is tilted towards the Sun in summer and away from it in winter. This applies in varying degrees to the other planets; the length of their years increases with mean distance from the Sun.

Z

ZEEMAN EFFECT. The phenomenon of the splitting of spectral lines formed in a magnetic field, which was predicted, before its discovery, by Zeeman in 1896.

Magnetic fields have been detected upon the Sun and some stars by observing the Zeeman effect. The **magnetograph** enables separations of the lines, which are very small compared with the widths of the lines themselves, to be measured with accuracy, with the aid of special filters.

If a star has a magnetic field and no filters are used when its spectrum is photographed, the components of each line merge together; the overall effect is to broaden the lines.

ZENITH. The point on the **celestial sphere** vertically above the observer.

ZERO, ABSOLUTE. See **Absolute Zero.**

ZERO GRAVITY. The state of **free fall,** or the total force acting at a point where gravitation and *all other* accelerations cancel out. Zero gravity has nothing to do with distance from the centre of any gravitational attraction, since the gravitational field of any body, no matter how small, extends infinitely far in all directions.

ZODIAC. The belt of sky lying within eight or ten degrees of the **ecliptic**, within which

Year	*Mean Solar Days*	d	h	m	s
Tropical	365·242195	365	5	48	45
Sidereal	365·256360	365	6	9	10
Anomalistic	365·259643	365	6	13	53
Eclipse	346·620050	346	14	52	52
Gregorian	365·2425	365	5	49	12
Julian	365·25	365	6		

the Sun, Moon and most of the planets appear to travel. Starting from the *First Point of Aries*, it is divided into twelve *Signs of the Zodiac*, each represented by a symbol and a character (usually an animal). There are twelve *Zodiacal Constellations* which have the same names as the signs but do no longer coincide with them. Since the naming of the First Point of Aries 2,300 years ago, precession has carried the signs westward 30°, or a whole sign, with respect to the constellations.

The Signs of the Zodiac played an important part in Astrology, and were studied with great care. We owe the existence and preservation of many astronomical records of the greatest antiquity to this otherwise often harmful and baseless superstition.

ZODIACAL CLOUD. The cloud of meteoric bodies responsible for the **Zodiacal Light.**

ZODIACAL LIGHT. This is a faintly luminous band extending along the whole length of the **ecliptic**. The brightest part is nearest the Sun; another, but much fainter, maximum occurs in the direction diametrically opposite to the Sun. This secondary maximum is named the *Gegenschein* or Counterglow. In temperate latitudes only the part of the Light near the Sun is often seen: it appears as a diffuse luminous area extending along the ecliptic from the position of the Sun, being widest and brightest near the horizon and becoming narrower and fainter away from the Sun. The most favourable times for observation are clear, moonless nights, after dusk in spring and before dawn in autumn. The advantage of these times is that the ecliptic then intersects the horizon at a steep angle, and parts of it quite close to the Sun are therefore observable when the Sun is sufficiently below the horizon to give a dark sky. In the tropics the ecliptic always makes a large angle with the horizon, and the Zodiacal Light is often quite conspicuous. (Exactly similar considerations apply to the visibility of the planet Mercury, which can never appear more than 28° from the Sun.)

Spectroscopic examination shows it to be reflected light, which indicates that it comes from solid or quasi-solid bodies which are concentrated along the plane of the planetary orbits. Whether they are directly connected with comets or meteors or genetically related in some way is a question not yet fully answered.

The *Gegenschein*, or Counterglow, found at the anti-solar point, was discovered independently on at least four occasions: by Jones 1853, Brorsen 1855, Backhouse 1868 and Barnard 1883. It is roughly elliptical in shape, measuring approximately 10° by 20° and elongated along the ecliptic. Although a number of ideas have been put forward to explain its origin and continual existence, the most likely explanation is the one which considers it to be a cloud of fine dust particles situated at a gravitational libration point; the result is a concentration of dust at the point which can then reflect sunlight.

This latter explanation is likely to be fully accepted in view of the now confirmed localized faint clouds of micrometeoric material associated with the **Lagrangian points** of the Earth–Moon system – which although reported often in the past was only positively confirmed by direct observation in 1968. Observers were flown to a height of 12 km. on separate occasions and saw clouds about 2° to 4° across just detectable at the limit of naked-eye visibility. These observers also saw the Gegenschein which was reported to be rather brighter than any of the Earth–Moon clouds.

THE ZODIACAL LIGHT, from a painting by E. L. Trouvelot.

APPENDIX

APOLLO 15. James Irwin standing by the rover parked near the lunar module, with Mount Hadley in the background.

APOLLO, SUMMARY. The *Apollo* lunar programme has now been completed. Orbital details of *Apollo 8* and *11* are given on p. 38, and the following is a brief summary of the last seven missions.

APOLLO 11:

July 16–24, 1969. First landing on Moon on July 20 in position 0·69° N, 23·43° E, in the SW corner of the Sea of Tranquillity. (See frontispiece, and for details of crew p. 19.) Some 35 kg of lunar rock samples were collected during an EVA lasting 2½ hours. The American flag was planted, and the landing and EVA were televised live to Earth. A screen for measuring solar wind, a seismometer and a laser reflector were set up. Aldrin's first impression: 'The lunar surface looks like a collection of every variety of shape, angularity, granularity . . . just about every kind of rock. Their colour depends on what angle you're looking at.' Lunar module left in selenocentric orbit during return, service module burnt up on re-entry after separation from command module. Crew wore germ-proof clothing and entered three weeks' quarantine after landing.

APOLLO 12:

November 14–24, 1969. Landing on November 19 in 3·20° S, 23·39° W, in the

THE APOLLO 12 MOONWALKS covered some 3 km. The positions of the lunar module *Intrepid*, the old *Surveyor 3* and the nuclear powered instruments package are indicated by white crosses.

Ocean of Storms, 600 m from *Surveyor 3*. Two EVAs totalling over 7 hours, during which 80 kg of rock samples were collected; a nuclear-powered 'surface instruments package' was part of the equipment. A seismometer, solar wind detector and a device for measuring micro-meteorite impacts were set up. The ground was much softer than at Tranquillity Base and there was much glassy material. Pieces from *Surveyor 3* were taken back. The lunar module ascent stage was deliberately crashed on the surface after use, and the seismic reverberations were measured.

APOLLO 13:

April 11–17, 1970. The explosion of an oxygen tank and serious electrical failures at 330 000 km from the Earth caused the mission to be abandoned. After swinging round the Moon the astronauts returned to Earth, shutting down all systems in the command module and transferring to the lunar module as a life-raft for four days. The mission triumphantly demonstrated the flexibility and multiple redundancies of the design, and the ability of ground control to analyse and overcome important failures.

APOLLO 14:

January 31–February 9, 1971. Landing on February 5 in 3·67° S, 17·49° W, in the uplands region north of Fra Mauro. First use of modularized equipment transporter ('golf-cart'). Chief experiments included seismic recordings of small explosions detonated later from Houston to help in mapping the geological structure and to detect water that might be trapped below the surface, and magnetic measurements. Two EVAs totalling 9½ hours.

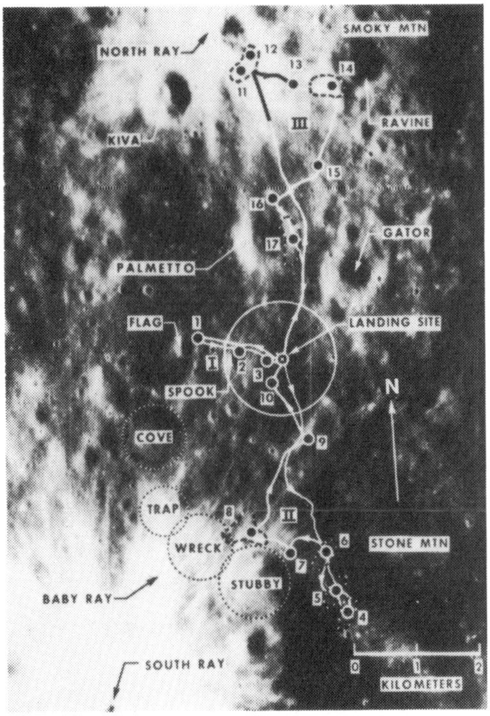

APOLLO 16 TRAVERSES and stopping places, superimposed on an *Apollo 14* reconnaissance photograph of the Descartes highlands.

APOLLO 15:

July 26–August 7, 1971. Landing on July 30 in 26·08° N, 3·66° E, close to Hadley rille and near the Apennines at the SE edge of Mare Imbrium. First use of lunar rover; three EVAs totalling over 18 hours. New activities included the extraction of 3-metre cores, sampling of the very old, crystalline 'Genesis' rock, and the release from lunar orbit of a sub-satellite to transmit scientific data from the Moon for over a year.

TWO POSSIBLE TYPES OF MOON BASE. The upper one is basically a string of different, soft-landed lunar modules linked to each other and to a common power source by 'umbilical cords'. The units can assist each other in the case of failure. The lower design by Lockheed is considerably more ambitious. Power is derived from a remote nuclear source, and the base is equipped with spherical transporters mounted within wheels.

APPENDIX

APOLLO 16:

April 16–27, 1972. Landing on April 21 in 9·0° S, 15·5° E, in Descartes region. Three EVAs with lunar rover ranging up to 4 km from the landing site. Mineral specimens included glassy spheres and white soil beneath a thin, grey surface layer. A small sub-satellite was released for particle and field measurements in selenocentric orbit.

APOLLO 17:

December 7–19, 1972. Landing on December 11 in 20·15° N, 30·73° E, in Taurus-Littrow region near the SE rim of the Sea of Serenity, in moutainous terrain. Investigations centred on sub-surface characteristics, electrical properties of the surface, possible mechanisms to account for the spreading of material, detection of minute traces of atmosphere and measurements of gravity, tidal surface movements and heat flow. Three EVAs totalling 22 hours; collection of 120 kg of specimens, including some unusual, orange material.

SPLIT LUNAR BOULDER near the Taurus-Littrow landing site being examined by *Apollo 17* scientist-astronaut Jack Schmitt.

APOLLO APPLICATIONS:

Apollo service and command modules can dock directly on to **Skylab** and are ferrying crews to and from it. For rendezvous with the Russian **Soyuz** spacecraft a special Docking Module has been developed; it is basically an adapter, fitted at one end with the American and at the other with the Russian docking system, and serves as an air-lock in which the astronauts can become acclimatized to the different atmospheres of their respective craft.

The practical application of the *Apollo* and *Skylab* missions range over the fields of meteorology, communications, crop and forest surveys, land use patterns, location of fish and identification of minerals by infra-red photography (see picture facing p. 121) and hydrographic studies.

SKYLAB, with Apollo ferry linked to it at docking port *A* (*left*), and Apollo and a Soyuz spacecraft joined by means of the special docking module (*right*).

SCIENTIFIC RESULTS:

It is impossible in a short article to give a meaningful summary of the vast amount of experimental and theoretical results that are based on examinations of the lunar material brought back by the *Apollo* missions and on the instrumental readings obtained through them. Most of the work published so far may be found in the six volumes of the *Proceedings of the Second* and *Third Lunar Science Conference* (published by the M.I.T. Press, Cambridge, Mass.) In them, some

360 principal investigators report on their findings in a total of 228 papers, and there is much more to come. Over 1 600 samples from the Moon have been examined for every aspect of their mineralogy and petrology, their chemical composition and their physical properties. Refined seismic, electrical, magnetic, thermal and gravitational observations have been made near the landing sites and from the command module in lunar orbit.

The most obvious question to ask about the *composition* of the lunar material is: how does it compare with terrestrial rocks? How does it fit in with theories of the origin of the Moon and of the solar system, and the history of the surface? Every technique known to Man has been applied to determine the chemical composition, including wet analysis, emission **spectroscopy,** X-ray fluorescence, neutron activation analysis and electron probe micro-analysis. Searches were made, with negative results, for traces of microbial or other *biological material. Isotope analysis* gave data on the age of individual rocks and on the Moon's heat balance and thermal history, for the heat produced in the Moon is largely of radio-active origin. *Mineralogical examination* entailed the production of very thin, highly polished rock-sections for investigation under the microscope by transmitted, reflected and polarized light. The results indicate what minerals are present, how they were formed, and what treatment they have been subjected to since their formation.

For example, determination of the oxidation states of all minerals indicated that lunar rocks solidified under extremely reducing conditions (corresponding to a partial pressure of free oxygen of less than 10^{-13} atm), at temperatures between 1 100°–1 200° K. The distortions of individual grains suggested the nature of the forces that affected them in their present or past locations. Ironically, the microscope has proved to be of greater importance than the telescope in deducing the evolution of the lunar surface.

A single 3-metre core sample contained 57 separate layers of soil, which between them illustrate 2 400 million years of lunar history. They are from 1 to 12 cm thick, and each probably represents a different meteorite impact which deposited a new layer of loose top soil. The bottom of the core consists of coarser material and may belong to the original lunar surface that is buried beneath the impacted soil.

Meteorites bombard the Moon constantly, and most are tiny pellets or even smaller particles. But about once a month a meteorite as large as a tennis ball strikes the surface and gouges out a sizeable crater. The duration of the resulting seismic reverberations indicates great crustal strength and depth, with no evidence of a sharp boundary between crust and mantle.

Magnetic Studies were complicated by the fact that the **solar wind** and the electrical conductivity of parts of the interior combine to create an induced magnetic field, but the permanent magnetisation found in some lunar rocks suggests that these rocks were formed in a weak magnetic field of internal lunar origin. This field was not strong enough to require a liquid lunar core for its explanation, nor is the present field, which is about 1 000 times weaker than that of the Earth and which was probably originally induced by the Sun's magnetic field. Two heat-flow probes drilled 2 metres apart during the *Apollo 15* mission showed that the temperature increases by 1° C for each 60 cm of depth, which is about $\frac{1}{5}$ of the Earth's heat flow, and suggests a local abundance of radio-active minerals higher than it can be in the interior. The same phenomenon (i.e. excess concentration of surface radioactivity) has been found also by α-particle spectrometry from the orbiting vehicles (*Apollo 15* and *16*) to an even greater extent than is true in the Earth's crust, and may be due to the vagaries of nucleogenesis in the last stages of the formation of the solar system. As for past surface temperatures, we can draw some conclusions by plotting the *glow curves* from readings obtained during heating lunar samples in the laboratory. These lead to the maximum temperatures to which the samples have been heated in the natural state, and thus to the maximum temperature to which the lunar surface has been subjected since the rocks were formed, provided we are quite clear about the past locations of the sample, i.e. whether it was formed where it was found or transported by impact or other means since its formation.

Minute traces of an *atmosphere* and of water vapour have been recorded, but they can be accounted for by the solar wind and by contamination from space vehicles, whose exhaust gases can amount to 9 tonnes or more per landing. There is a delicate ionosphere; it consists of fleeting ion clouds fed in by the Sun or by the occasional escape of ancient gases like helium trapped in a lunar rock.

It is probably true to say that during the last few years lunar material has been subjected to examinations as sophisticated and as thorough as those which have been devoted to terrestrial materials in a generation. In so far as the results permit general conclusions to be drawn, these will be found in Professor Kopal's article on the **Moon.** But it will be some time before the present flow of information has been digested and fitted into a coherent theory of the history of the Moon within the solar system.

COMET KOHOUTEK, 1973 *f*. This comet was discovered by Dr. L. Kohoutek at the Hamburg Observatory on March 7, 1973. Best available orbital details at the time of going to press are:

Perihelion distance in A.U.	0·142
Inclination	14°·3
Eccentricity	nearly 1·0
Predicted magnitude for Dec. 29, 1973	−2·5

This means that it should be a very bright and prominent object towards the end of 1973, rivalling **Halley's Comet**, with a tail stretching across one third of the night sky. It is hoped that **Skylab** will be manned at the appropriate time to make the first-ever detailed observations of a comet from outside the Earth's atmosphere. The orbit is highly eccentric and not yet distinguishable from a parabolic orbit, so that the period of the comet remains at present undetermined.

MARS, RECENT CONJECTURES. In November 1971 the complex robot spacecraft *Mariner 9* fired its braking rocket and was captured in orbit around Mars, thus becoming the first man-made satellite of another planet.

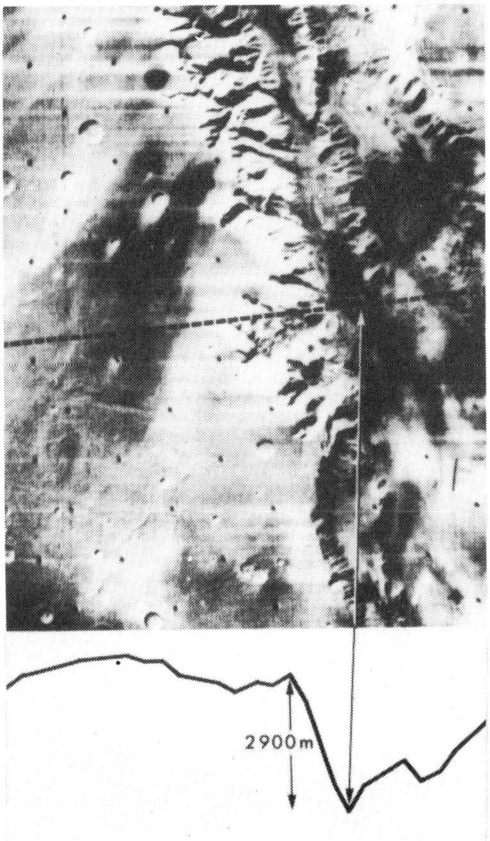

A CHASM ON MARS, 140×450 km, in Tithonius Lacus, photographed by *Mariner 9.* The resemblance to a river canyon system is probably superficial, as many of the 'tributaries' are closed depressions. Subsidence along lines of crustal weakness and the erosive action of winds may have shaped this feature. A cross-section along the black broken line is shown with the vertical scale exaggerated.

For nearly a year it sent back a steady stream of pictures and scientific data, which provided about 100 times the information accumulated by all previous missions to Mars. It also decisively changed our view of the planet. In our main article on **Mars,** Patrick Moore sums up the more firmly established part of our knowledge; what follows below is based chiefly on the speculations of Professor Bruce C. Murray and his colleagues.

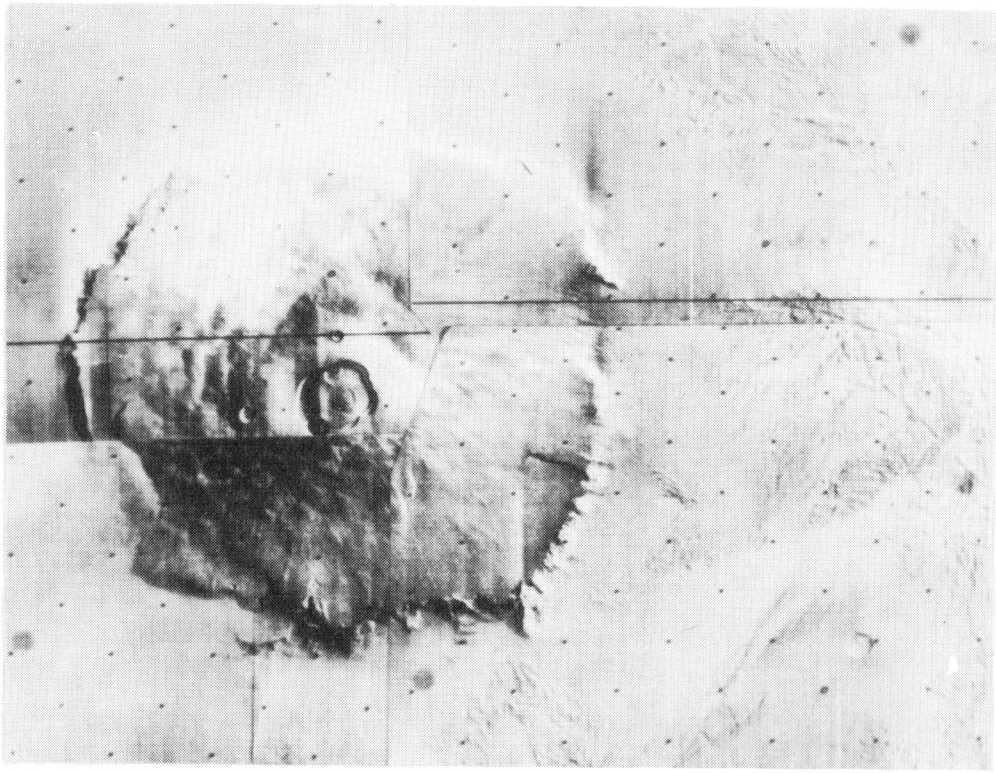

NIX OLYMPICA, a huge volcano on Mars, and the largest feature of its kind known anywhere. The mountain rises from a great plain, is 500 km wide at the base, and has a central caldera 65 km wide. The height is probably about 25 km. The picture is a composite of four *Mariner 9* photographs. Nix Olympica is part of the long, volcanic Tharsis Ridge. Long tongues of lava flow and honeycombs of lava channels are visible.

One question that arises stems from the fact that, whereas earlier reconnaissances photographed large areas showing scarcely any evidence of internal activity on Mars, *Mariner 9* revealed the existence of a chain of enormous volcanoes, and much other evidence of surface igneous activity. The explanation may be that Mars is just beginning to 'boil' internally, that the Olympica–Tharsis Ridge area is the first to show these effects, and that they will ultimately spread to the planet as a whole. If Mars had the same original composition and proportion of radio-active material as the Earth, one would expect it to heat up more slowly because it has only a tenth of the mass of the Earth. Nix Olympica suggests that deep convection currents may be churning which, in some hundreds of millions of years, could lead to something resembling the slow drift of crustal plates on the Earth.

Some theories have been put forward that account for the canyons and river-like channels in terms of lava flows and other tectonic events and the erosive action of winds. Others conjecture that liquid water was trapped in underground reservoirs below a layer of permafrost, and that on one or perhaps many periodic occasions this water was released during volcanic episodes, causing strong floods or a cycle of heavy rainfall. All these theories face serious difficulties.

CRATERS ON MARS, photographed by Mariner IV from an altitude of about 6,000 miles. The electronically enhanced picture reveals details about a hundred times smaller than the best telescopes can obtain under seeing conditions from the Earth. A number of craters similar to those on the Moon are apparent; the two large ones in the centre and at upper left seem to be completely filled in.

The present atmsophere may, in a large measure, be due to outgassings from the Tharsis Ridge area. The existence of widespread blankets of rock material and other evidence of earlier volcanism and sedimentation suggest that rather large volumes of volatile substances issued from the planet's interior; and that they provided an atmosphere which in turn produced a laminated terrain and provided the wind transport and erosion mechanisms to form the channels and the great canyons.

The answers to these questions and to the vexed problem of the past and present distribution of water on the planet, with its implications for the existence of life, must wait for sophisticated chemical and mineralogical analysis of the surface material. New data from the Russian *Mars* landers launched late in 1973, from second-generation Russian landers and from the complex U.K. *Viking* lander programme of 1976 can tell us much about organic and inorganic surface materials and their past chemical evolution, including whether or not they have reacted with water. The climax to the great search for life on Mars may come when we receive, probably by unmanned means, a sample of Martian soil for refined analysis in terrestrial laboratories. The successful return to the Earth of Moon material by unmanned *Luna* probes makes it reasonable to hope that this can be achieved within two or three decades.

SKYLAB. The first experimental space station, i.e. an orbiting workshop in which men can live and carry out useful work for periods of up to two months. Its hull consists of a modified Saturn 5 rocket, and it was launched on May 14, 1973. (See table on p. 38 and diagram on p. 258).

Its basic objectives include: the application of modern techniques in studies of the Earth's resources; the study of the effect of extended space flight on humans; the use of sensors for gathering data on the distribution of snow, clouds and water, and of heat and dust patterns near volcanoes; the investigation of the effects of zero gravity on the melting, moulding and welding of metals, on plants and bacteria, and on the coordination of spiders when constructing webs; and a variety of other projects, which may include spectroscopic observations of Comet Kohoutek ; the latter would give far better results than those which could be obtained from Earth.

Skylab's biggest and most important instrument is the Apollo Telescope Mount, which effectively makes it a manned astronomical observatory. Orbiting clear of the atmosphere, the astronauts use it to study the Sun under perfect observing conditions across the full spectrum of its radiation.

Medical surveillance focuses on bones, muscles, lungs and blood circulation, and on the balance of minerals in body fluids. Skylab is designed to give its occupants a considerable measure of domestic comfort and means of taking regular exercise. There is, for instance, a 'bicycle' exerciser which provides data on the user's respiration, heart rate and pulse. The astronauts can cook and eat their meals as they sit at table in the wardroom. There are toilet and washroom facilities; a shower within a cylindrical water-proof curtain; cards, books, a vacuum cleaner, refrigerators and a large supply of clothes. For sleeping, the crew members zip themselves into sleeping bags which are secured to the bulkheads.

Skylab circles the Earth every 93 minutes, and is visible from the ground as a fast-moving 'star'. With its orbit inclined at 50° to the equator, Skylab passes at different times over virtually all land areas outside the polar regions, including the Soviet Cosmodrome of Tyuratam, the Zvyezdograd rocket city, the Chinese missile testing centre at Shuang-Ch'eng-tze and the atomic test centre at Lop Nor.

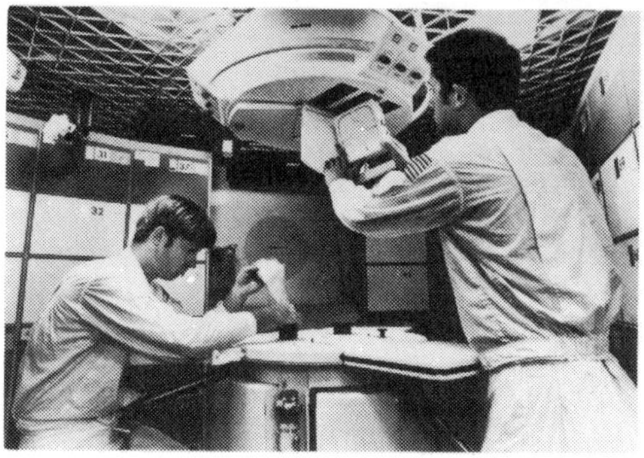

SKYLAB KITCHEN. It includes a cooker suspended from the ceiling and a water gun for reconstituting dehydrated foods. To avoid drifting across the cabin, the man on the left pushes against a metal brace.

Skylab is well equipped to ride out most emergencies. If life support systems failed, the crew could retreat to their **Apollo** module and return to Earth. If the Apollo module were not usable, it could be discarded and the crew could stay long enough to be rescued by another Apollo module docking in its place or in an alternative docking port. An Apollo ferry can quickly be modified to carry five men, i.e. two from the rescue team and three from Skylab. Finally, if docking should be impossible, the crew can 'space-walk' to the rescuing ferry.

Various faults developed during the Skylab missions which made it seem likely that these rescue facilities would be invoked: a heat shield failed to deploy properly before the first crew arrived, and prevented the solar panels that provide power for the workshop from opening fully, while the temperature inside Skylab rose to tropical levels. Nevertheless the crew was sent, and was able to carry out makeshift repairs and to occupy the station for the planned stay. Sometime after their return, the second crew was despatched with a new substitute heat shield; during their occupancy a number of other faults developed, including fuel leakages and faults in the control jets of the Apollo module, as well as a leak of coolant from Skylab itself, together with minor electrical faults. At the time of writing it seems probable that in spite of these difficulties no rescue mission will have to be launched.

382479

The New space encyclopaedia.